Preparative Centrifugation

The Practical Approach Series

SERIES EDITORS

D. RICKWOOD

Department of Biology, University of Essex
Wivenhoe Park, Colchester, Essex CO4 3SQ, UK

B. D. HAMES

Department of Biochemistry and Molecular Biology,
University of Leeds, Leeds LS2 9JT, UK

Affinity Chromatography
Anaerobic Microbiology
Animal Cell Culture (2nd edition)
Animal Virus Pathogenesis
Antibodies I and II
Biochemical Toxicology
Biological Data Analysis
Biological Membranes
Biomechanics—Materials
Biomechanics—Structures and Systems
Biosensors
Carbohydrate Analysis
Cell-Cell Interactions
Cell Growth and Division
Cellular Calcium
Cellular Neurobiology
Centrifugation (2nd Edition)
Clinical Immunology
Computers in Microbiology
Crystallization of Nucleic Acids and Proteins
Cytokines
The Cytoskeleton
Diagnostic Molecular Pathology I and II
Directed Mutagenesis
DNA Cloning I, II, and III
Drosophila
Electron Microscopy in Biology
Electron Microscopy in Molecular Biology
Electrophysiology
Enzyme Assays
Essential Molecular Biology I and II
Eukaryotic Gene Transcription
Experimental Neuroanatomy
Fermentation
Flow Cytometry
Gel Electrophoresis of Nucleic Acids (2nd Edition)
Gel Electrophoresis of Proteins (2nd Edition)
Genome Analysis
Growth Factors
Haemopoiesis
Histocompatibility Testing
HPLC of Macromolecules
HPLC of Small Molecules
Human Cytogenetics I and II (2nd edition)
Human Genetic Diseases
Immobilised Cells and Enzymes
Immunocytochemistry
In situ Hybridization
Iodinated Density Gradient Media
Light Microscopy in Biology
Lipid Analysis
Lipid Modification of Proteins

Lipoprotein Analysis
Liposomes
Lymphocytes
Lymphokines and Interferons
Mammalian Cell Biotechnology
Mammalian Development
Medical Bacteriology
Medical Mycology
Microcomputers in Biochemistry
Microcomputers in Biology
Microcomputers in Physiology
Mitochondria
Molecular Genetic Analysis of Populations
Molecular Neurobiology
Molecular Plant Pathology I and II
Monitoring Neuronal Activity
Mutagenicity Testing
Neural Transplantation
Neurochemistry
Neuronal Cell Lines
Nucleic Acid and Protein Sequence Analysis
Nucleic Acids Hybridization
Nucleic Acids Sequencing
Oligonucleotide Synthesis
PCR
Peptide Hormone Action
Peptide Hormone Secretion
Photosynthesis: Energy Transduction
Plant Cell Culture
Plant Molecular Biology
Plasmids
Pollination Ecology
Post-implantation Mammalian Embryos
Preparative Centrifugation
Prostaglandins and Related Substances
Protein Architecture
Protein Engineering
Protein Function
Protein Phosphorylation
Protein Purification Applications
Protein Purification Methods
Protein Sequencing
Protein Structure
Protein Targeting
Proteolytic Enzymes
Radioisotopes in Biology
Receptor Biochemistry
Receptor–Effector Coupling
Receptor–Ligand Interactions
Ribosomes and Protein Synthesis
RNA Processing
Signal Transduction
Solid Phase Peptide Synthesis
Spectrophotometry and Spectrofluorimetry
Steroid Hormones
Teratocarcinomas and Embryonic Stem Cells
Transcription Factors
Transcription and Translation
Virology
Yeast

Preparative Centrifugation

A Practical Approach

Edited by

DAVID RICKWOOD

Department of Biology, University of Essex, Wivenhoe Park, Colchester

OXFORD UNIVERSITY PRESS
Oxford New York Tokyo

Oxford University Press, Walton Street, Oxford OX2 6DP

Oxford New York Toronto
Delhi Bombay Calcutta Madras Karachi
Petaling Jaya Singapore Hong Kong Tokyo
Nairobi Dar es Salaam Cape Town
Melbourne Auckland

and associated companies in
Berlin Ibadan

Oxford is a trade mark of Oxford University Press

A Practical Approach is a registered trade mark
of the Chancellor, Masters, and Scholars of the University of Oxford
trading as Oxford University Press

Published in the United States
by Oxford University Press, New York

A catalogue record for this book is available from the British Library

Library of Congress Cataloging in Publication Data

Preparative centrifugation : a practical approach / edited by D. Rickwood.
p. cm. -- (Practical approach series)
Includes bibliographical references.
1. Centrifugation. 2. Biochemistry--Methodology. 3. Cytology--Methodology. I. Rickwood. D. (David) II. Series.
QP519.9.C44P74 1992 574.19'285--dc20 91-47023
ISBN 0 19 963208 1 (h/b)
ISBN 0 19 963211 1 (p/b)

Typeset by Dobbie Typesetting Limited, Tavistock, Devon
Printed and bound in Great Britain by Information Press, Ltd., Oxford, England

Preface

To the casual observer it could appear that there have been no real changes in centrifugation in the last ten years. Centrifuges may now have instrument panels with LEDs for digital output instead of wavering needles but, superficially at least, centrifugation does not seem to have changed in recent years. In fact, apart from the theory of centrifugation and its associated equations, there have been large changes in both the instrumentation and the experimental protocols used for various separation methods. Over the years ultracentrifuges have ceased to be the exclusive preserve of the specialist and now they are used widely by most laboratory personnel. At the same time there has been a dramatic increase in the sophistication of centrifuges as a result of the availability of cheap microprocessors, and there has been a parallel increase in the availability of small but powerful microcomputers in the laboratory. These have opened up new areas of calculation and simulation of centrifugal separations to essentially everyone working with centrifuges. It is for this reason that the use of computer programs in centrifugation is emphasized. Experimental protocols have also become more advanced, driven by the need to isolate increasingly pure sub-cellular fractions. At the same time many quite complex separations are carried out on a routine basis by novices in the centrifugation field. This book sets out both to provide instant solutions for a very wide range of separations as well as providing the reader with insights to the bases of the different types of separations in the expectation that these will be of help in developing new types of fractionation procedures.

Colchester D.R.
November 1991

Health warning

Some of the chemicals used for procedures described in this book may be associated with biological hazards. In addition, the reader should be aware of the hazards associated with the handling of animal tissue samples. While the editor and authors have tried to indicate the hazards associated with the different procedures, it is the responsibility of the reader to ensure that, in following the procedures described in this book, they use safe working practices.

Contents

Contributors

ADRIAAN BROUWER
TNO-IVVO, Zernikedreef 9, 2333 CK Leiden, The Netherlands.

MILOSLAV DOBROTA
School of Biological Sciences, University of Surrey, Guildford, Surrey GU2 5XH, UK

W. HOWARD EVANS
National Institute for Medical Research, The Ridgeway, Mill Hill, London NW7 1AA, UK.

TERRY C. FORD
Department of Biology, University of Essex, Colchester, Essex CO4 3SQ, UK.

JOHN GRAHAM
MIC Medical Ltd, Merseyside Innovation Centre, 131 Mount Pleasant, Liverpool L3 5TF, UK.

HENK F. J. HENDRIKS
TNO Institute for Experimental Gerontology, PO Box 5815, 2280 HV Rijswijk, The Netherlands.

RICHARD H. HINTON
School of Biological Sciences, University of Surrey, Guildford, Surrey GU2 5XH, UK.

STEVEN HUMPHRIES
Department of Biology, University of Essex, Colchester, Essex CO4 3SQ, UK.

DONALD L. KNOOK
TNO Institute for Experimental Gerontology, PO Box 5815, 2280 HV Rijswijk, The Netherlands.

DAVID RICKWOOD
Department of Biology, University of Essex, Colchester, Essex CO4 3SQ, UK.

HANS ROTH
Nissei Sangyo GmbH, Berliner Strasse 91, 4030 Ratingen 3, Germany.

S. PETER SPRAGG
Department of Chemistry, University of Birmingham, Edgbaston, Birmingham B15 2TT, UK.

JENS STEENSGAARD
Institute for Medical Biochemistry, University of Aarhus, DK 8000 Aarhus C, Denmark.

Abbreviations

BSA	bovine serum albumin
DAB	diaminobenzidine tetrahydrochloride
DEPC	diethylpyrocarbonate
DFP	di-isopropylfluorophosphate
DMSO	dimethyl sulphoxide
DTT	dithiothreitol
EDTA	ethylenediamine tetraacetic acid
EGTA	ethyleneglycobis(β-aminoethyl)ether tetraacetic acid
GBSS	Gey's balanced salt solution
NP-40	Nonidet P-40
PAGE	polyacrylamide gel electrophoresis
PBS	Phosphate-buffered saline
PEG	polyethylene glycol
PET	polyethyleneterephthalate
PMSF	phenylmethylsulphonyl fluoride
POPOP	1,4-bis-(5-phenyloxazol-2-yl)benzene
PPO	2,5-diphenyloxazole
PVP	polyvinylpyrrolidone
RCF	relative centrifugal force
RNP	ribonucleoprotein
mRNP	messenger ribonucleoprotein
r.p.m.	revolutions per minute
SDS	sodium dodecyl sulphate
SV40	simian virus 40
TCA	trichloroacetic acid

1

Theoretical aspects of practical centrifugation

S. PETER SPRAGG and JENS STEENSGAARD

1. Introduction

This chapter provides the reader with the theory and numerical bases upon which the practices of centrifugation are based. It particularly emphasizes the quantitative features of practical aspects of centrifugation such as sedimentation and determination of the efficiency of rotors. An in-depth knowledge of the various mathematical aspects of centrifugation is not an essential prerequisite for the average person. Hence the reader can omit this chapter if the protocols required are given in subsequent chapters of the book. However, if it is necessary to devise new centrifugation procedures, then an understanding of this chapter will provide an insight into the different parameters which should be modified to define the optimum conditions for centrifugal separations. For the sake of clarity the number of equations in the text has been kept to a minimum and, where derivations of equations are new or not readily available, these have been included in Appendix A of this chapter; derivations of equations that have been widely published (e.g. derivation of the Lamm equation) are not given here but readers are provided with appropriate references where these derivations can be found.

The centrifuge was introduced to biochemistry during the 1920s, largely through the efforts of Swedish workers at the University of Uppsala (1). These researchers developed an instrument which had an optical system for observing sedimentation of the sample during centrifugation and this sedimentation was recorded on film using either refractive index changes or optical absorption to visualize the boundaries of sedimenting macromolecules. After this time the analytical centrifuge continued to provide quantitative data on the physical chemistry of macromolecules. This instrument reached its zenith during the 1970s but then declined in importance as various new methods for defining the characteristics of macromolecules were developed. These developments have not displaced the ultracentrifuge as an instrument for separating and characterizing macromolecules and other biological material. The theory which was developed to describe sedimentation of macromolecules in analytical machines and which is described in this chapter can be applied to guide experiments carried out using preparative centrifuges.

2. Units of variables and constants

When applying the equations for sedimenting particles it must be remembered that this is a physical process and the more familiar definition of molecular mass does not apply. Molar masses must be given in kilograms and not as relative masses, nor as Daltons. Now the internationally recognized dimensions collectively called SI units are used. However, most of the data on solution properties of molecules were tabulated before SI units were introduced. In order to avoid confusion over the use of this older data *Table 1* lists factors for converting from the old cgs system to SI together with accepted symbols. Care is necessary in using units since it is all too easy to arrive at answers which are 1000-fold too large or small.

3. Theoretical basis of centrifugation

3.1 Basic centrifugal equations

Centrifugation techniques can be divided into three main methods, namely differential pelleting, rate-zonal centrifugation, and isopycnic centrifugation. However, irrespective of the technique, there is just one mathematical equation which describes all centrifugal systems, the Lamm partial differential equation (2). Basically, this is an equation describing the movement of particles in the presence of an external force, the centrifugal force, and it has been the subject of intensive study by many theoreticians; this work is summarized by Fujita (3). It might be thought that integration of this equation would produce a formula which completely describes the sedimentation of any particle under actual experimental conditions, but this ideal has eluded the best mathematicians. Instead there are several analytical solutions which depend on the initial mathematical boundary conditions and, hence, on the type of experiment and particle. Even these equations do not allow for all the practical conditions found in real centrifugal experiments (e.g. interactions of the gradient medium with samples), and these can only be included as approximations. Despite these major weaknesses the analytical equations can be applied to preparative centrifugal experiments to guide the choice of separation conditions.

Centrifugal separations can be carried out in two ways; velocity and equilibrium. In velocity separations particles move in the direction of the centrifugal force and separate according to size and density, while in equilibrium separations particles move to a position in the gradient which reflects their solvated densities. The experimental conditions used to achieve these separations are not fixed but depend on whether the sample is uniformly distributed throughout the solution or is introduced as a narrow zone supported on top

Table 1. Factors for converting cgs into SI units

Unit	cgs to SI	Dimensions	Symbol	Value (SI)
Gas constant	$\times 10^7$	ML^2t^{-2}	R	8.31434
Density	$\times 10^3$	ML^{-3}		
Partial specific volume	$\times 10^{-3}$	L^3M^{-1}	$\bar{v}$	
Viscosity	$\times 10^{-1}$	$ML^{-1}t^{-1}$		
Molecular mass	$\times 10^{-3}$	M	M	
Temperature	$+273.12$	K	T	
Diffusion	$\times 10^{-4}$	L^2t^{-1}	D	
Sedimentation coefficient	$\times 1$	t	S	
Avogadro's number	$\times 10^3$	molecules/ mole	N	6.022169×10^{26}

M = mass(kg); L = length (metres); t = time (sec)

of a density gradient. The term 'differential pelleting' is used to describe experiments in which particles are distributed throughout the solution and separated according to a combination of mass and density. For such separations there is no density gradient and particles never reach or approach equilibrium.

Before dealing with the special conditions routinely used in preparative centrifugation it is important to outline the factors which control the velocity of the sedimenting particle. As the rotor accelerates to its constant running speed, particles begin to sediment towards the bottom of the tube. It is generally assumed this acceleration time is small compared with the time at the uniform speed. The centrifugal force (RCF) relative to the Earth's gravitational force (g) can be calculated from the square of the speed, ω, given as radians/sec and the centrifugal radius, r (cm):

$$\text{RCF} = \omega^2 r/g$$

The speed in radians/sec (ω) can be converted to revolutions/min (r.p.m.) by the expression:

$$\text{r.p.m.} = \pi\omega/30$$

When the speed of rotation is expressed as revolutions/min (Q) then the original equation closely approximates to:

$$\text{RCF} = 11.18\, r(Q/1000)^2$$

Thus, for a mixture of particles, those furthest away from the centre of rotation are subjected to greater centrifugal forces and the system can be formulated using Newtonian principles in which, at uniform velocity, the sedimenting particles meet an opposing force. The magnitude of this opposing force depends on the shape of the particle. A filamentous particle presents more friction than a solid sphere of the same mass. Thus the shape of the particle in solution must be considered when interpreting its true velocity of sedimentation and this hydrodynamic shape is expressed as a frictional coefficient, Nf, where N is Avogadro's number and f is the frictional coefficient of a single particle. The value for this coefficient cannot be explicitly calculated if the shape is unknown but it can be calculated for spherical particles of radius a in a solution of viscosity η using Stokes' equation (Equation 1):

$$f = 6\pi\eta a \qquad [1]$$

In order to convert this frictional coefficient to a force it must be multiplied by the velocity of the particle ($\mathrm{d}r/\mathrm{d}t$), where t is the time which, when equated with the sedimenting force, gives the equation for sedimentation of a single particle (Equation 2).

$$\text{effective mass} = M(1 - \bar{v}\rho_{\mathrm{m}}) \qquad [2a]$$

$$\text{force} = M(1 - \bar{v}\rho_{\mathrm{m}})\omega^2 r \qquad [2b]$$

$$\text{opposing force} = Nf(\mathrm{d}r/\mathrm{d}t) \qquad [2c]$$

$$M = Nf(\mathrm{d}r/\mathrm{d}t)/(1 - \bar{v}\rho_{\mathrm{m}})\omega^2 r \qquad [2d]$$

The bracketed term in the denominator of Equation 2d is the difference in density between that of the solvent (ρ_{m}) and that of the particle expressed as the partial specific volume, $\bar{v}$ (the solvated volume of a unit mass of particle). Although $\bar{v}$ is equal to the reciprocal of the density of the solvated particle, it is necessary to remember that for solvated particles or molecules the density term of Equation 2 is dependent on the composition of the solvent.

Examination of Equation 2d shows that $(\mathrm{d}r/\mathrm{d}t)/\omega^2 r$ is the velocity of the particle divided by the centrifugal acceleration. Since this ratio is an experimental variable it can be normalized to give velocity per unit acceleration and called the sedimentation coefficient (s, which is defined mathematically in Chapter 5). At the molecular level f is a complex coefficient which is related to Brownian motion and inversely proportional to the diffusion coefficient, D (see Equation 3).

$$D = RT/Nf \qquad [3]$$

Thus, introducing Equation 3 into Equation 2d gives Equation 4, the full Svedberg equation for a thermodynamically ideal particle:

$$M = RTs/[(1 - \bar{v}\rho_{\mathrm{m}})D] \qquad [4]$$

Equation 4 does not describe the distribution of the particles during sedimentation but refers to the movement of a single particle. Practically, this movement corresponds to the velocity of the sample peak in zonal experiments. The sample moves to the bottom of the tube provided the centrifugal forces are greater than the back-flow caused by molecular diffusion.

A maximum value for D can be calculated by combining Equation 3 with Stokes' equation (Equation 1) and replacing a with the volume of a sphere ($4\pi a^3/3$), multiplied by Avogadro's number to convert to moles of particles. Substituting for a in Equation 1 gives for a molar frictional coefficient ($F = Nf$):

$$F_{\text{min}} = 6\pi N\eta\,(3M\bar{v}/4\pi N)^{1/3} \qquad [5]$$

This represents the minimum frictional coefficient for a spherical molecule, but particles or molecules are not spheres. A convenient way to accommodate reality with theory is to compare the experimental diffusion or frictional coefficient with that calculated from Equation 5 to give F/F_{min}. This ratio increases as the shape deviates from a sphere; for example, with DNA, F/F_{min} is about 15 because it behaves in solution like a filamentous molecule, while for globular proteins it approximates to 1.2 (see also ***Figure 7*** of Chapter 5).

Having two opposing forces acting on the particles suggests that an experimental condition can be described where sedimentation equals diffusion and no net flux of material occurs with time. This is called sedimentation equilibrium when Equation 4 converts to:

$$M = RT(\mathrm{d}C/\mathrm{d}r)/[(1 - \bar{v}\rho_{\text{m}})\omega^2 Cr] \qquad [6]$$

Equations 4 and 6 are the basic equations which describe all types of sedimentation but they can be modified to include differing experimental conditions. Their simplicity belies their ability to describe practical results from a centrifugal experiment under many varying conditions. The sedimentation coefficient, s, as defined in Chapter 5, does not equate directly with experimental values because solvents have varying viscosities and densities which alter the velocity of the particles. This can be allowed for by calculating an equivalent coefficient for sedimentation in water at 20 °C and quoting this value in the literature as $s_{20,\text{w}}$ (the formulae are given in Chapter 5 together with the inverse for converting $s_{20,\text{w}}$ to operational coefficients). A further factor concerns the concentration of the particle. When a particle sediments it displaces the solvent and since any solute affects the organization of adjacent solvent molecules this acts as a drag on the moving solute. The influence of the solute molecule extends out from its surface and as the concentration of the solute is increased there comes a point where the volume of free solvent is very limited; it is almost as though the solute molecules 'knock' into one another. Allowance for these effects on the value of the sedimentation coefficient can be made using the empirical expressions:

$$s = s^0(1 - gC) \quad [7a]$$

$$s = s^0/(1 + kC) \quad [7b]$$

where s^0 is the coefficient at infinite dilution. Equation 7a applies to solutions of globular particles at concentrations not exceeding a few per cent; for filamentous particles Equation 7b is more appropriate. For very high concentrations (greater than ~100 mg/ml) the dependence departs from simple linear relationships and terms containing higher powers of C become necessary. There is a similar dependence of D on concentration although this is usually less marked for particles and only in special cases do allowances for finite concentrations affect calculations. There are occasions when the concentration of the gradient medium is very high (e.g. CsCl gradients) and under these conditions it is not the weight concentrations which should be used in Equation 7 but the thermodynamic activities. These features have been introduced into equations, particularly for isopycnic separations, through a thermodynamic β coefficient for systems at true equilibrium (summarized in ref. 4).

These non-ideal effects resulting from the concentration of particles mean that observed parameters must be corrected before they can be compared. One example where the concentration dependence of the sedimentation coefficient, s, becomes important is the shape of sample bands. Higher concentrations move with a lower velocity than lower concentrations and since concentrations within a sample peak vary from virtually zero to the maximum concentration this causes the leading edge of zones to sharpen. Naturally sharp sample zones always occur with large particles because the diffusion is small ($D \propto 1/M^{1/3}$) so sharpening caused by concentration dependence is only noticeable when sedimenting relatively small macromolecules. It does mean, however, that concentration dependence must be taken into account when comparing simulated and experimental peaks. Most analytical solutions for the Lamm equation do not include allowance for the concentration effects although approximations can be included in numerical solutions.

3.2 Practical aspects of sedimentation in tubes

3.2.1 Wall effects

Up until this point sedimentation has been treated without any real consideration of the way in which the sample solution is contained. Basically, samples are centrifuged in three different types of container. In the analytical centrifuge a sector-shaped cell is used, in zonal rotors the rotor itself is the container, and samples can be considered to be in a sector-shaped vessel. In the majority of cases samples are centrifuged in tubes, and since these are usually narrow and parallel-sided they do pose a number of problems from the point of view of sedimentation.

The centrifugal force is radial in direction which means that the sedimenting particles try to move radially away from the centre of rotation. To allow for this, analytical cells are sector-shaped with the angle subtended to the centre of rotation. It would not be very practical to make individual containers in this shape for preparative work and so usually all centrifugation in preparative centrifuges is carried out in tubes of varying sizes. Because of the radial direction of the centrifugal force, the particles moving down the tube tend to collide with the wall and can disrupt the sedimentation of the particles; this is known as wall effect.

The magnitude of the wall effects depends on the type of the rotor and the shape of the tube. Zonal rotors, described in Chapter 8, are available and, as particles in these rotors sediment in sector-shaped chambers of the rotor rather than tubes, there are no wall effects; however, it is usually inconvenient to use these for most gradient separations. In the case of swing-out rotors (*Figure 1a*) then only those particles contained within a cone reach the bottom without colliding with the walls. It can be calculated that for a tube 3 cm long of radius 1 cm, placed in a rotor with the end 6 cm from the centre, then about 42% of the particles will strike the sides. This effect can be formulated for the following system:

- radius of tube r_T
- length of tube L + radius of the hemispherical bottom (r_b)
- bottom at radius R
- volume of solution which is not affected by the walls V

Then, for the system shown in *Figure 1a*, and neglecting the small contribution from a curved bottom, the total volume of the cone subtended from the centre to the bottom of the tube can be calculated by:

$$\pi r_T^2 (R/3)$$

This includes the volume outside the tube which extends from the meniscus to the centre of the rotor and this is calculated as:

$$\pi [(r_T - (r_T/R)]^2 \, [R - (R + r_b)]/3$$

Subtracting the volume outside the tube from the total volume and rearranging gives:

$$V = \pi r_T^2 \{3 - [3(L + r_b)/R] + [(L + r_b)/R)]\}/3 \qquad [8]$$

The terms within the curly brackets represent the proportion of particles that hit the walls. Increasing the radial position of the tube without increasing its diameter or length decreases the proportion of particles that collide with

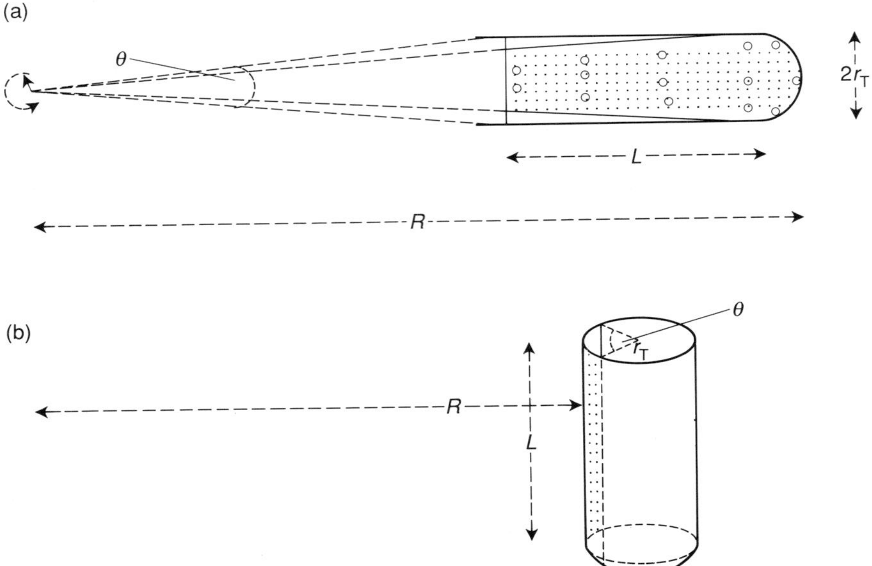

Figure 1. Diagram of the forces (dotted lines) on particles in a cylindrical tube. (a) Proportion of particles hitting the sides in a swing-out rotor. Particles within the shaded region sediment without colliding with the walls. The radius of the tube is r_T, L is the length, R is the maximum radius (r_{max}) of the rotor. (b) Geometrical arrangement of the solution in a vertical tube. R is the radial distance from the centre of rotation to the centripetal edge of the tube (i.e. r_{min}), L is the cylindrical length of the tube extending from the top to the point where the hemispherical bottom starts, r_T is the radius of the tube, and Θ is the angle subtended by the two ends of the chord defining the width of a plane in the tube.

the wall; for example, when R is equal to 7 cm, r_T is equal to 1 cm and L is equal to 3 cm, and the proportion hitting the side drops to about 37%.

Particles which reach the walls of the tube are subjected to less centrifugal force than those within the cone; their centrifugal acceleration is equal to:

$$M\omega^2 R \cos(\Theta)$$

where the angle Θ is defined as $\cos(r_T/R)$ and their velocity would be correspondingly reduced. It is likely, however, that a proportion of those particles hitting the wall would stick and leave a smear down the tube. Another possibility is that they would accumulate as unfractionated aggregates having increased effective mass, producing an increased force which would make them sediment rapidly to the bottom. These abnormal consequences cannot be included in simulations but should be allowed for when interpreting results from centrifugation in tubes; they emphasize the fact that only the particles near the top which do not sediment are likely to be the purest. Since wall effects enhance the rate at which particles pellet in tubes, it follows that increasing wall effects should enhance the sedimentation rate of particles. This can be done by adding beads (1–3 mm diameter) to the tubes, although in practice this is rarely done.

In the case of vertical rotors, the particles sediment across the tube and so while they are sedimenting over the first half of the tube there are no wall effects (*Figure 1b*); however, once past the halfway mark the particles begin to sediment against the wall of the tube. Wall effects are severest in fixed-angle rotors where the majority of the particles will sediment against the wall of the tube. This can lead to rapid pelleting of the sample (see Section 4.3.2 of Chapter 2).

While simple trigonometric calculations can determine the proportion of the sample particles that are likely to hit the wall, there is no quantitative measure as to the degree of interaction of the particles with the wall in terms of the expected amount of pelleting and the extent to which sample bands are disrupted by such interactions. This is a very complex problem since the degree of interaction depends on the nature of the sample; for example, the wall effects do not appear to affect the rate-zonal sedimentation of small ($\leqslant$ 10S) proteins in a fixed-angle rotor while under similar conditions polysomes will pellet on to the wall of the tube.

3.2.2 Hydrostatic pressure

It is usually not appreciated that a spinning rotor generates not only a centrifugal force which is used to separate the particles, but also hydrostatic pressure within the medium in which the particles are suspended. The hydrostatic pressure generated is dependent on the speed of centrifugation and the height of the liquid column (3). The hydrostatic pressure generated (P) in excess of atmospheric at any distance r (cm) from the centre of rotation can be calculated from the equation:

$$P = \rho\,\omega^2(r^2 - r_{men}^2)/2$$

converting from angular velocity to revolutions/min (r.p.m.) gives:

$$P = 5.48 \times 10^{-3}\,\rho\,Q^2(r^2 - r_{men}^2)$$

where r_{men} is the distance of the meniscus from the centre of rotation and ρ is the density of the solution. If the solution is uniform then the density is simply that of the solution but if the tube contains a gradient then it is simplest to take the average value at the median radius, that is $(r + r_{men})/2$. For more accurate work it is necessary that the density is integrated between r and r_{men}. The pressures generated can be very large indeed; for example, the pressure at the bottom of a 1 cm column of water 8 cm from the centre of rotation centrifuged at 50 000 r.p.m. can be calculated ($r_{men} = 0.07$ m, $r = 0.08$ m, $\rho = 10^3$ kg/m^{-3}) as 112×10^6 N m^2, equivalent to about 120 atmospheres.

Obviously, the pressure in denser solutions would be proportionately greater. The high pressures generated during centrifugation can have very significant effects on altering the permeability of membranes and can dissociate complexes such as ribosomes and cytoskeletal proteins. This occurs because sum of the partial specific volumes of the complex is less than that of the complex. Studies have been done using hydrostatic pressure to determine changes in the partial specific volume of associating systems (18).

4. Differential pelleting

In this type of separation the particles are initially distributed uniformly throughout the solution at the start of the run and as time progresses particles pellet to the bottom with the particles of the sample moving with varying velocities down the tube (*Figure 2*). The position reached by the particles at any given time can be calculated from knowing their sedimentation coefficients, s, and the speed of the rotor using Equation 10, the integral of Equation 4 given in Chapter 5, which gives:

$$r = r_{\text{men}} \exp(s\,\omega^2\,t)$$

The value chosen for s must be corrected for varying densities and viscosities of the solvent, which may vary during sedimentation (see Chapter 5). This calculation is useful for optimizing centrifugal conditions for pelleting the largest component in a mixture (r should be chosen so it equals the maximum radius). It does not give any indication of the overlap of different particles; to estimate this it is necessary to simulate the shape of the particle distribution using the Lamm equation (see Section 7).

Equation 10 shows that doubling the time of a run does not double the distance travelled by the particle or the separation of components; for example, a 20S particle ($s = 20 \times 10^{13}$ sec) centrifuged at 60 000 r.p.m. and starting at 3.00 cm from the centre of rotation would be at 3.80 cm after 3000 sec and 4.82 cm from the centre of rotation after 6000 sec. Using the same centrifugal variables but for a 7S particle the corresponding radial positions would be 3.26 cm and 3.54 cm. Thus, the separation between the two particles would have increased from about 0.54 cm to about 1.28 cm as a result of doubling the time of the run but diffusion increases zone widths by about 40% during this extra time. This separation would give poor purity of the smaller particle by differential pelleting. These simple calculations emphasize that, in order to get good separations of two particles, it is desirable to have a large difference in size, at least a 10-fold difference in sedimentation coefficients, of the two particles. It can also be seen that such differences in sedimentation coefficient would be sufficient to obtain satisfactory rate-zonal separations on a sucrose gradient. This fact explains why, despite the additional experimental inconvenience of rate-zonal separations, zonal experiments have become widely used as preparative procedures.

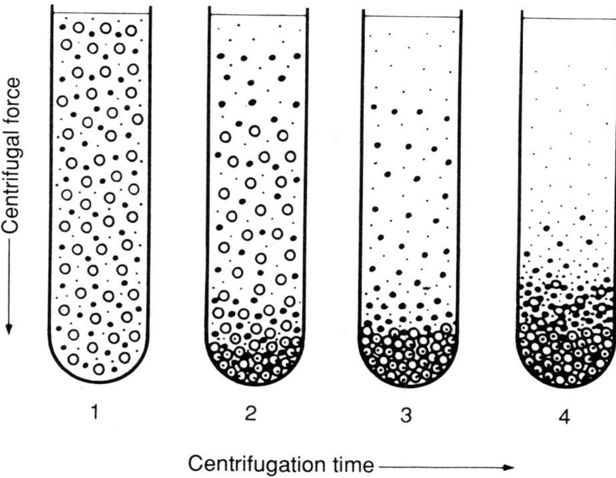

Figure 2. Diagrammatic representation of the fractionation of particles by differential pelleting. (Reproduced from reference 19 with permission.)

5. Gradient separations

Gradient separations involve separation of the sample as a zone in the gradient and practical examples of these types of separation are given in other chapters. Most calculations relate to the maximum concentration of the zone, and it moves with a velocity determined by the sedimentation coefficient of the particles and the difference in density between the particles and the surrounding medium.

Gradient separations can be divided into two types:

(a) rate-zonal runs where separation depends on the relative velocities of the zones

(b) isopycnic equilibrium where separation depends on the buoyant densities of the particles

Irrespective of the type of separation, it is important to know the position of the sample band in the tube. This is not easily measured as a distance but it can be calculated after fractionation of the gradient into fractions of known volumes provided that the maximum distance from the centre of rotation and the relationship between volume and radius is known. The relationship between volume and radius will vary depending on the shape of tube and the type of rotor. Calculations for zonal rotors are given in Section 5.2 of Chapter 5. Here the volume to radius relationship for swing-out and vertical rotors will be described.

5.1 Calculation of radii from fraction volumes

5.1.1 Volume–radius relationships for tubes of swing-out rotors

Converting volumes to radii is often necessary when relating fractions collected from the tubes. Assuming the cylindrical tube has a hemispherical bottom of internal radius r_b (usually obtained from the manufacturer's specification) and whose internal length of the cylindrical part is L, then using the same symbols as for Equation 8 the total radius, R, can be calculated from the volume and the radius at the meniscus (r_{men}) by:

$$R = r_{men} + (V/\pi r_T^2) \qquad [9a]$$

The volume, V, is the total volume of the fractions, that is the summed volumes up to the one being considered. If samples are taken from the hemispherical bottom of the tube then the relationship between volume and radius is less easily calculated. It can be shown that in this case the volumes are related to the radius by:

$$V = (\pi r_b h^2) - (1/3\pi h^3) \qquad [9b]$$

where h is the internal radial length within the hemispherical bottom. Thus, to calculate the true radius, Equation 9b must be solved first to find h and this is added to the radius calculated up to the final sample before the rounded bottom.

5.1.2 Volume radius relationships for tube in vertical rotors

When samples are centrifuged in tubes of vertical rotors they migrate across the diameter of the tube, and when the rotor stops the bands rotate through 90° as gravity causes the liquid to reorientate. In this case the distance moved by the sample is not directly related to the accumulated volume of the fractions. The situation during centrifugation is shown diagrammatically in *Figure 1b*. Note that the tubes are always run completely full and so the total volume of the solution (gradient plus sample) is the volume of the tube. From the summed volumes which have been taken from the reorientated solution, one calculates the sample radius within the solution allowing for the contributions from the hemispherical top and bottom of the tube (5). The calculation involves an intermediate step of estimating Θ, the angle subtended by the two ends of a chord across the tube. This chord represents the level of solution in the rotating tube either before sampling has started (meniscus) or after taking the samples. The procedure is complicated by the inclusion of trigonometrical functions within a cubic equation (Equation 10) and solutions can only be found using numerical procedures. The equations and method of solving them is given in Section 1.3 of Appendix A of this chapter.

$$V_s = [(1/2)r^2 L] - [(1/2)r^2 L \sin\Theta] + [r^2 \sin^2(\Theta/2)] - [(1/3)r^3 \sin^3(\Theta/2)] \qquad [10]$$

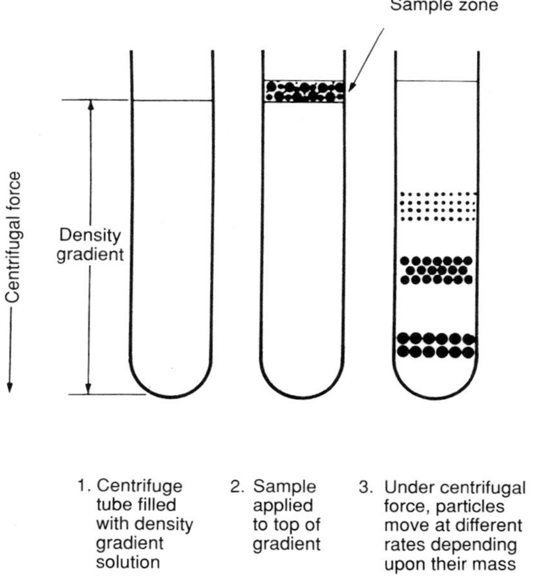

Figure 3. Diagrammatic representation of the fractionation of particles by rate-zonal centrifugation. This diagram shows how particles with differing rates of sedimentation form bands in the gradient. (Reproduced from reference 19 with permission.)

The limitation of the accuracy of the conversion of volumes to radii is determined by the manufacturer's tolerances for the tubes; these can be variable in terms of wall thickness. Under these circumstances it is a good idea to check each batch of tubes and calibrate them using the above equations, and either use a graph to convert volumes to radii or fit the calibration results to a polynomial which reproduces the relationship numerically.

5.2 Rate-zonal separations

In the case of rate-zonal separations (*Figure 3*) the radial position of the sample peak is related to the time of sedimentation and can be used to calculate the value of the particle's sedimentation coefficient.

Subsequent chapters in this book describe the practical aspects of rate-zonal separations so for present purposes all that is necessary to note is that the sample solution is loaded on to a denser gradient of inert solute (usually sucrose) as a narrow band. The density at the temperature of the run for any position within the zone, $\rho_{(r)\mathrm{T}}$, is given by:

$$\rho_{(r)\mathrm{T}} = \rho_\mathrm{m} + [C(r)_\mathrm{m}(1 - \rho_\mathrm{m}\bar{v}_\mathrm{m})] + [C_{(r)\mathrm{p}}(1 - \rho_\mathrm{m}\bar{v}_\mathrm{p})] \qquad [11]$$

where, ρ_m is the density of solvent, $\bar{v}_m$ is the partial specific volume of the small solute (subscript m) and $\bar{v}_p$ is the partial specific volume of the particle (subscript p). For a mixture, the right-hand side would be expanded to include further contributions from all the components. As with any system under the influence of gravitational forces a zone will only float if its total density is less than that of the underlying solution. The role of the centrifugal force is to drive the heavier particles through the gradient. In the case of the gradient solute, such as sucrose, the gradient will remain essentially unchanged by centrifugation. As the sample zone moves through this gradient it meets increased concentrations of the small molecules. Because the small molecules have a higher rate of diffusion than that of the sample particles their contribution to the density gradient within the zone will be reduced with time. Provided that the density of the particle is greater than that of the surrounding solution and the total density of the zone (density of particles and gradient solute) remains less than the density of the supporting gradient below it, then normal sedimentation will occur. This describes the ideal, stable gradient in which sample zones broaden only by diffusion of the particles to produce a shape which can be calculated from Equation 12 (6):

$$C_{x,t} = [C_0(\delta/2)(\pi^{1/2}\sigma)]\exp[-(x-x_0)^2/2\sigma^2] \qquad [12]$$

In this equation δ is the width of the zone at the start, x is the radial coordinate for a rectangular cell, x_0 is the radial position of the maximum concentration, C_0 is the loading concentration of the sample at zero time, and σ is the width of the zone at time t, where σ is given by:

$$\sigma = 2(DZt)^{1/2}$$

$$\text{where } Z = [\exp(2s\omega^2 t) - 1]/2s\omega^2 t$$

Equation 12 applies to a zone sedimenting down a gradient; the width of the zone may be constant or it may increase with radius. Experience shows that often the observed width of the zone does not fit this model and the sedimentation coefficient calculated from the sedimentation of x_0 does not equal that expected. One common reason for this discrepancy is the instability caused by inversion of the density gradient across the zone. A zone can only move with the velocity determined by the coefficients of its components if the density at any point within the zone is less than that immediately below this point. Mathematically this condition is described as:

$$\rho_{(r)\mathrm{T}} \leqslant \rho_{(r+\mathrm{d}r)\mathrm{m}} \qquad [13]$$

Since small molecules diffuse quicker into zones towards the bottom of the gradient where the concentration of the gradient medium is much greater than that of the particles, this diffusion process disrupts the gradient within the zone. Under extreme conditions the concentrations of gradient medium within the zone will equal those of the supporting gradient adjacent to the leading edge. There remains, however, a residual density from the particles which, in the absence of any contribution from the supporting molecules, produces a zonal density which is greater than the density of the solution below the zone. In other words the leading edge of the zone tumbles towards the bottom until a stable condition (Equation 13) is obtained. Convective broadening caused by density instabilities has been the subject of several investigations and these have shown that the effect is proportional to the ratio of the diffusion coefficients of the particles and small molecules raised to a power greater than two (7, 8). Examples of this phenomenon are described in detail in Section 2.1.3 of Chapter 3.

This unwanted broadening of bands reduces the effectiveness of the separation and can readily be detected by estimating sedimentation coefficients for zones in a series of experiments where only the initial zonal concentrations of the particles are varied. As described in Section 2.1.3 of Chapter 3 and Section 9.3 of Chapter 5, overloading the gradients will then result in skewing of the peaks and a marked increase in the observed sedimentation coefficients with increasing concentrations, the opposite to that expected from hydrodynamic concentration phenomena. This is one good reason why it is good practice to estimate sedimentation coefficients from rate-zonal experiments when optimizing the centrifugal variables prior to routine runs.

5.3 Isopycnic separations

5.3.1 Basic equations for isopycnic separations

Whereas rate-zonal separations will often yield unstable zones, a similar problem does not occur in isopycnic runs. This is because it is a true equilibrium procedure where the concentrations of all the molecules will be determined by a balance of the external centrifugal field with diffusion (*Figure 4*). The problem is that it takes an infinite time to reach true equilibrium! The implication of this is that all 'equilibrium runs' use conditions which only approach equilibrium. However, the deviation from true equilibrium is very small and unimportant for the interpretation of the experiment. These practical limits do not reduce the usefulness of the theoretical considerations of isopycnic sedimentation.

The shape of a zone at equilibrium for a homogeneous zone resembles a Gaussian shape, expressed as:

$$C_{(r)} = C_{(r,\Theta)} \exp[-\omega^2 r_\Theta M\bar{v}(\mathrm{d}\rho_\mathrm{m}/\mathrm{d}r)_r (r - r_\Theta)^2/2RT] \qquad [14a]$$

Here r_Θ is the radial coordinate of the isopycnic point of the zone which corresponds to maximum concentration (C_r, Θ) of the particles and C_r is the concentration at radius r. By analogy with the Gaussian equation, the width at two-thirds height is equivalent to the variance (σ^2) given in Equation 14b.

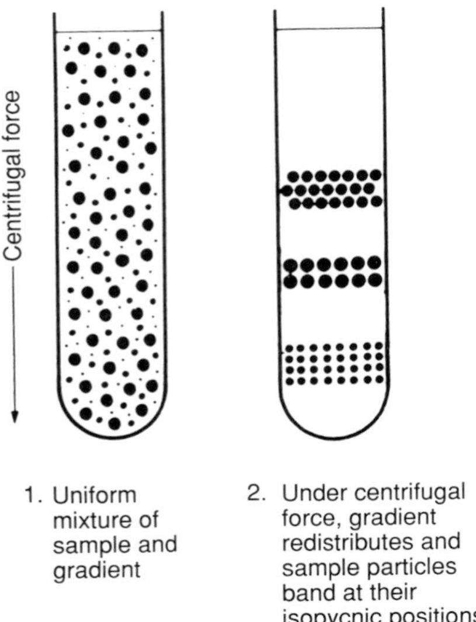

Figure 4. Diagrammatic representation of the fractionation of particles by isopycnic centrifugation in a self-forming gradient. A similar technique can also be used with preformed gradients. (Derived from reference 19.)

$$\sigma^2 = RT/\omega^2 r_\Theta M\bar{v}(\mathrm{d}\rho_\mathrm{m}/\mathrm{d}r)_r \qquad [14b]$$

Examination of Equation 14b shows that the width of the zone decreases as the speed of the rotor and the slope of the gradient increase. The property of particles which determines the radial separation of the peaks is their solvated densities (expressed as partial specific volumes, $\bar{v}$) while the width of the zones determines the effectiveness of the separation. The width of the zones is determined by the diffusion of the particle which in turn is dependent on the total amount of material within a zone. The amount can be calculated by integrating the concentrations divided by volumes over the whole zone and this shows the mass is proportional to the linear width, σ (see Section 1.1 of Appendix A) of this chapter. The volume of the band is proportional to the area of the plane of a zone and so, as the area increases, so does the capacity of the gradient. Doubling the speed and/or gradient slope will allow four times the quantity to be contained for the same width of the zone. These considerations suggest that ideal runs should be made at maximum rotor speeds and using steep density gradients. There are, however, counter factors which offset this idea and possibly the most important are the limits set by the solubilities of the particles and the density-gradient medium. Increased speeds lead to high concentrations of the gradient medium at the bottom of the tube making it essential to check that these concentrations do not exceed the solubility of the molecules, otherwise they will precipitate and possibly produce forces which exceed the design specifications of the rotor.

5.3.2 Time to approach equilibrium

Practical considerations mean that experimental times need to be limited and it would be useful if some guide could be given for calculating reasonable experimental times. Many formulae have been proposed to calculate quasi-equilibrium times, all derived from true equilibrium conditions (see Section 5.7.1 of Chapter 3 and Section 4.2.3 of Chapter 4). Their use suggests that the centrifugation times required for an acceptable separation are longer than the times determined empirically. An example of one of these equations (9), re-derived elsewhere (10), is:

$$(r_{\Theta}-r)/r=[(r_{\Theta}-r_{\text{men}})/r]\exp[\omega^2 r(\mathrm{d}s/\mathrm{d}r)t] \qquad [15]$$

where the calculated time is that required for any solute at r to sediment to r_{Θ}, the true equilibrium coordinate. Theoretically the left-hand ratio should be equal to zero but this is numerically unusable so it is suggested that it is set to 0.001. The rate of change of s with radius ($\mathrm{d}s/\mathrm{d}r$) can be calculated by replacing it by $(\mathrm{d}s/\mathrm{d}\rho_{\mathrm{T}})(\mathrm{d}\rho_{\mathrm{T}}/\mathrm{d}r)$, ignoring the effect of changing viscosities (as in the case of CsCl gradients). The term $\mathrm{d}s/\mathrm{d}\rho_{\mathrm{T}}$ is equal to $-K\bar{v}$ where K is equal to MD/RT and $\mathrm{d}\rho_{\mathrm{T}}/\mathrm{d}r$ is the linear total density gradient. Using Equation 15 and inserting reasonable parameters for biological macromolecules and gradients gives times in the order of tens of hours for biological particles. If the equation is used to calculate equilibrium times of gradient media like CsCl then it will take even longer to approach equilibrium, possibly more than a hundred hours. These long times occur because of the long gradient pathlengths used in many preparative gradients. Vertical rotors have been introduced and these have shorter pathlengths. In these tubes sedimentation occurs across the diameter of the tube, and the tube length does not affect the time required to reach equilibrium but only provides reasonable gradient volumes for supporting the total quantities of the sample particles.

Calculating useful experimental times for quasi-equilibrium isopycnic runs can only be done by simulating sedimentation of the components. This can be done if the simulation calculates the time required to generate a density profile which embraces the density of the particle. In this case it is not necessary for the gradient to reach quasi-equilibrium and the changes of density with time can be calculated by using analytical solutions to the Lamm equation (10). In order to calculate experimental sedimentation and diffusion coefficients of particles at increasing radial coordinates it is necessary to know the densities and viscosities of the solvent at the chosen radial coordinates. These can only be calculated if the concentrations of the gradient medium are known. Hence, these must be calculated first for a discrete set of radial coordinates and at fixed time intervals using the sedimentation formula. Using these values the sedimentation and diffusion coefficients of the particle can be corrected and the concentrations of particles calculated for the same coordinates and times using the same formula but different coefficients. This iteration can be repeated with increasing times until there is negligible change in the position of the particle's zone.

At the beginning it is not necessary to calculate the concentrations of the particles and gradient solutes for all the radial positions at each time increment, and only the density (hence, concentration) of the solute for the bottom region is required. Only when this density is greater than that of the particle is it necessary to calculate concentrations and densities of the solutes for all the radial positions. Calculations can be done using a formula derived by Yphantis and Waugh (11). This equation applies to a cylindrical tube in which the gradient material and particles are uniformly distributed at the beginning of the run. Using this procedure the time for an ideal particle to reach its isopycnic position was 22 h using the following conditions:

- density of particle = 1.3 g/ml
- radius of meniscus = 3.00 cm
- column length = 4.00 cm
- temperature = 20 °C
- rotor speed = 30 000 r.p.m.
- particle molecular mass = 1×10^6 daltons
- gradient material = CsCl
- initial CsCl concn. = 30% (w/v)

The isopycnic position of the particle is at 6.66 cm from the centre of rotation and the width of the zone is about 0.6 cm. The assumptions made in this calculation are that CsCl is the only solute responsible for the densities and ignores any hydrodynamic and thermodynamic non-ideality of the coefficients. This means that the final positions of particles may not correspond exactly with that found experimentally but the procedure does yield useful information for determining the optimum centrifugation times. The source code for this program is given in Appendix B of this chapter.

One feature of this type of simulation is that it is easy to determine if the initial concentration of gradient material was sufficient to form a density gradient within a reasonable time and if it includes the isopycnic density of the particle. If the speed and initial concentrations of gradient material were too low then error messages arising from numerical instabilities (producing 'overflow') will be produced. If the speed is too high the concentrations of the gradient material near the bottom will exceed their solubilities and this is noted. The source code for an alternative program (20) for predicting the shape of gradients at true equilibrium is given in Appendix C of this chapter.

5.3.3 Separation and resolution of samples

Separation and resolution of zones in an isopycnic experiment are interconnected properties. Separation is defined as the distance between the peaks of two zones and is equal to:

$$(r_\Theta^{(2)} - r_\Theta^{(1)}) = [RT/(\mathrm{d}\rho/\mathrm{d}r)\,\omega^2]\,[(M_{(2)}\bar{v}_{(2)}\sigma_{(2)}{}^2)^{-1} - (M_{(1)}\bar{v}_{(1)}\sigma_{(1)}{}^2)^{-1}] \qquad [16]$$

Equation 16 shows that the radial separation is proportional to the inverse of the product of two experimental variables ($\mathrm{d}\rho/\mathrm{d}r$) and ω^2 as well as with the difference between the products of the particulate masses and partial specific volumes. Thus, decreasing one or both of the experimental variables increases the separation for a given pair of particles.

Resolution is a more complex problem since it is defined through zonal overlap and this is defined as $|\sigma_{(1)} + \sigma_{(2)}|$. This difference is proportional to the reciprocal of $\omega(\mathrm{d}\rho/\mathrm{d}r)^{1/2}$. In other words, changing the centrifugal parameters affects resolution less than the separation of the peaks. Finding optimum conditions which give maximum resolution and separation as well as including allowances for the maximum capacity of particles contained in the zones is not easy (the width of an isopycnic zone increases as more material accumulates). Decreasing the width by increasing the speed (Equation 14b) influences the quantity of material contained in the zone as well as reducing the separation. These factors have to be considered when designing individual isopycnic experiments.

6. Calculation of rotor parameters

The efficiency of all centrifuge rotors is traditionally expressed in terms of k-factors (k, k', or k^*). Using water as centrifuge medium the time in hours (t) required to pellet a particle is given by:

$$t = k/s \qquad [17]$$

where s is the sedimentation coefficient of the particle in Svedberg units. The simple k-factor is based on the following logic. From the definition of the sedimentation coefficient it can be seen that:

$$s\omega^2 t = k(\ln r_{max} - \ln r_{min}) \qquad [18]$$

Hence, k, with t expressed in hours and the radii in centimetres, will be:

$$k = \frac{\ln r_{max} - \ln r_{min}}{(2\pi Q/60)^2\, 3600 \times 10^{-13}} \qquad [19]$$

or

$$k = \frac{2.53 \times 10^{11}\,(\ln r_{max} - \ln r_{min})}{Q^2} \qquad [20]$$

k-factors are usually calculated on the basis of the shape of the rotor pocket (neglecting wall thickness of the tubes) and from the maximum speed of the rotor. Time needed for acceleration and deceleration of the rotor is also ignored in this calculation. Most rotors accordingly perform slightly better than indicated by their k-factor. Everything else being equal, it is best to choose a rotor with the lowest k-factor since this will give the fastest separation.

Manufacturers always give the k-factors of rotors for the maximum speed of the rotor and this is normally given in the instruction manual. However, if derating is required because the solutions have a density higher than the design density of the rotor (usually 1.2 g/ml), or if the centrifuge available is not able to run the rotor at maximum speed, then a new k-factor (k_n) can be calculated from the original (k_o), the original and the current maximum speeds (Q_o and Q_n), respectively as follows:

$$k_n / k_o = (Q_o / Q_n)^2 \qquad [21]$$

The simple k-factor reflects the sedimentation processes in a homogeneous centrifugation medium with the density and viscosity of water. In the case of sucrose gradient centrifugation separations the increasing density and viscosity of the centrifugation medium act as a brake on the sedimentation processes. Several related constants have been devised to characterize sedimentation of particles through gradients. Constants designated k', k^*, or k^* (1.3) seek to quantitate the sedimentation processes in a 5–20% (w/w) sucrose gradient at 5 °C for particles with a density 1.3 g/ml, a reasonable value for proteins and DNA in sucrose solutions, whereas k^* (1.7) covers particles like RNA with a particle density of 1.7 g/ml in sucrose. The rate of sedimentation of particles through a 5–20% sucrose gradient is about half as fast as through water, that is the k' or k^*-factor is about twice as large as the k-factor.

The values of k^*-factors are more complicated to calculate, though the logic is essentially the same. In sucrose gradient centrifugation sedimentation coefficients are evaluated by numerical integration of the following expression:

$$s_{20,w}\,\omega^2 t = \frac{\rho_p - \rho_{20,w}}{\eta_{20,w}} \int_{r_{min}}^{r_{max}} \frac{\eta_{T,m}}{\rho_p - \rho_{T,m}} \times \frac{dr}{r} \qquad [22]$$

Then k', k^* (1.3), and k^* (1.7) factors can be obtained by dividing the right-hand side of Equation 22 by ω^2. The key problem then is to divide the path through the centrifuge tube into a number of segments. Density and viscosity in each segment are calculated by use of Barber's polynomials (see Section 5.3 of Chapter 5) for each segment, and a summation procedure is completed when passing through the segments. The complete program is given in Appendix D of Chapter 5. In this program the tube of the rotor is divided into 100 segments as an example, and sucrose concentrations are evaluated for the middle of each segment. While these factors have been devised primarily for sedimentation in 5–20% sucrose gradients they can be used for estimating sedimentation of the sample under other conditions. As an example, many separations are carried out in 0.25 M sucrose solution which is isotonic; 0.25 M sucrose is 8.55% and so to predict sedimentation in this solution it is more accurate to use the k^*-factor rather than the k-factor. By careful application of k-factors and k^*-factors it is possible to get a good estimate of the time required to pellet particles within any sample. The source code for a program which calculates k-factors is given in Appendix D of this chapter.

It is also possible to use k-factors to calculate centrifugation conditions when changing from one type of rotor to another. The relationship between the k-factor of a rotor and centrifugation time required is very simple as follows:

$$t_1 k_1 = t_2 k_2 \qquad [23]$$

Hence, if it is necessary to centrifuge a sample in a different type of rotor or at a different speed then the new centrifugation time required for pelleting particles can be calculated from this equation.

7. Simulation of centrifugation experiments

Probably the most frequently asked question when developing a new application is 'what centrifugal conditions are likely to give me the best separation?' Hence, before embarking on a series of experiments, it is useful to know how changing the experimental conditions is likely to affect the outcome of the centrifugal separations. A simple way of doing this is to use the k-factor or k^*-factor to estimate likely optimal conditions for the separation as described in the previous section. However, a more accurate approach is to simulate the sedimentation process. This involves calculating the positions and widths of the sample zones for chosen conditions using the Lamm equation to formulate the computations. For differential boundaries and macromolecules the integral solutions to the Lamm equation (3) often suffices, while for isopycnic zonal experiments Equation 14a can provide useful guides to the width of zones at equilibrium.

It is more difficult to simulate zonal velocity experiments because the zones move along gradients of density and viscosity which themselves can change with time. An approximate solution can be obtained by assuming a constant gradient and calculating radial positions of zones from the reverse procedure to that used to correct sedimentation coefficients (see Chapter 5). This only gives values for the positions of the sample peaks and does not yield information about resolution of zones because it does not give any guide to the widths of zones at different times. To calculate positions and shapes of zones means using numerical methods to integrate the Lamm equation. Despite the difficulties associated with numerical integration of partial differential equations, Cox (12) successfully used the explicit method to integrate the Lamm equation numerically. Programs that carry out the simulation of zonal experiments have been devised (13) as well as descriptions of the influences that changing intervals have on the results (14).

A second method can be employed which is more stable and faster to execute than the explicit method, and this is called the implicit method. It depends on the use of sets of simultaneous equations which contain projections of concentrations to the next time increment. This gives for coordinate points simultaneous equations containing the known and unknown projected variables; this is called the Crank–Nicholson method. There are many variations on this method and the procedures are well described in most books on numerical methods, and since the equations are specific to the problem it is not described here. The advantage with this method is it is stable and convergent for all finite values of the time increments Δt whereas the explicit method will only work for very small values of Δt. This procedure has been used successfully for integrating the Lamm equation over a wide range of experimental conditions (15).

References

1. Svedberg, T. and Pederson, K. (1940). *The ultracentrifuge*. Oxford University Press, Oxford.
2. Lamm, O. (1929). *Z. physik. Chem. (Leipzig)* **A143**, 177.
3. Fujita, H. (1962). *Mathematical theory of sedimentation analysis*. Academic Press, New York.
4. Birnie, G. D. (1978). In *Centrifugal separations in molecular and cell biology* (ed. G. D. Birnie and D. Rickwood), p. 169. Butterworths, London.
5. Young, B. D. and Rickwood, D. (1981). *J. Biochem. Biophys. Methods* **5**, 95.
6. Vinograd, J. and Bruner, R. (1966). *Biopolymers* **4**, 131.
7. Sartory, W. K. (1969). *Biopolymers* **7**, 251.
8. Sartory, W. K., Halsall, H. B., and Breillat, J. P. (1976). *Biophys. Chem.* **5**, 107.
9. Baldwin, R. L. and Shooter, E. M. (1963). In *Ultracentrifugal analysis* (ed. J. T. Williams), p. 143. Academic Press, New York.
10. Spragg, S. P. (1978). In *Centrifugal separations in molecular and cell biology* (ed. G. D. Birnie and D. Rickwood), p. 7. Butterworths, London.
11. Yphantis, D. A. and Waugh, D. F. (1956). *J. Phys. Chem.* **60**, 623.
12. Cox, D. J. (1971). *Arch. Biochem. Biophys.* **146**, 181.
13. Steensgaard, J., Funding, L., and Meuwissen, A. T. P. (1974). In *Methodological developments in biochemistry* (ed. E. Reid), Vol. 4, p. 67. Longmans, London.

14. Steensgaard, J. and Funding, L. (1974). In *Methodological developments in biochemistry* (ed. E. Reid), Vol. 4, p. 55. Longmans, London.
15. Dishon, M., Weiss, G. H., and Yphantis, D. A. (1966). *Biopolymers* **1**, 449.
16. Barber, E. J. (1966). In *The development of zonal centrifuges and ancillary systems for tissue fractionation and analysis* (ed. N. G. Anderson), p. 219. NCI Monograph 21, National Cancer Institute, Bethesda, USA.
17. *Handbook of chemistry and physics*, 53rd edn. CRC Press, Ohio.
18. Marcum, J. M. and Borisy, G. G. (1978). *J. Biol. Chem.* **253**, 2852.
19. Griffith, O. M. (1979). In *Ultracentrifuge rotors: a guide to their selection*. Beckman Instruments Inc, Palo Alto, CA, USA.
20. Steensgaard, J. and Rickwood, D. (1985). In *Microcomputers in biology: a practical approach* (ed. C. Ireland and S. P. Long), p. 241. IRL Press Ltd, Oxford.

Appendix A: Derivations of equations

1.1 Shape of equilibrated zones

At physical and chemical equilibrium in a centrifuge Equation 2d can be written as:

$$(\mathrm{d}C/\mathrm{d}r) \times 1/C = [\omega^2 r M(1 - \bar{v}_\mathrm{m}\rho_\mathrm{s})]/RT \qquad [24]$$

Assuming a linear and constant density gradient across the zone then the density at any position is given as:

$$\rho_{(\mathrm{r})} = (1/\bar{v}_\mathrm{m}) + (r - r_\theta)(\mathrm{d}\rho/\mathrm{d}r) \qquad [25]$$

and when substituted in Equation 24 gives:

$$(\mathrm{d}C/\mathrm{d}r) \times 1/C = -\omega^2 r_\theta M\bar{v}(r - r_\theta)(\mathrm{d}\rho/\mathrm{d}r)/RT \qquad [26]$$

For small deviations from equilibrium when r equals r_θ and $(r - r_\theta)\mathrm{d}r$ is equal to $0.5(r - r_\theta)^2$, then Equation 26 becomes:

$$\mathrm{dln}C = -\omega^2 r_\theta M\bar{v}(\mathrm{d}\rho/\mathrm{d}r)_{r=\theta}\, \mathrm{d}(r - r_\theta)^2/2RT \qquad [27]$$

which when integrated gives:

$$C(r) = C(r_\theta)\exp[-\omega^2 r_\theta M\bar{v}(\mathrm{d}\rho/\mathrm{d}r)_{r=\theta}(r - r_\theta)^2/2RT] \qquad [28]$$

Comparing Equation 28 with the probability integral shows that an equilibrated zone assumes a Gaussian shape with two equal half-width at two-thirds height,

$$\sigma_\mathrm{eq} = (RT/[\omega^2 r_\theta M\bar{v}(\mathrm{d}\rho/\mathrm{d}r)_{r=\theta}]^{1/2} \qquad [29]$$

1.2 Total quantity of particles in a zone

The total quantity of particles in a zone, m, can be calculated by:

$$m = A \int_{r=r_{\text{men}}}^{r=r_b} C_{\text{r}} r^n \, \text{d}r \qquad [30]$$

where A is the area (constant for a cylindrical tube) and n is equal to 1 for a tube and 2 for a sector-shaped vessel. Substituting in Equation 24 gives:

$$m = A\,(RT/\bar{v}M\omega^2) \int_{r_{\text{men}}}^{r_b} [r/(1-\bar{v}\rho_{(\text{r})})]\,\text{d}C \qquad [31]$$

and replacing

$$RT/(M\omega^2\bar{v}) \text{ by } \sigma^2 r_\theta (\text{d}\rho/\text{d}r)$$

from Equation 29 gives:

$$m = A\,\sigma^2 r_\theta \int_{r_{\text{men}}}^{r_b} \text{d}C \qquad [32]$$

Here the integral approximates to C and the total mass is given by:

$$m = A\,\sigma\, r_\theta\, C \qquad [33]$$

showing how the quantity in a zone determines its width.

1.3 Estimating radii from volumes sampled from vertical rotor tubes

The geometrical arrangement for a solution in the rotating tube is given in *Figure 1*. From this it is possible to formulate a relationship between operational radii and volumes of the tube taken after the solution has reorientated. The planes which define the limits of these volumes form chords across the tube and the edges of these chords subtend an angle θ (radians) at the centre. The general expression for any radius will be given assuming the radius of the cylindrical part of the tube equals that of the hemispherical parts. While tubes in vertical rotors are almost always run full, the mathematical treatment here does include part-filled tubes.

It is assumed that the volume of solution, V_{s}, in the tube is equal to the total volume of the tube, V_{T}, and so can be calculated from the manufacturer's specification which are known

$$V_{\text{T}} = \pi r^2 L + (2\pi r^3)/3$$

where r is the radius of the cylindrical and hemispherical parts of the tube having a length L. Hence, the volume of air above the solution is V_{T} less the volume of solution V_S. The area of the segment of air within the cylindrical part of the tube is equal to:

$$(1/2)\,r^2\,(\theta - \sin\theta)$$

and this must be added to the area of the segment that extends into the hemispherical top and bottom to give the total volume of air. The total volume of these two segments

$$= \pi\,(3r - h)\,h^2/3$$

(h is the half width of the chord). This width of the chord across the tube equals $2h$ giving for half width, equal to $r\sin(\theta/2)$. Thus, the volume of the segment within the rounded top and bottom is given by substituting for h to give Equation 34:

$$= \pi\;[3\,r - r\sin(\theta/2)]\;[\mathrm{r}\sin(\theta/2)]^2/3 \qquad [34]$$

The volume of the air equals the sum of the volume of the segment in the cylindrical part (area $\times\,L$) plus that in the hemispherical top and bottom of the tube giving Equation 35:

$$\text{Vol. of air} = (1/2)\,r^2\,L\,[\theta - \sin(\theta)] + (1/3)\,\pi\,[3\,r - r\sin(\theta/2)]\;[r\sin(\theta/2)]^2 \qquad [35]$$

But the volume of air remaining after sampling is given by

$$V_{\mathrm{air}} = \pi\,r^2\,L + (2/3)\,\pi\,r^3 - V_{\mathrm{s}}$$

Equating this with Equation 35 and rearranging gives the relationship $f(x)$ between the volume of the solution added to the tube and the unknown angle:

$$f(x) = (1/2)\,r^2\,L\,\theta - (1/2)\,r^2\,L\sin(\theta) + r^2\sin^2(\theta/2)\,\pi r/2 - (1/6)\,\pi\,r^3\sin^3(\theta/2) - V_{\mathrm{s}} - [\pi r^2\,L + (2/3)\,\pi\,r^3] \qquad [36]$$

If the tube was completely full and had no meniscus then the part of Equation 36 in square brackets is omitted. When programming the equation, V_{s} starts at zero and increases as radii equivalent to successive samples are calculated. The only unknown in Equation 36 is the angle and this must be found before the radius can be calculated. From the geometrical arrangement (*Figure 1*) it is apparent the radius of the sample (r_s) is:

$$r_s = R + (r - Z)$$

where

$$Z = r\cos(\theta/2)$$

and R is the radial distance of the centripetal edge of the tube from the centre of the rotor. Combining these two expressions shows that the radius becomes:

$$R + r\,[1 - K\cos(\theta/2)]$$

The cosine is subtracted from 1 when collecting samples from the top ($K = 1$) and added when sampling from the bottom ($K = -1$) of the stationary tube.

The functions $f(x)$ and Z are used to calculate radii from all volumes. In this general case V_s (Equation 36) is replaced by the volume of the air (when sampling from the top but not when collecting from the bottom) plus the total volume collected up to the ith sample. If all the samples are of equal volume then V_s is replaced for top—sampling by $V_{air} + nV_i$ where n is the sample number in the series $n = 1,2,\ldots,i,\ldots, \leqslant V_s$.

The calculation hinges on estimating a value for the angle using Equation 36. This cubic equation can be solved numerically using the well known Newton–Raphson procedure. This is an iterative calculation in which values for θ are continually updated by a correction term in the manner shown in Equation 37.

$$\theta_{(j)} = \theta_{(j-1)} - f(x)/f'(x) \qquad [37a]$$

$$f'(x) = (1/2)\,r^2 L - (1/2)\,r^2 L\cos(\theta) +$$
$$r^3 \pi \sin(\theta/2)\cos(\theta/2) - r^3 \sin^2(\theta/2)\cos(\theta/2)/2 \qquad [37b]$$

The subscripts refer to the values for the jth and $(j-1)$th cycles of the iteration while $f'(x)$ is the derivative of $f(x)$ (Equation 36) and given in Equation 37b. The iteration is continued until the ratio is less than a fixed value (say 10^{-6}) when (j) is the estimated angle. Only then can Z be calculated before completing the calculation of the radius.

1.4 Density and viscosity equations

Calculating viscosity and density from given concentrations of supporting materials implies they can be related either from molecular properties or empirically from polynomial expressions. A set of coefficients for orthogonal polynomials relating densities and viscosities of sucrose to its concentration

Table 2 Coefficients for calculating relative density and relative viscosity (η/η_0) at 20 °C for concentration given in % w/w [g(anhydrous solute)/g(solution)]. Calculated from data (17) and for Equation 38a or Equation 38b. Maximum errors are given for the ranges

Substance	A_1	A_2	A_3	Error
(a) Relative density				
CsCl	1.009985	5.723211×10^{-3}	1.202414×10^{-4}	6%
Glycerol	0.9988583	2.449206×10^{-3}	2.317415×10^{-6}	1%
KI	1.001713	6.768155×10^{-3}	7.752044×10^{-5}	2%
KBr	1.001379	6.681938×10^{-3}	6.702068×10^{-5}	2%
NaBr	1.001567	7.222810×10^{-3}	7.720270×10^{-5}	1%
(b) Relative viscosity				
CsCl	1.021052	-8.770035×10^{-3}	1.711131×10^{-4}	3%
Glycerol (Equation 38a)				
0–40%	1.120447	-1.035501×10^{-2}	1.964791×10^{-3}	6%
(Equation 38b)				
40–90%	5.263194	-1.530875×10^{-1}	1.732818×10^{-3}	6%
KI	1.002583	-6.767929×10^{-3}	1.008861×10^{-4}	4%
KBr	1.005648	-6.117955×10^{-3}	1.483215×10^{-4}	4%
NaBr	1.001567	-4.991555×10^{-3}	6.062503×10^{-4}	5%

at a range of temperatures has been compiled by Barber (16) and are repeated in Chapter 5. Similar coefficients have not been reported for other materials and a curvilinear set has been compiled for glycerol, CsCl, KI, KBr, NaBr but only for data at 20 °C (17). The coefficients given in *Table 2* were estimated for Equation 38a using a regression procedure.

The densities and viscosities (equivalent to y, Equation 38a) can be calculated for concentrations up to 64% w/w for CsCl, 40% w/w for KI, KBr, NaBr, and 90% w/w for the density of glycerol. The viscosity of solutions of glycerol give a more complicated relationship over this range of concentrations than can be described by a simple quadratic equation. For concentrations 0–40% w/w the quadratic equation is adequate (Equation 38a) but for concentrations 40–90% w/w an exponential relationship is required (Equation 38b).

$$y = A_1 + A_2 C + A_3 C^2 \quad [38a]$$

$$y = \exp(A_1 + A_2 C + A_3 C^2) \quad [38b]$$

The concentrations for these calculations must be in g(anhydrous solute)/g(solution), % w/w (w), and in order to convert to the more usual units of concentration [g(anhydrous solute)/litre solution, q] it is necessary to know the partial specific volume of the solute ($\bar{v}$, expressed in cgs units). The conversion can be made using Equation 39a, while the inverse of calculating q from w requires Equation 39b.

$$w = q\,100/[1000 + q\,(1 - \bar{v})] \qquad [39a]$$

$$q = 1000\,w/[100 - w\,(1 - \bar{v})] \qquad [39b]$$

Unlike those for sucrose, the coefficients only apply to one temperature and because they are empirical they will not give realistic values outside the measured range of concentrations.

Appendix B. Calculation of isopycnic centrifugation using the Yphantis–Waugh equation

```
10 REM Calc. C/CO for rectangular cell using
20 REM Yphantis-Waugh equation (J.Phys.Chem.60,623,1956)
30 PI = 3.14159
40 DEFDBL Z
50 DIM D20(2), S20(2), SOM(2), TAU(2)
60 DIM ZSUD(8), ZSUG(2), ZSUD1(7), ZSUD2(5), ZSUE1(7), ZSUE2(5)
70 DIM ZA(2), ZB1(2), ZB2(2), VBA(2), BMOL(2)
80 DIM RPAR(100), CONMOL(100), CONMAT(100), DENS(100)
90 AVNO = 6.0222E+23: RG = 8.31434
100 IFLAG% = 0: CLS : ATST = .02286: MATST = -.02286: DTST = .01
110 COLOR 7, 0: PRINT "C/CO versus radius at incrementing times"
120 PRINT "for supporting medium (e.g.Sucrose or CsCl)"
130 PRINT " using Yphantis -Waugh equation (J.Phys.Chem.60,623,1956)"
140 INPUT "Radius of meniscus (cms.)"; RME
150 INPUT "Length of column (cms.)"; ALEN
155 INPUT "Speed of rotor(RPM)"; RPM
160 INPUT "Temp. of run (C)"; TEM
170 INPUT "Initial, uniform conc.gradient material (g/l)"; COINI
180 INPUT "Mol. wt. of macromol. (Daltons)"; AMM: BMOL(2) = AMM * .001
190 INPUT "Density of macromol. (g/cm^3)"; RHOMM: VBA(2) = .001 / (RHOMM)
200 INPUT "Is this mol. filamentous (Y/N)"; ANS$
210 IF ANS$ = "Y" THEN DMUL =1/15 ELSE DMUL = 1
220 GOSUB 1720: VBA(1) = VBAR * .001
230 BMOL(1) = AMOL * .001: SGTES = RHOMM - (.04 * RHOMM): TTES = 1000000!
240 RME = RME * .01: ALEN = ALEN * .01
250 COINI = COINI * 100 / (1000 + (COINI * (1 - VBAR)))
260 TEM = TEM + 273.12: VBAR = VBA(2): YPTE = 85: YPTE1 = -YPTE: RSTA = 0
270 PRINT "it is possible to find a (possibly)";
275 print "hypothetical density which forces"
280 PRINT "the conc. of macromol. to appraoch zero at the bottom. Please enter"
290 INPUT "the rel. bottom-conc. you wish to reach (>=.1,<=1)"; BOTCO
300 PRINT "Assumes water is the solvent & solutes are thermodynamic. ideal"
310 VISW = .001: RHO = 1000: CONC = COINI: GOSUB 2460
320 GOSUB 2840: STARS = SS20
330 REM Calc equiv Diff.coef. of particle :spherical
340 FE2 = 3 * BMOL(2) * VBA(2) / (4 * PI * AVNO)
350 FE2 = LOG(FE2) / 3: FE2 = EXP(FE2): FE2 = 6 * PI * AVNO * VISC * FE2
360 FE2 = FE2 * DMUL: D20(2) = RG * TEM / FE2
370 PRINT " Experimental diffusion coef. of mat.(M^2/sec)=";
```

```
380 PRINT USING "##.####^^^^"; D20(1)
390 PRINT "Spherical dif. coef. macromol (m^2/sec)=";
400 PRINT USING "##.####^^^^"; D20(2)
410 REM Calc. S.
420 S20(1) = BMOL(1) * (1 - (VBA(1) * RHO)) * D20(1) / (RG * TEM)
430  PRINT "Equiv. Sed. Coef.mat. (secs^-1)=";
440 PRINT USING "##.####^^^^"; S20(1)
450 S20(2) = BMOL(2) * (1 - (VBA(2) * RHO)) * D20(2) / (RG * TEM)
460 PRINT "Equiv. Sed. Coef.(W,T).of macromol.=";
470 PRINT USING "##.####^^^^"; S20(2)
480 REM calc. constants for equation
490 RBOT = RME + ALEN: RDIF = ALEN
500 RBAR = RME * RME + RBOT * RME + RBOT * RBOT
510 RBAR = RBAR / 3: RBAR = SQR(RBAR)
520 OM2 = PI * RPM / 30: OM2 = OM2 * OM2
530 EQUL = BMOL(1) * (1 - (VBA(1) * RHO)) * OM2 / (2 * RG * TEM)
540 R2 = (RBOT * RBOT - RBAR * RBAR)
550 R1 = (RME * RME - RBAR * RBAR): CEM = EQUL * R1: CEM = EXP(CEM)
560 CEB = EQUL * R2: CEB = EXP(CEB)
570 CONC = CEB * COINI: GOSUB 2460: GOSUB 2840
580 R2R1 = LOG(RBAR / RME)
590 TSEC = R2R1 / (OM2 * S20(2) * STARS): TSEC = ABS(TSEC)
600 IF TSEC > 10800! THEN TSEC = 10800!: GOTO 610 ELSE 610
610 IF SG < RHOMM THEN 620 ELSE 640
620 PRINT "Bottom SG at inf. time="; SG; " Cannot conc. macro."
630 'STOP
640 PSG = RHOMM - (.01 * RHOMM): PSG2 = RHOMM + (.03 * RHOMM): ETST = 1000
650 LOCATE 22, 10, 0: MREP% = 0
660 COLOR 0, 4: PRINT "WAIT:Computing time to pseudo-equilibrium"
670 TPR% = 0: MES% = 10: LCT% = 0: SGTES2 = RHOMM + (.04 * RHOMM)
680 SOM(1) = S20(1) * OM2 * RBAR: SOM(2) = S20(2) * RBAR * OM2
690 REM TAU must be *TIME before use
700 TAU(1) = 2 * S20(1) * OM2: TAU(2) = 2 * S20(2) * OM2
710 RINC = RME * .02: TINC = TSEC * .1: PAS% = 0
720 TIME = TSEC: CTES = .25: RINC2 = RINC * .1
730 LOOPY% = 1: R = RBOT - (2 * RINC): DD20 = 1: VBAR = VBA(1): SS20 = 1
740 GAM = RBAR / (2 * RDIF): GOSUB 2950
750 CONC = CCO * COINI: GOSUB 2460
760 IF SG < SGTES2 THEN 770 ELSE 790
770 LOOPY% = 1: SS20 = 1: DD20 = 1: TIME = TIME + TINC
780 IF TIME < TTES THEN GOSUB 2950: GOTO 750 ELSE 1360
790 LOOPY% = 2: VBAR = VBA(2): GOSUB 2840: TPR% = 0: GOSUB 2950
800 IF ABS(CCO) < BOTCO THEN RATC = CCO: GOTO 810 ELSE 770
810 PRINT "Relative conc.macromol @ "; R; " m. approx. "; CCO;
820 PRINT " at (hrs)"; TIME / 3600
830 INPUT "Is this satisfactory?(Y/N)"; ANS$
840 IF ANS$ = "Y" THEN
850      CLS
860      COLOR 7, 0
870      R = RME
880      GOTO 930
890 ELSE
900      SGTES2 = SGTES2 + .001
910      GOTO 770
920 END IF
930 LOOPY% = 1: SS20 = 1: DD20 = 1: GOSUB 2950
940 CONC = CCO * COINI: GOSUB 2460: GOSUB 2840: LOOPY% = 2: GOSUB 2950
950 IF ABS(CCO) < .001 THEN 980 ELSE 960
960 IF RSTA = 0 THEN RSTA = R * 100: RINC = RINC * .4: GOTO 970 ELSE 970
970 ISC% = ISC% + 1: CONMOL(ISC%) = CCO: CONMAT(ISC%) = CONC
971 DENS(ISC%) = SG: RPAR(ISCC%) = R * 100
980 R = R + RINC
990 IF R >= RBOT THEN 1000 ELSE 930
1000 CLS : MPR% = 0: PRINT TAB(3); "R"; TAB(12); "Grad.Conc.(wt%)";
```

```
1010 PRINT TAB(29); "Macro C/C(0)"; TAB(44); "Sp.Gr."
1020 IBGN% = 1: II% = 1
1030 IF RPAR(II%) < RME * 100 THEN 1040 ELSE 1070
1040 IBGN% = IBGN% + 1: II% = II% + 1: GOTO 1050
1050 IF RPAR(II%) < RME * 100 THEN 1040 ELSE 1060
1060 IF IBGN% > ISC% THEN IBGN% = 1: GOTO 1070 ELSE 1070
1070 INO% = ISC% - IBGN% + 1
1080 IF INO% > 15 THEN
1090     IST% = CINT(ISC% / 15)
1100     GOTO 1150
1110 ELSE
1120     IST% = 1
1130     GOTO 1150
1140 END IF
1150 FOR II% = IBGN% TO ISC% STEP IST%
1160 PRINT USING "##.##"; RPAR(II%);
1170 PRINT TAB(14); : PRINT USING "##.###"; CONMAT(II%);
1180 PRINT TAB(29);
1190 PRINT USING "+##.###^^^^"; CONMOL(II%);
1200 PRINT TAB(43); : PRINT USING "##.###"; DENS(II%);
1210 IF CONMAT(II%) > SAT THEN
1220         PRINT TAB(55);
1230         PRINT "Mat. crystallises"
1240         GOTO 1280
1250    ELSE
1260         PRINT : GOTO 1280
1270 END IF
1280 NEXT II%
1290 PRINT "Compare with true equil. for gradient mat. "; GR$
1300 PRINT USING "Meniscus= #.###"; CEM * COINI;
1310 PRINT USING " ,Bottom   = #.###"; CEB * COINI
1320 PRINT "Centrifuge pseudo-equilibrium  time (hrs)"; TIME / 3600
1330 PRINT "C/C(0) for macromol. was 0 for radii below"; RSTA; " cm."
1340 INPUT "Satisfied? (Y/N)"; ANS$
1350 IF ANS$ = "N" THEN ISC% = 0: GOTO 1400 ELSE STOP
1360 PRINT "May never reach isopycnic;Time="; TIME / 3600; " Sp.Gr.="; SG
1370 SGTES2 = SGTES2 * .95: TTES = TTES * 1.2
1380 INPUT "Continue iterating?(Y/N)"; BANS$
1390 IF BANS$ = "Y" THEN 1400 ELSE COLOR 7, 0: STOP
1400 LOOPY% = 0: PAS% = 0: RINC = RME * .05
1410 R = RBOT - RINC: CLS : MREP% = 1: SGTES2 = 1.05 * SGTES2
1420 LOCATE 21, 10, 0: COLOR 0, 4: PRINT "WAIT;Repeating computation"
1430 GOTO 770
1440 REM Sub to calc. C/C(0) using Yphantis-Waugh equation
1450 REM J.Phys.Chem.,60,623,1956
1460 ESUM = 0!: M% = 1
1470 YR = (R - RME) / RDIF
1480 A = M% MOD 2: IF A = 0 THEN M1% = 1 ELSE M1% = -1
1490 ANUM = (1 - (M1% * EBRAK)) * M% * EMCON
1500 EMUL = M% * M% * ECON * TIME
1510 IF EMUL > 80 THEN EMUL = 80: GOTO 1530 ELSE 1520
1520 IF EMUL < -80 THEN EMUL = 0: GOTO 1540 ELSE 1530
1530 EMUL = EXP(-EMUL)
1540 ANUM = ANUM * EMUL: DENOM = 1 + (M% * M% * EINV)
1550 DENOM = DENOM * DENOM
1560 EMR = ANUM / DENOM
1570 BT = M% * PI * YR: ST = SIN(BT): CT = COS(BT) * 2 * PI * M% * ALPHA
1580 EMR = EMR * (ST + CT): ESUM = ESUM + EMR
1590 IF M% > 1 THEN 1600 ELSE 1610
1600 DIFS = ESUM - EMRO: IF ABS(DIFS) < 1E-10 THEN 1650 ELSE 1610
1610 EMRO = ESUM
1620 M% = M% + 1: IF M% > 100 THEN 1650 ELSE 1480
1630 GOTO 1480
1640 REM calc C/C(0)
```

```
1650 A1 = YR - (.5 * GAM * TAU1 * TIME): A3 = A1 / (2 * ALPHA)
1660 IF A3 > 88 THEN A3 = 88: GOTO 1670 ELSE 1670
1670 A4 = EXP(A3)
1680 A4 = A4 * ESUM
1690 YZ = YR / ALPHA: A2 = COMU * EXP(YZ)
1700 CCO = A2 + A4
1710 RETURN
1720 REM Sub to retrieve data for gradient material
1730 REM Sucrose coeffs.;Barber, NCI Mono.21,219,1966.
1740 PRINT " Coeficients for Visc. & Density for following:-"
1750 RESTORE: READ NAM$: PRINT NAM$
1760 INPUT "Enter name of gradient material(IN CAPS.)"; GR$
1770 RESTORE 2080: IP% = 1
1780 READ NAM$: IF NAM$ = GR$ THEN 1810 ELSE IP% = IP% + 1: GOTO 1790
1790 IF IP% > 6 THEN 1800 ELSE 1780
1800 PRINT "Are you sure? Repeat input": GOTO 1760
1810 REM Got name ,now get molecular data
1820 LP% = 0: RESTORE 2100
1830 READ AMOL, VBAR, D20(1): LP% = LP% + 1
1840 IF LP% = IP% THEN 1880 ELSE 1850
1850 IF LP% > 6 THEN 1860 ELSE 1830
1860 PRINT "Fault reading Mol.Params": STOP
1870 REM Now get sp.gr. & density coefficients
1880 LP% = 0: RESTORE 2140
1890 READ SAT, ALIM: LP% = LP% + 1: IF LP% = IP% THEN 1900 ELSE 1890
1900 IF IP% = 6 THEN 1980 ELSE LP% = 0: RESTORE 2160: GOTO 1910
1910 IF IP% > 2 THEN RESTORE 2230: LP% = 2: GOTO 1920 ELSE 1920
1920 READ ZA(0), ZA(1), ZA(2), ZB1(0), ZB1(1), ZB1(2): LP% = LP% + 1
1930 IF LP% = 6 THEN 1860 ELSE IF LP% = IP% THEN 1940 ELSE 1920
1940 REM Got 1st. coeffs. but see if glycerol & read extra coefs.
1950 IF IP% = 2 THEN 1960 ELSE 1970
1960 READ ZB2(0), ZB2(1), ZB2(2)
1970 GOTO 2060
1980 RESTORE 2320
1990 REM Restore to line for sucrose data
2000 FOR KC% = 0 TO 8: READ ZSUD(KC%): NEXT KC%
2010 FOR KC% = 0 TO 2: READ ZSUG(KC%): NEXT KC%
2020 FOR KC% = 0 TO 7: READ ZSUD1(KC%): NEXT KC%
2030 FOR KC% = 0 TO 5: READ ZSUD2(KC%): NEXT KC%
2040 FOR KC% = 0 TO 7: READ ZSUE1(KC%): NEXT KC%
2050 FOR KC% = 0 TO 5: READ ZSUE2(KC%): NEXT KC%
2060 RETURN
2070 DATA "CSCL, GLYCEROL, KI, KBR, NABR, SUCROSE"
2080 DATA "CSCL","GLYCEROL","KI","KBR","NABR","SUCROSE"
2090 REM molecular data (g,c,s),MOL.WT + VBAR + DIFF. CONST.(SI)
2100 DATA 168.37,0.266,1.7E-9,72.11,0.793,7.7E-10
2110 DATA 166.03,0.291,1.8E-9,119.01,0.319,1.8E-9,102.91,0.268,1.0E-9
2120 DATA 342.3,0.637,4.1E-10
2130 REM Limiting concs. (WT%):-SAT + Fitted Limit
2140 DATA 75.,64.,100.,100.,81.,40.,40.,40.,51.,40.,76.,76.
2150 REM Coeffs. in above sequence
2160 DATA 1.009985#,5.723211D-3,1.202414D-4
2170 DATA 1.021052,-8.770035D-3,1.711131D-4
2180 REM Glycerol
2190 DATA 9.98853D-1,2.449206D-3,2.317415D-6
2200 DATA 1.1204471#,-1.035501D-2,1.964791D-3
2210 DATA 5.263194#,-1.530875D-1,1.732818D-3
2220 REM Potassium iodide, KI
2230 DATA 1.001713#,6.768155D-3,7.752044D-5
2240 DATA 1.002583#,-6.767929D-3,1.008861D-4
2250 REM Potassium bromide KBR
2260 DATA 1.001379#,6.681938D-3,6.702068D-5
2270 DATA 1.005648#,-6.117955D-3,1.483215D-4
2280 REM Sodium bromide NABR
```

```
2290 DATA 1.001567#,7.22281D-3,7.72027D-5
2300 DATA 1.032189#,-4.991555D-3,6.062503D-4
2310 REM Sucrose data
2320 DATA 1.0003698#,3.9680504D-5,-5.8513271D-6
2330 DATA 3.8982371D-1,-1.0578919D-3,1.2392833D-5
2340 DATA 0.17097594#,4.7530081D-4,-8.9239737D-6
2350 DATA 146.06635#,25.251728#,7.0674842D-2
2360 DATA -1.5018327#,9.4112153#,-1.1435741D3
2370 DATA 1.0504137D5,-4.6927102D6,1.0323349D8
2380 DATA -1.1028981D9,4.5921911D9
2390 DATA -1.0803314#,-2.0003484D1,4.6066898D2
2400 DATA -5.9517023D3,3.5627216D4,-7.8542145D4
2410 DATA 2.1169907D2,1.6077073D3,1.6911611D5
2420 DATA -1.4184371D7,6.0654775D8,-1.2985834D10
2430 DATA 1.3532907D11,-5.4970416D11
2440 DATA 1.3975568D2,6.6747329D3,-7.8716105D4
2450 DATA 9.0967578D5,-5.538083D6,1.2451219D7
2460 REM Routine to calc. relative viscs & sp.gr.
2470 REM must enter with wt. fraction in CONC
2480 DEF FNSPG (ZW) = ZA(0) + ZA(1) * CONC + ZA(2) * CONC * CONC
2490 DEF FNVIS1 (ZW) = ZB1(0) + ZB1(1) * CONC + ZB1(2) * CONC * CONC
2500 DEF FNVIS2 (ZW) = ZB2(0) + ZB2(1) * CONC + ZB2(2) * CONC * CONC
2510 AMW = 18: IF IP% > 6 THEN 2520 ELSE 2530
2520 PRINT "No gradient has been defined": RETURN
2530 IF IP% = 6 THEN 2600 ELSE 2550
2540 REM cscl & glycerol calcs:assumes CONC in WT.%
2550 ZCON = CDBL(CONC): SG = FNSPG(ZCON)
2560 IF IP% = 2 THEN 2570 ELSE VISC = FNVIS1(ZCON): GOTO 2820
2570 IF CONC > 48 THEN 2580 ELSE VISC = FNVIS1(ZCON): GOTO 2820
2580 VISC = FNVIS2(ZCON): VISC = EXP(VISC): GOTO 2820
2590 REM Sucrose calcs.,1st. convert to mol. fraction
2600 CO = CONC * .01
2610 ZWF = CDBL(CO): AMFR = CO / AMOL
2620 AMFR = AMFR / (AMFR + ((1 - CO) / AMW))
2630 ZCON = CDBL(AMFR): TEMC = TEM - 273.12
2640 TEMC2 = TEMC * TEMC: ZW = ZWF * ZWF
2650 Z1 = ZSUD(0): Z1 = Z1 + (ZSUD(1) * TEMC + ZSUD(2) * TEMC2)
2660 Z2 = ZSUD(3) + ZSUD(4) * TEMC + ZSUD(5) * TEMC2: Z2 = Z2 * ZWF
2670 Z3 = ZSUD(6) + ZSUD(7) * TEMC + ZSUD(8) * TEMC2: Z3 = Z3 * ZW
2680 SG = Z1 + Z2 + Z3
2690 ZC = ZCON / ZSUG(2): ZC = (ZC * ZC) + 1
2700 ZC = SQR(ZC): ZC = ZSUG(0) - (ZSUG(1) * ZC): ZC = ZC + TEMC
2710 IF CONC > 48 THEN 2770 ELSE 2720
2720 Z1 = ZSUD1(0) + ZSUD1(1) * ZCON
2730 FOR KC% = 2 TO 7: Z1 = Z1 + (ZSUD1(KC%) * (ZCON ^ KC%)): NEXT KC%
2740 Z2 = ZSUE1(0) + ZSUE1(1) * ZCON
2750 FOR KC% = 2 TO 7: Z2 = Z2 + (ZSUE1(KC%) * (ZCON ^ KC%)): NEXT KC%
2760 GOTO 2810
2770 Z1 = ZSUD2(0) + ZSUD2(1) * ZCON
2780 FOR KC% = 2 TO 5: Z1 = Z1 + (ZSUD2(KC%) * (ZCON ^ KC%)): NEXT KC%
2790 Z2 = ZSUE2(0) + ZSUE2(1) * ZCON
2800 FOR KC% = 2 TO 5: Z2 = Z2 + (ZSUE2(KC%) * (ZCON ^ KC%)): NEXT KC%
2810 ZV = Z1 + (Z2 / ZC): ZV = 2.3025851# * ZV: VISC = EXP(ZV)
2820 VISC = VISC * .001
2830 RETURN
2840 REM Sub. to correct S & D for visc & Sp.Gr.,returns S/S(20),D/D(20)
2850 REM Enters knowing VBAR,TEM,SG,VISC
2860 REM IFLAG=0 then calcs. constants for water
2870 IF IFLAG% <> 0 THEN 2900
2880 DEN1 = (1 - (VBAR * 1000)): DEN1 = 1 / DEN1
2890 DRAT = TEM * .001 / 293.12: IFLAG% = 1
2900 NUM1 = (1 - (VBAR * SG * 1000))
2910 SS20 = .001 * NUM1 * DEN1 / VISC
```

```
2920 DD20 = DRAT / VISC
2930 REM SS20=S/S(20,W),DD20=D/D(20,W)
2940 RETURN
2950 ALPHA = D20(LOOPY%) * DD20 / (RDIF * SOM(LOOPY%) * SS20): ZMUL = -1
2960 IF ALPHA < 0 THEN 2980 ELSE 2970
2970 IF ALPHA < .012 THEN ALPHA = .012: GOTO 2990 ELSE 2990
2980 IF ABS(ALPHA) < .012 THEN ALPHA = -.012: GOTO 2990 ELSE 2990
2990 EMCON = 16 * PI * ALPHA * ALPHA: TAU1 = TAU(LOOPY%) * SS20
3000 EBRAK = ZMUL / (2 * ALPHA)
3010 IF EBRAK < -88 THEN EBRAK = 0: GOTO 3020 ELSE EBRAK = EXP(EBRAK)
3020 ECON = PI * PI * ALPHA * GAM * TAU(LOOPY%) * SS20
3030 EINV = 4 * PI * PI * ALPHA * ALPHA
3040 REM IF ALPHA < 0 THEN 4184 ELSE 4190
3050 REM COMU=1-EXP(1/ALPHA):COMU=1/(ALPHA*COMU):GOTO 4200
3060 COMU = EXP(1 / ALPHA) - 1: COMU = 1 / (ALPHA * COMU)
3070 TPR% = TPR% + 1: IF TPR% > MES% THEN 3080 ELSE 3110
3080 LOCATE 22, 5, 0: TPR% = 0
3090 PRINT "Still computing"
3100 PRINT "Time (hrs) so far="; TIME / 3600; " Dens. @ "; R; " m. ="; SG
3110 GOSUB 1440: RETURN
```

Appendix C. Program for the prediction of the shape of self-forming gradients at equilibrium

```
1     ON ERROR GOTO 40000
5     LET COL = 3
90    DIM D(1, 20), CS(2, 20), C4(3, 20), NA(4, 20)
      DIM RB(5, 20), DI(20), RI(20)
100   SCREEN 2: KEY OFF: CLS
110   LOCATE 3, 25: PRINT "CALCULATION OF GRADIENT PROFILE"
120   LOCATE 8 + COL, 6
      PRINT "This program calculates the profiles of self forming"; ""
130   LOCATE 10 + COL, 22: PRINT "gradients and the equilibration times"
150   LOCATE 12 + COL, 27: PRINT "of  gradients  and  sample."
160   LOCATE 22, 4: PRINT "Press the space bar to continue."
165   GOSUB 10000
170   Q$ = INKEY$: IF Q$ = "" THEN 170
180   IF Q$ = "q" OR Q$ = "Q" THEN GOTO 50000
190   IF Q$ <> " " THEN 170
240   CLS
245   LOCATE 3, 25: PRINT "CALCULATION OF GRADIENT PROFILE"
247   MEN = 0
250 GOSUB 10000
255 LOCATE 24, 46: PRINT "                              ";
260 LOCATE 22, 4: PRINT "
270 INPUT "MAXIMUM RADIUS in cm = ", RX$
271 LOCATE 8, 24
    PRINT "                              "
    LOCATE 10, 33
    PRINT "              ":
    LOCATE 13, 27
    PRINT "                              "
272 IF LEN(RX$) > 15 THEN GOTO 240
275 IF RX$ = "Q" OR RX$ = "q" THEN CLS : SYSTEM: ELSE RX = VAL(RX$)
276 FOR Z = 1 TO LEN(RX$): LET CHAR = ASC(MID$(RX$, Z, 1))
      IF CHAR = 46 THEN GOTO 278
277   IF CHAR < 48 OR CHAR > 57 THEN GOTO 260
278 NEXT Z
```

```
279 LOCATE 17, 10
    PRINT "RMAX="; USING "#####.##"; VAL(RX$);
    PRINT "                                        "
283 MEN = 1
284 GOSUB 10000
285 LOCATE 22, 4
290 PRINT "MINIMUM RADIUS in cm =                                                    ";
295 LOCATE 22, 26: INPUT "", RN$
296 IF LEN(RN$) > 15 THEN GOTO 240
297 IF RN$ = "L" OR RN$ = "l" THEN 240 ELSE RN = VAL(RN$)
298 IF RN$ = "Q" OR RN$ = "q" THEN RUN
299 FOR Z = 1 TO LEN(RN$)
         LET CHAR = ASC(MID$(RN$, Z, 1))
         IF CHAR = 46 THEN GOTO 301
300      IF CHAR < 48 OR CHAR > 57 THEN GOTO 285
301 NEXT Z
302 LOCATE 17, 45: PRINT "RMIN="; USING "####.##"; VAL(RN$);
305 IF RX > RN GOTO 360
310 LOCATE 8, 24: PRINT "MINIMUM RADIUS MUST BE LESS THAN"
320 LOCATE 10, 33: PRINT "MAXIMUM RADIUS"
330 LOCATE 13, 27: PRINT "T R Y   A G A I N   ! ! !"
340 GOTO 245
360 CLS
361 LOCATE 3, 25: PRINT "CALCULATION OF GRADIENT PROFILE"
365 RC = SQR((RN ^ 2 + RN * RX + RX ^ 2) / 3)
370 GOSUB 10000
380 LOCATE 23, 4: PRINT "(1 < Dens <= 1.9)"
390 LOCATE 22, 4: PRINT "                                                                           "
395 IF DO$ = "L" OR DO$ = "l" THEN 240 ELSE SDO = VAL(DO$)
396 IF DO$ = "Q" OR DO$ = "q" THEN RUN
397 FOR Z = 1 TO LEN(DO$): LET CHAR = ASC(MID$(DO$, Z, 1))
       IF CHAR = 46 THEN GOTO 399
398    IF CHAR < 48 OR CHAR > 57 THEN GOTO 390
399 NEXT Z
400 IF SDO <= 1.9 THEN 421
410 LOCATE 10, 24: PRINT "DENSITY MUST BE LESS THAN 1.8 g/ml"
415 LOCATE 15, 29: PRINT "T R Y   A G A I N   ! ! !"
420 GOTO 380
421 IF SDO > 1 THEN 430
422 LOCATE 10, 24: PRINT " DENSITY MUST BE MORE THAN 1 g/ml"
423 LOCATE 15, 29: PRINT "T R Y   A G A I N   ! ! !"
425 GOTO 380
430 CLS : GOSUB 10000
435 LOCATE 3, 25: PRINT "CALCULATION OF GRADIENT PROFILE"
440 LOCATE 22, 4: PRINT "                                                                     "; :
441 IF LEN(RP$) > 15 THEN GOTO 430
450 IF RP$ = "l" OR RP$ = "L" THEN 370
460 IF RP$ = "q" OR RP$ = "Q" THEN RUN:  ELSE RP = VAL(RP$)
461 FOR Z = 1 TO LEN(RP$)
       LET CHAR = ASC(MID$(RP$, Z, 1))
       IF CHAR = 46 THEN GOTO 463
462    IF CHAR < 48 OR CHAR > 57 THEN GOTO 440
463 NEXT Z
485 PI = 4 * ATN(1)
490 W = 2 * RP * PI / 60
500 PRINT
510 GOTO 4000
520 CLS : PRINT : PRINT : PRINT : PRINT : PRINT : PRINT : PRINT "
525 PRINT "                              CODE  MEDIUM"
527 PRINT : PRINT
530 PRINT "                              1.....CESIUM CHLORIDE"
535 PRINT
540 PRINT "                              2.....CESIUM SULFATE"
```

```
545 PRINT
550 PRINT "                              3.....SODIUM IODIDE"
555 PRINT
560 PRINT "                              4.....RUBIDIUM CHLORIDE"
562 LOCATE 3, 25: PRINT "CALCULATION OF GRADIENT PROFILE"
565 GOSUB 10000
570 LOCATE 22, 4
    PRINT "                                                 ";
    LOCATE 22, 4
    INPUT "Enter code for medium used : ", M$
575 IF M$ = "l" OR M$ = "L" THEN RESTORE: GOTO 4010:  ELSE M = VAL(M$)
576 IF M$ = "q" OR M$ = "Q" THEN RUN
577 IF LEN(M$) > 15 THEN GOTO 520
578 FOR Z = 1 TO LEN(M$)
       LET CHAR = ASC(MID$(M$, Z, 1))
       IF CHAR = 46 THEN GOTO 580
579    IF CHAR < 48 OR CHAR > 57 THEN GOTO 570
580 NEXT Z
585     GOSUB 10000
590     H = 10 * SDO
600     IF M = 1 THEN M = 120: B = CS(2, H): GOTO 700
610     IF M = 2 THEN M = 130: B = C4(3, H): GOTO 700
620     IF M = 3 THEN M = 140: B = NA(4, H): GOTO 700
630     IF M = 4 THEN M = 150: B = RB(5, H): GOTO 700
640     GOTO 570
700     IF B = 0 THEN
            CLS
            LOCATE 12, 29:
            PRINT "THAT'S IMPOSSIBLE!!"
            GOTO 361
         END IF
701     BO = B * 1E+09
702     GOTO 2010
710     CLS : LOCATE 9, 22
715     PRINT "      THE ISOCONCENTRATION POINT"
720     LOCATE 11, 15:
        PRINT "(WHERE DENSITY ="; INT(SDO * 10) / 10; "g/ml) IS AT A RADIUS OF"
730     LOCATE 11, 57: PRINT USING "####.##"; RC; : PRINT "cm"
740     LOCATE 16, 14
        PRINT "Please input the radius for which a density is required!"
760     LOCATE 3, 31: PRINT "Gradient Profile"
790     MEN = 0
791     GOSUB 10000
792     MEN = 1
795     IF GG$ = "n" OR GG$ = "N" THEN
          IF RAPH = 0 THEN LOCATE 24, 24: PRINT "                    ";
        END IF
796     LOCATE 19, 16: PRINT "MIN RADIUS="; USING "###.##"; RN; : PRINT "
800     LOCATE 22, 4: PRINT "                                                        "
801     LOCATE 23, 2: FOR EDW = 1 TO 38: PRINT " "; : NEXT
802     IF RI$ = "q" OR RI$ = "Q" THEN
           RUN
        ELSE
           RAPH = RAPH + 1: RI(RAPH) = VAL(RI$)
        END IF
803     IF RI$ = "q" OR RI$ = "Q" THEN
           IF RAPH = 0 THEN MONI = 1
           GOTO 2100
        ELSE
           IF RAPH = 0 THEN RAPH = 1: RI(1) = VAL(RI$): GOTO 805
        END IF
```

```
804     IF RAPH > 6 THEN
            FOR DOR = 1 TO 6
                RI(DOR) = RI(DOR + 1)
                DI(DOR) = DI(DOR + 1)
            NEXT
        RAPH = 6
805     IF RI(RAPH) < RN THEN
            LOCATE 23, 2
            PRINT "VALUE SMALLER THAN MINIMUM RADIUS."
            RAPH = RAPH - 1
            GOTO 790
        END IF
806     IF RI(RAPH) > RX THEN
          LOCATE 23, 2
          PRINT "VALUE LARGER THAN MAXIMUM RADIUS."
          RAPH = RAPH - 1
          GOTO 800
        END IF
830     DI(RAPH) = SDO + (W ^ 2 / (2 * BO)) * (RI(RAPH) ^ 2 - RC ^ 2)
840     CLS : FOR DOR = 1 TO RAPH
850     LOCATE 5 + DOR * 2, 19
        PRINT "Density at"; USING "###.##"; RI(DOR);
        PRINT " cm equals "; USING "###.##"; DI(DOR);
        PRINT " g/ml"
860     NEXT
865     LOCATE 3, 25: PRINT "CALCULATION OF GRADIENT PROFILE"
870     MEN = 0
880     GOSUB 10000: MEN = 1: GOTO 796
2010    CLS : LOCATE 12, 14
2015    PRINT "Do you wish to calculate the approximate time required"
2017    PRINT
2020    PRINT "                   for gradient formation and sample equilibration?"
2032    PRINT
2040    GOSUB 10000
2041    LOCATE 3, 25: PRINT "CALCULATION OF GRADIENT PROFILE"
2042    LOCATE 22, 4: INPUT "YES/NO (Y/N)", GG$
2045    '
2047    IF GG$ = "L" OR GG$ = "l" THEN 520
2048    IF GG$ = "Q" OR GG$ = "q" THEN RUN
2050    IF INSTR("Nn", GG$) <> 0 THEN 710
2060    IF INSTR("Yy", GG$) = 0 THEN 2010
2065    CLS : LOCATE 15, 14
2070    PRINT "Use an estimate if you do not know the  correct value."
2080    LOCATE 13, 20: PRINT " Please input the s-value of your sample"
2081    LOCATE 3, 25: PRINT "CALCULATION OF GRADIENT PROFILE"
2082    GOSUB 10000: LOCATE 22, 4: PRINT "
2083    IF SW$ = "l" OR SW$ = "L" THEN 2010 ELSE SW = VAL(SW$)
2084    IF SW$ = "q" OR SW$ = "Q" THEN RUN
2085    IF LEN(SW$) > 15 THEN GOTO 2065
2086    FOR Z = 1 TO LEN(SW$)
            LET CHAR = ASC(MID$(SW$, Z, 1))
            IF CHAR = 46 THEN GOTO 2088
2087        IF CHAR < 48 OR CHAR > 57 THEN GOTO 2082
2088    NEXT Z
2089    LOCATE 18, 26: PRINT "....and the buoyant density": GOSUB 10000
2090    LOCATE 23, 4: PRINT "            "; : LOCATE 23, 4: INPUT "g/ml = ", BD$
2095    IF BD$ = "L" OR BD$ = "l" THEN LOCATE 18, 26: PRINT "
2096    IF BD$ = "Q" OR BD$ = "q" THEN RUN
2097    FOR Z = 1 TO LEN(BD$)
            LET CHAR = ASC(MID$(BD$, Z, 1))
            IF CHAR = 46 THEN GOTO 2099
2098        IF CHAR < 48 OR CHAR > 57 THEN GOTO 2090
2099    NEXT Z
```

```
2100    CLS
        IF GG$ = "n" OR GG$ = "N" THEN RAPH = RAPH + 1
           GOTO 805
        ELSE
          LOCATE 7, 4
          PRINT "The time for the gradient to form will be approximately ";
        END IF
2102    IF MONI = 1 THEN 2110
2105    TG = INT(5.6 * ((RX - RN) ^ 2) * 100) / 100
2110    '
2115    PRINT TG; " hrs."
2116    IF MONI = 1 THEN MONI = 0: GOTO 2120
2118    IF BO = 0 OR W = 0 OR RP = 0 OR SW = 0 THEN 20000
2120    DT = SDO + (W ^ 2 / (2 * BO)) * (RN ^ 2 - RC ^ 2)
2130    RQ = SQR(RN ^ 2 + ((2 * BO) / W ^ 2) * (BD - DT))
2140    SQ = (9.83E+13 * BO * (BD - 1)) / (RP ^ 4 * RQ ^ 2 * SW)
2150    LOCATE 10, 10
        PRINT "The time required for the sample to reach equilibrium will be"
2170    LOCATE 11, 26
        PRINT "approximately"; INT(SQ * 100) / 100; " hrs."
2180    LOCATE 14, 20: PRINT "At equilibrium the sample will be at an";
2190    LOCATE 15, 24
        PRINT "approximate radius of"; : PRINT INT(RQ * 100) / 100; " cm."
2300    LOCATE 18, 16: PRINT "These figures are only a guide and to calculate"
2330    LOCATE 19, 28: PRINT "the exact gradient profile."
2340    LOCATE 23, 4: PRINT "Press the space-bar to continue"
2342    LOCATE 3, 29: PRINT "Gradient Profile"
2345    MEN = 0
2346    GOSUB 10000
2347    LOCATE 24, 17: PRINT "                                    ";
2348    MEN = 1
2350    ZZ$ = INKEY$: IF ZZ$ <> " " THEN 2350
2360    GOTO 710
3999    END
4000    IF DFUIO = 1 THEN RESTORE: GOTO 4010
4002    DFUIO = 1
4010    DATA 0.00,2.04,1.55,1.33,1.22,1.17,1.14,1.12
4020    DATA 0.00,1.06,0.76,0.67,0.64,0.66,0.69,0.74
4030    DATA 5.10,3.19,2.82,0.00,0.00,0.00,0.00,0.00
4040    DATA 0.00,3.42,2.76,2.25,0.00,0.00,0.00,0.00
4050    READ CS(2,11), CS(2,12), CS(2,13), CS(2,14), CS(2,15), CS(2,16)
4055    READ CS(2,17), CS(2,18)
4060    READ C4(3,11), C4(3,12), C4(3,13), C4(3,14), C4(3,15), C4(3,16)
4065    READ C4(3,17), C4(3,18)
4070    READ NA(4,11), NA(4,12), NA(4,13), NA(4,14), NA(4,15), NA(4,16)
4075    READ NA(4,17), NA(4,18)
4080    READ RB(5, 11), RB(5,12), RB(5,13), RB(5,14), RB(5,15), RB(5,16), RB(5,17)
4085    READ RB(5, 18)
4145    P$ = "##.##        ##.##      ##.##     ##.#       ##.##"
4160    DATA 1.1,1.2,1.3,1.4,1.5,1.6,1.7,1.8
4170    READ D(1,11), D(1,12), D(1,13), D(1,14), D(1,15), D(1,16), D(1,17), D(1,18)
4190    FOR I = 11 TO 18
4210    NEXT
4220    MEN = 0
4230    GOTO 520
4999    END
5000    LOCATE 22, 4: PRINT "Press the space bar to continue";
5010    MAR$ = INKEY$: IF MAR$ = "" THEN 5010
5015    IF MAR$ = "q" OR MAR$ = "Q" THEN 360
5017    IF MAR$ <> " " THEN 5010
5020    RETURN
5100    RETURN
10000 ' window
```

```
10010 PI = 103.638 / 32.989
10195 IF MEN = 1 THEN
         LOCATE 23, 53
         PRINT "Press Q then ENTER to quit";
         GOTO 10205
      END IF
10200   LOCATE 23, 53: PRINT "Press Q then ENTER to quit";
10202   RETURN
10205   LOCATE 24, 46: PRINT "Press L then ENTER for last value";
10210   RETURN
20000   CLS
20010   LOCATE 10, 22: PRINT "     S O M E T H I N G   W E N T"
20020   LOCATE 12, 22: PRINT "            W R O N G ! ! "
20030   LOCATE 15, 32: PRINT "Please try again..."
20040   LOCATE 22, 4: PRINT "Press the space bar to restart..."
20045   GOSUB 10000
20050   BAAL$ = INKEY$: IF BAAL$ = "" THEN 20050
20060   IF BAAL$ = "q" OR BAAL$ = "Q" THEN RUN
20065   IF BAAL$ <> " " THEN 20050
20070   RUN 240
30000 REM LETTER CHECK FOR RI$
30005   IF RI$ = "Q" OR RI$ = "q" THEN GOTO 801
30010   IF LEN(RI$) > 15 THEN CLS : GOTO 710
30020   FOR Z = 1 TO LEN(RI$)
           LET CHAR = ASC(MID$(RI$, Z, 1))
           IF CHAR = 46 THEN GOTO 30040
30030      IF CHAR < 48 OR CHAR > 57 THEN GOTO 800
30040   NEXT Z
30050   GOTO 801
40000 REM ERROR HANDLER
40005 BEEP: BEEP
40010 CLS
40015 LOCATE 12, 16
      PRINT "AN ERROR WHICH WAS NOT EXPECTED BY THE WRITERS OF"
      LOCATE 13, 24
      PRINT "THIS SOFTWARE HAS OCCURRED"
      LOCATE 17, 21
      PRINT "Press ENTER to restart the program"
40020 LET A$ = INKEY$: IF A$ = "" THEN GOTO 40020
40030 IF ASC(A$) = 13 THEN RUN ELSE GOTO 40020
50000 REM ENTER CHECK
50005 LOCATE 24, 76: PRINT "Q";
50010 LET A$ = INKEY$: IF A$ = "" THEN GOTO 50010
50020 IF ASC(A$) = 13 THEN CLS : SYSTEM
50025 IF ASC(A$) = 8 THEN LOCATE 24, 76: PRINT " "; : GOTO 170
50030 GOTO 50010
```

Appendix D. A program for the calculation of rotor parameters

```
10      ON ERROR GOTO 15000
100     SCREEN 2: KEY OFF: CLS
109     GOSUB 110: GOTO 140
110       LOCATE 3, 25: PRINT "CALCULATION OF ROTOR PARAMETERS"
135     RETURN
140     REM
150     REM      MEANING OF SOME IMPORTANT VARIABLES :
160     REM         R1=RMIN                   R2=RMAX
170     REM         M1=MAX RPM                G1=MAX RCF
```

```
180     REM       G2=AVER. RDF          G3=MIN RCF
190     REM       K1=K-FACTOR           K2=K*-FACTOR
200     REM       T=ROTOR TEMPERATURE (5 D.C.)
210     REM       S1=SUCROSE CONCENTRATION
220     REM       P1=PARTICLE DENSITY (1.35 G/ML)
230     REM       B=GRADIENT IMPACT ON SEDIMENTATION
240     REM       O=OMEGA
250     REM       P2=PI
260     REM
270     I1$ = " RMIN      =  ##.## cm      RMAX     = ##.## cm         "
280     I2$ = " MAX RPM   =  #####         "
290     I3$ = " RCF MIN   = ###### xg      "
300     I4$ = " RCF AVG   = ###### xg      "
310     I5$ = " RCF MAX   = ###### xg      "
312     I6$ = " K-VALUE   =  ####. h       K*VALUE = #####. h          "
318       LOCATE 7, 14
319       PRINT "This program provides some commonly used factors for"
320       LOCATE 9, 16
321       PRINT "comparisons between and evaluations of different"
322       LOCATE 11, 18: PRINT "          ultracentrifuge rotors."
325     LOCATE 23, 3
330     INPUT "INPUT ROTOR NAME:", R$
331     MEN = 1
332     LOCATE 7, 3: PRINT R$; "                                    "
333     GOSUB 11000
335     LOCATE 23, 3
        INPUT "RMIN in cm = ", R1$
        IF R1$ = "Q" OR R1$ = "q" OR R1$ = "L" OR R1$ = "l" THEN GOTO 11010
336 IF VAL(R1$) <= 0 AND LEN(R1$) < 15 THEN LOCATE 23, 2: PRINT "
337 IF LEN(R1$) <= 15 THEN LOCATE 23, 2: PRINT "                                     "
338 IF LEN(R1$) > 15 THEN
         LET EDWI = EDWI - 1
         BEEP
         CLS
         GOSUB 11000
         GOTO 335
     END IF
340 LOCATE 23, 3: INPUT "RMAX in cm = ", R2$
    IF R2$ = "q" OR R2$ = "Q" OR R2$ = "L" OR R2$ = "l" THEN GOSUB 11010
341 IF VAL(R2$) <= 0 AND LEN(R2$) < 15 THEN
        LOCATE 23, 2
        PRINT "                                    "
        LET EDWI = EDWI - 1
        GOSUB 11010
        GOTO 340
     END IF
342 IF LEN(R2$) > 15 THEN
        LET EDWI = EDWI - 1
        BEEP
        CLS
        GOSUB 11000
        GOTO 340
     END IF
343 IF LEN(R2$) <= 15 THEN
          LOCATE 23, 2
          PRINT "                                    "
          LET EDWI = 2
          GOSUB 11010
    END IF
345  LOCATE 23, 3: INPUT "MAX RPM = ", M1$:
     IF M1$ = "Q" OR M1$ = "q" OR M1$ = "L" OR M1$ = "l" THEN GOSUB 11010
346 IF VAL(M1$) <= 0 AND LEN(M1$) < 15 THEN
         LOCATE 23, 2
```

```
        PRINT "                                         "
        LET EDWI = EDWI - 1
        GOSUB 11010
        GOTO 345
    END IF
347 IF LEN(M1$) > 15 THEN
        LET EDWI = EDWI - 1
        BEEP
        CLS
        GOSUB 11000
        GOTO 345
    END IF
348 IF LEN(M1$) <= 15 THEN
       LOCATE 23, 2
       PRINT "                                                       "
       LET EDWI = 3
       GOSUB 11010
    END IF
349 LET EDWI = 3
350 MEN = 0
353     GOSUB 11000
370     REM
380     REM CALCULATION OF RELATIVE CENTRIFUGAL FORCES.
390     P2 = 4 * ATN(1!)
400     G1 = (M1 * P2 / 30) ^ 2 * R1 / 981
410     G2 = (M1 * P2 / 30) ^ 2 * ((R1 + R2) / 2) / 981
420     G3 = (M1 * P2 / 30) ^ 2 * R2 / 981
430     REM
440     REM CALCULATION OF NORMAL K-VALUE
450     O = M1 / 60 * 2 * P2
460     K1 = (LOG(R2) - LOG(R1)) * 1E+13 / O ^ 2 / 3600
470     REM
480     REM CALCULATION OF K*-VALUES
490     R3 = (R2 - R1) / 100
500     T = 5!
510     P1 = 1.35
520     K2 = 0
530     O = M1 / 60 * 2 * P2
540     FOR R = R1 + .5 * R3 TO R2 - .5 * R3 STEP R3
550     S1 = 5! + (R - R1) * (15 / (R2 - R1))
560     C = S1 / 100
570     Y = (C / 342.3) / (C / 342.3 + (1 - C) / 18.032)
580     GOSUB 700
590     B = (1.002 / V) * ((P1 - D) / (P1 - 1!))
600     K2 = K2 + R3 / (R * O ^ 2 * B)
610     NEXT R
620     K2 = K2 * 1E+13 / 3600
629 GOSUB 630: GOTO 690
630     LOCATE 7, 3: PRINT R$
635     LOCATE 9, 2: PRINT USING I1$; R1; R2
640     LOCATE 11, 2: PRINT USING I2$; M1
645     LOCATE 13, 2: PRINT USING I3$; G1
650     LOCATE 15, 2: PRINT USING I4$; G2
655     LOCATE 17, 2: PRINT USING I5$; G3
659     LOCATE 19, 2: PRINT USING I6$; K1; K2
660 REM
680 REM
685 RETURN
690     LOCATE 23, 3
        PRINT "Press space-bar to continue"
        LOCATE 24, 64
        PRINT "               ";
```

```
691     Y$ = INKEY$: IF Y$ <> " " THEN 691 ELSE RUN
693     SYSTEM
700     REM DENSITY FOR 0<T<30 AND 0<C<.75
710     ED$ = INKEY$
        IF ED$ = "q" OR ED$ = "Q" THEN CLS : GOSUB 10010: RUN 109
720     D1 = 1.0003696# + .000039680504# * T - .0000058513271# * T ^ 2
730     D2 = .38982371# - .0010578919# * T + .000012392833# * T ^ 2
740     D3 = .17097594# + .000475300081# * T - .0000089239737# * T ^ 2
750     D = D1 + D2 * C + D3 * C ^ 2
760     REM
770     REM VISCOSITY FOR 0<T<60 AND 0<C<.48
780     G4 = SQR(1! + (Y / .070674842#) ^ 2)
790     C2 = 146.06635# - 25.251728# * G4
800     A = -1.5018327# + 9.4112153# * Y
810     A = A - 1143.5741# * Y ^ 2 + 105041.37# * Y ^ 3
820     A = A - 4692710.2# * Y ^ 4 + 103233490# * Y ^ 5
830     A = A - 1102898100# * Y ^ 6 + 4592191100# * Y ^ 7
840     B = 211.69907# + 1607.7073# * Y
850     B = B + 169116.11# * Y ^ 2 - 14184371# * Y ^ 3
860     B = B + 606547750# * Y ^ 4 - 12985834000# * Y ^ 5
870     B = B + 135329070000# * Y ^ 6 - 549704160000# * Y ^ 7
880     V = 10 ^ (A + B / (T + C2))
890     RETURN
900     REM
910     END
10000 ' window
10010 PI = 103.638 / 32.989
10195 IF MEN = 1 THEN
          LOCATE 23, 52
          PRINT "Press Q then ENTER to quit";
          GOTO 10205
      END IF
10200 IF EDWI <> 3 THEN
          LOCATE 24, 53
          PRINT "Press Q then ENTER to quit";
        ELSE
          LOCATE 24, 64
          PRINT "Press Q to quit";
       END IF
10202 RETURN
10205 LOCATE 24, 45: PRINT "Press L then ENTER for last value";
10210 RETURN
11000 CLS : GOSUB 110: GOSUB 10000
11010 EDWI = EDWI + 1
11012 ON EDWI GOSUB 11050, 11060, 11070, 11080
11015 IF EDWI = 4 THEN
        GOSUB 660
        LOCATE 23, 3
        PRINT "I'm calculating!!!"
      ELSE
        GOSUB 680
      END IF
11020 '
11030 RETURN
11050 IF R$ = "q" OR R$ = "Q" THEN CLS : GOTO 693
11056   LOCATE 7, 3: PRINT R$; "                                        "
11057   RETURN
11060 IF R1$ = "q" OR R1$ = "Q" OR R1$ = "l" OR R1$ = "L" THEN
        CLS
        GOSUB 10010
        RUN 109
      ELSE
        R1 = VAL(R1$)
      END IF
```

```
11065  IF R1 = 0 THEN IF RMIN <> 0 THEN R1 = RMIN ELSE R1 = 2
11066   LOCATE 7, 3: PRINT R$; "
11067   LOCATE 9, 2: PRINT USING LEFT$(I1$, 21); R1
11068   LOCATE 9, 23: PRINT "
11069 RETURN
11070 IF R2$ = "l" OR R2$ = "L" THEN RETURN 12010 ELSE R2 = VAL(R2$)
11072 IF R2$ = "q" OR R2$ = "Q" THEN CLS : GOSUB 10010: RUN 109
11075  IF R2 = 0 THEN IF RMAX <> 0 THEN R2 = RMAX ELSE R2 = 2
11076   LOCATE 7, 3: PRINT R$; "
11077   LOCATE 9, 2: PRINT USING I1$; R1; R2
11078 IF R1 > R2 THEN
        LOCATE 22, 3
        PRINT "RMIN CANNOT BE HIGHER THAN RMAX. Try again."
        EDWI = EDWI - 1
        RETURN 12010
      END IF
11079 RETURN
11080 IF M1$ = "l" OR M1$ = "L" THEN RETURN 12020 ELSE M1 = VAL(M1$)
11082 IF M1$ = "q" OR M1$ = "Q" THEN CLS : GOSUB 10010: RUN 109
11085  IF M1 = 0 THEN M1 = .1
11086,  LOCATE 7, 3: PRINT R$; "                                          "
11087   LOCATE 9, 2: PRINT USING I1$; R1; R2
11088   LOCATE 11, 2: PRINT USING I2$; VAL(M1$)
11089 RETURN
12010   LOCATE 7, 3: PRINT R$; "                                          "
12012   LOCATE 9, 2: PRINT USING LEFT$(I1$, 21); R1
12013   RMIN = R1
12015   EDWI = EDWI - 2: GOTO 335: '       GOTO RMIN
12020   LOCATE 7, 3: PRINT R$; "                                          "
12022   LOCATE 9, 2: PRINT USING I1$; R1; R2
12023   LOCATE 11, 2: PRINT USING I2$; VAL(M1$)
12024   RMIN = R1: RMAX = R2
12025   EDWI = EDWI - 2: GOTO 340: '       GOTO RMAX
15000 REM ERROR HANDLING
15005 BEEP: BEEP
15010 CLS
15020 LOCATE 12, 16
      PRINT "AN ERROR WHICH WAS NOT EXPECTED BY THE WRITERS OF"
      LOCATE 13, 24
      PRINT "THIS SOFTWARE HAS OCCURRED"
      LOCATE 17, 21
      PRINT "Press  ENTER  to restart the program"
15030 LET A$ = INKEY$: IF A$ = "" THEN GOTO 15030
15040 IF ASC(A$) = 13 THEN RUN ELSE GOTO 15030
```

2

Centrifuges and rotors

HANS ROTH and DAVID RICKWOOD

1. Classification of centrifuges

Centrifuges can be classified on the basis of a number of different criteria. However, centrifuges are usually classified on the basis of their speed since this is a fairly good indicator of the centrifugal force (RCF) which can be generated by the machine and hence also indicates the range of possible applications that the centrifuge can be used for. *Table 1* gives a summary of some of the principal classes of centrifuge, namely low-speed, high-speed, and ultracentrifuge. While these classes used to be quite distinct, over the years the boundaries have become somewhat more blurred in terms of the size, costs, and applications of the different machines. The primary interest of the user is simply an answer to the question 'what can I use it for?'. The answer to this question is more complicated than it might seem. As an example, plasmid DNA can be prepared on isopycnic CsCl–ethidium bromide gradients (see Section 3.3.1 of Chapter 4) using one of the high-performance high-speed machines generating 100 000*g*. However, using such machines gradients need to be centrifuged for at least 30 h while using an ultracentrifuge the same isolation could be done in 2–5 h depending on the type of rotor used. Hence the time available to the researcher can also be a crucial factor in choosing the type of centrifuge for a particular application.

For historical reasons, and for the purposes of this chapter, centrifuges will be classified on the basis of their maximum speed. In cases where salient points need to be made about the applications of different classes of centrifuge then these will be made in the appropriate section.

2. Different classes of centrifuge

2.1 Low-speed centrifuges

These machines range from small bench centrifuges cooled to room temperature by drawing air through the bowl, and small fully-refrigerated compact centrifuges designed to fit into the kneewell under the laboratory bench when not in use, to the large refrigerated floor-standing machines capable of

Table 1. Classes of centrifuges and their applications

	Class of centrifuge		
	low-speed	**high-speed**	**ultracentrifuge**
Speed ranges (r.p.m. × 10^3)	2–6	18–30	35–120
Maximum RCF (× 10^3)	8	100	700
Refrigerator	some	yes	yes
Vacuum system	none	some	yes
Acceleration/braking	some	variable	variable
Applications			
Cells	yes	yes	yes
Nuclei	yes	yes	yes
Precipitates	some	most	yes
Membrane organelles	some	yes	yes
Membrane fractions	—	some	yes
Ribosomes/polysomes	—	some	yes
Macromolecules	—	some	yes

centrifuging up to nine litres at a time. These machines are nearly always used routinely for the initial processing of samples. This class of centrifuge is used for pelleting cells and the larger cell organelles such as nuclei and chloroplasts. Gradient separations of cells are possible using this type of machine as long as acceleration and deceleration characteristics are suitable. This type of machine has also been used for elutriator rotors (see Chapter 7). This category of centrifuge is widely used in many different types of laboratory. These centrifuges generate much less relative centrifugal force than the other classes and their design should ensure safe operating conditions at all times. However, it is important to check before each run whether the rotor and tube carrier configuration is compatible with the instrument (in some cases, it is possible to put a swing-out rotor into a centrifuge the bowl of which is too small for that rotor!) and can be used safely at the speed you require. Follow the operating instructions carefully, and especially obey load/density limits as specified to prevent overstressing the rotor which may lead to catastrophic failure. At least weekly, check the function of lid locks, lid support mechanisms, and whether the centrifuge is properly grounded; these checks become more important as the machine is used and components become worn. It is also important to ensure that there is no material within the safety envelope, usually of about 30 cm around the instrument, which could cause secondary damage in case of the centrifuge moving as result of a rotor disruption or sudden failure of the drive.

2.2 High-speed centrifuges

Right through up until the end of the 1980s these machines were classified as having a maximum speed of 21–24 000 r.p.m. and generating up to 80 000*g*.

With improvements in technology a new generation of higher-performance centrifuges has become available capable of maximum speeds of 28–30 000 r.p.m. and generating centrifugal forces in excess of 100 000*g*. High-speed centrifuges are the principal preparative laboratory centrifuges, being used for almost all types of both low- and high-speed applications. They are simple to operate like the low-speed machines and much cheaper to purchase and maintain than ultracentrifuges. The large amounts of heat that can be generated by these machines at maximum speed means that it is essential to have a refrigeration system. Some machines also have a vacuum pump which, by reducing the air pressure inside the bowl, reduces both the amount of heat and noise generated as well as allowing more accurate temperature control of the rotors. However, accurate temperature control, that is less than ± 2 °C, is seldom needed for most high-speed centrifuge applications. The presence of a vacuum pump is likely to increase maintenance costs and can be an inconvenience given the fact that most runs on this type of machine are 10–20 min in duration. Some centrifuges minimize this inconvenience by using software within the centrifuge to determine whether or not the vacuum pump is actually required for each particular run.

The principal use of these machines remains the same as it has always been, that is the preparation of subcellular fractions, usually by pelleting, for further processing or analysis. Large-capacity rotors capable of centrifuging up to three litres are available for these types of machines and almost all machines can generate more than 50 000*g* depending on the type of rotor. However, those machines capable of generating centrifugal forces of 100 000*g* or more and with more sophisticated control systems can be used for a range of ultracentrifuge applications, although often run times may be a little longer (1,2). The availability of vertical rotors has also enhanced the ability of these machines to carry out rate-zonal separations for particles such as polysomes (3).

2.3 Ultracentrifuges

These can be divided into two types, analytical and preparative ultracentrifuges; the former are designed to be able to measure the sedimentation velocity or density of macromolecules with great accuracy and are usually very specialized, dedicated machines with a price tag to match. Much of the work of the analytical centrifuge can now be carried out by using other techniques such as gel electrophoresis or even DNA sequencing. The decline of the range of applications for the analytical centrifuge has stunted its development; as an alternative some manufacturers produce add-on attachments which are able to provide the user with a reasonable analytical centrifuge. Even so, the setting up of such attachments is rather an enthusiast's hobby and it is not usually done on a routine basis. As described in Chapters 4 and 5, preparative ultracentrifuges can be used for doing some types of analytical studies including the calculation of sedimentation coefficients using sucrose gradients, but the degree of accuracy obtainable is significantly less than can be obtained using an analytical ultracentrifuge. Recognition of the trend towards micro-scale preparations has led to the development of micro-ultracentrifuges. Benchtop and free-standing ultracentrifuges have been developed to accommodate small-volume, low-mass rotors with low moments of inertia and tremendously rapid acceleration. The maximum speed of these rotors can be 100 000 r.p.m. or more. For ease of rotor/sample loading, and space-saving reasons, floor-standing model micro-ultracentrifuges have much to recommend them. If only a full-size floor-standing model ultracentrifuge is available, similar results can be obtained, provided that its maximum speed is 80 000 r.p.m. or greater, by using the special low-mass rotors manufactured by some companies. These low-mass rotors have tube sizes in the range of 1–5 ml giving rotor volumes of up to 40 ml. These small rotors have very small *k*-factors and so most separations take a much shorter time than when using conventional rotors.

The centrifugal force generated by centrifuges can be in excess of 600 000*g*. It is noticeable that, although maximum speeds have tended to increase from 75 000 to 120 000 r.p.m. the maximum centrifugal force generated by rotors has tended to level off at just over 600 000*g*, presumably reflecting the limits of the strength of titanium and other materials used for manufacturing rotors. Thus the limiting factor today is no longer the instrument/drive speed but the design of rotors, tubes, and tube-closure methods.

2.4 Miscellaneous centrifuges

The broad classes of preparative centrifuges described in the previous sections cover the main types likely to be met in most laboratories. However, there are also many other types of specialist centrifuges ranging from microcentrifuges developing 12–15 000*g* used for pelleting small (0.2–1.5 ml) samples to the specialized continuous-flow centrifuges, such as that shown in *Figure 1*, which are used for processing very large volumes of samples, perhaps as much as several hundred litres at a time. Continuous-flow centrifuges are widely used for downstream processing in biotechnology and can be used for pelleting, isopycnic separations, and phase separations.

3. Centrifuge technology

This section is not designed to give the reader detailed technical descriptions of the different systems available, but rather it surveys the different types of technology used for different systems to give the reader an insight as to the real advantages and disadvantages of each. This will at least allow the reader to appreciate the importance of the different technical information which is provided by manufacturers. One notable feature is the increase in the use of microprocessors and software to control centrifuges, thus increasing the precision of instruments, improving the operating safety, and facilitating the servicing of the equipment.

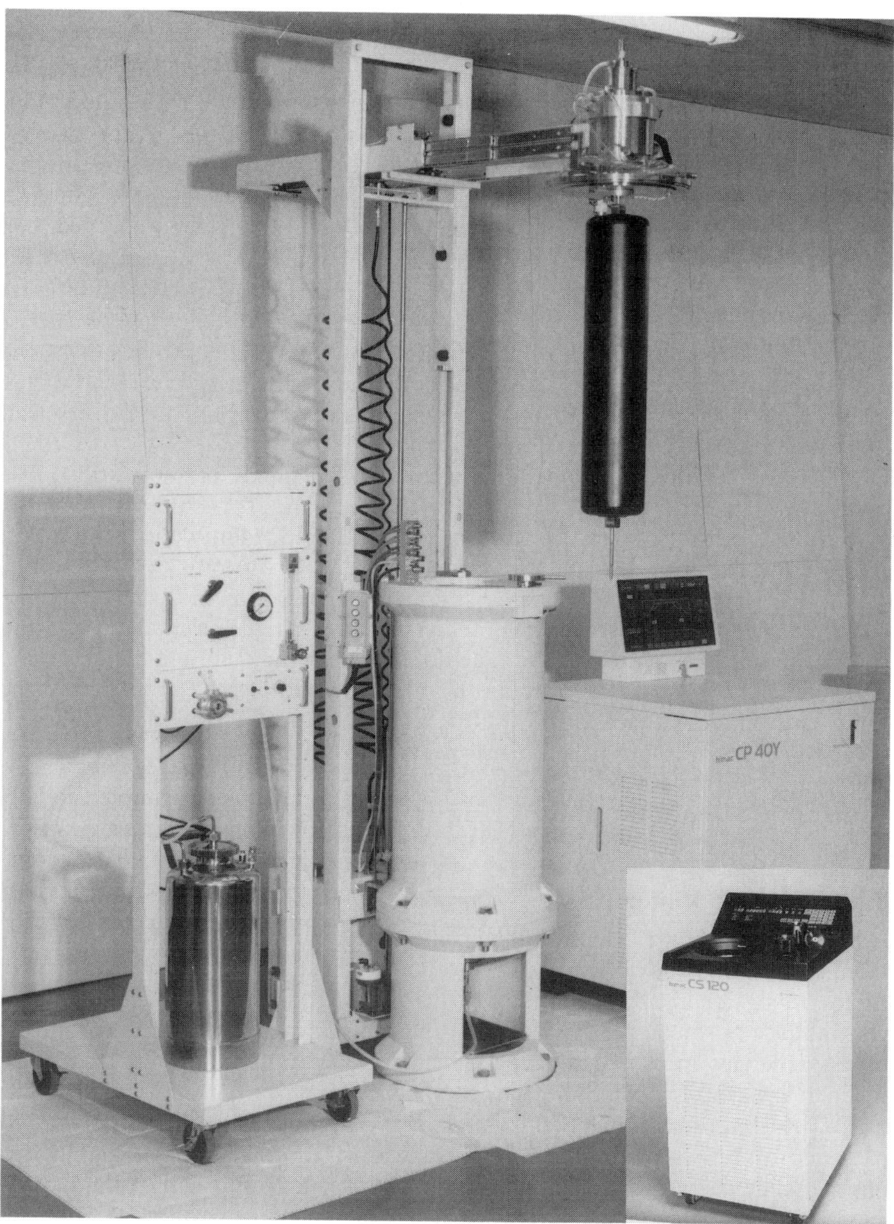

Figure 1. A Hitachi CP40Y continuous-flow ultracentrifuge showing the long, narrow cylindrical rotor suspended below the drive unit which is capable of 40 000 r.p.m. (118 000*g*). For comparison, the inset shows the relative size of a micro-ultracentrifuge. (Reproduced with kind permission of Hitachi Koki Co. Ltd.)

3.1 Drive systems

3.1.1 Turbine drive systems

It is of interest to note that in constructing some of the first ultracentrifuges Svedberg used the potential of the turbine for spinning rotors at very high speeds for many hours. Although centrifuges driven by air or steam turbines continued to be used for specialized machines they were neglected for a number of years. The introduction of the oil turbine drive by Dupont–Sorvall in the middle 1970s revived the concept in its OTD range of ultracentrifuges. The advantage of this system is that only the turbine and bearing housing are subjected to the stresses of high-speed rotation, tending to enhance the reliability of the system compared with the older gearbox-driven instruments. However, the complexities of the oil-pumping system and the large amounts of heat generated as a result of the inherent inefficiency of this drive arrangement tend to weigh against the overall reliability of this turbine drive system.

The acceleration control of turbine systems is good in terms of gentle acceleration of rotors at low speed, but the small size of the turbine limits the amount of torque generated by the system which in turn limits the acceleration rate for heavier rotors.

Another interesting turbine system is the Air-fuge (Beckman Instruments Inc) which has found its own specialist niche. This centrifuge has small aluminium rotors the bottoms of which are machined so that they can act as a turbine. Air jets from below the rotor act both to levitate and rotate the rotor. All the rotors are small and accelerate rapidly to 100 000 r.p.m. (165 000*g*); the actual speed of the rotor is controlled by the pressure of the air supply. There are five different sorts of rotor but the range of applications is limited by the restricted volume of the rotors. Even so, this centrifuge has proved popular for the routine separation of chylomicrons.

3.1.2 Electric motor drive systems

Although this is now the almost universally accepted drive system for centrifuges from the humblest low-speed centrifuge to the most advanced ultracentrifuge, there are significant differences in terms of the design of the drive systems. Originally, low-speed centrifuges had direct drive systems while in most of the ultracentrifuges and some high-speed machines the motor speed was geared up through a gearbox or belt drive. The motors of this type of machine were almost all DC brush motors controlled through a thyristor arrangement. However, the presence of brushes limits the performance of motors and involves added maintenance costs. Similarly, gearboxes and belt drives also introduce likely sources of mechanical deficiencies and a subsequent loss of reliability. The development of drive systems has been dominated by the introduction of induction motors and brushless DC motors. The induction motors are usually three phase motors but they have a non-linear speed/torque curve and the heat generated in these motors is considerable. Variable frequency synchronous motors have also been used and these have similar characteristics to induction motors. The DC brushless motor drives have magnets within the armature and generate much less heat, making them more efficient than induction motors. DC brushless motors also have very good torque characteristics, with the torque being linear with the applied current, it is easy to control acceleration and deceleration, and they exhibit good speed stability. These brushless motors are now featured in many types of different centrifuges from low-speed to ultracentrifuges.

The very low wear characteristics of the motors of direct drive centrifuges have allowed manufacturers to give very long guarantees on drive systems which would have been impossible for the older non-direct drive centrifuges.

3.1.3 Speed control of drive systems

One of the major features of the direct drive systems is the very high degree of accuracy of monitoring and maintaining the exact speed required. Whereas DC motors controlled by thyristors would maintain speeds within 1% of the desired value, that is to within 300 r.p.m. at a setting of 30 000 r.p.m., such accuracy is insufficient for accurate rate-zonal experimental work. In contrast, direct drives can be controlled to within 10 r.p.m. even at 80 000 r.p.m. as a result of the nature of the accurate electronic control systems used for these drives.

Control of the speed of a rotor is usually done by monitoring the speed of the motor, drive shaft, or rotor using a system based on either optical or magnetic impulses or electrical currents that are generated as the driveshaft turns. Speed control should be differentiated from the rotor overspeed protection devices which are based on either an optical disc system (as featured on many ultracentrifuge rotors) or by magnets inserted in a particular pattern in the base of the rotors. This latter method is finding increasing popularity in a number of rotor identification systems (see Section 3.4).

3.1.4 Out-of-balance detection systems

Out-of-balance detection systems are important not only in protecting the machine against damage but also to ensure that damage is not caused to the laboratory by a centrifuge running badly out of balance.

The sensitivity of drive systems to imbalance varies greatly between different classes of machine as well as between different manufacturers. Almost all systems monitor the degree of lateral displacement of the drive shaft in a variety of ways. The actual amount of lateral displacement is extremely dependent on the type of bushing and bearings used to stabilize the shaft. Since all rotors cause lateral displacement of the drive shaft as a result of precession (see Section 4.5), it is important that the out-of-balance system can differentiate between precession and serious imbalance of the rotor. The sensitivity of drive systems is very variable with some manufacturers developing imbalance tolerant drives. The recommendation, however, must be to try and ensure that rotors are always carefully balanced in order to obtain the best quality separation and to minimize wear on the drive system.

3.2 Temperature control

Rotors spinning in air generate a great deal of heat; even a small microcentrifuge rotor can heat up rapidly. It is possible to cool the rotor by drawing air at ambient temperature across the rotor, but most air-cooled machines run at 3–8 °C above ambient temperatures. An exception to this is the Air-fuge where the air driving the rotor is cooled by expansion and so the rotor runs about 6–8 °C below the temperature of the air. However, most people prefer to control the temperature more precisely to keep delicate biological samples intact and to reduce the possibility of artefacts arising from enzymatic digestion during centrifugation. Hence, most biomedical users prefer to use refrigerated, floor-standing low-speed machines for large volumes and refrigerated benchtop centrifuges for smaller samples. Since cooling with prechilled air is rather difficult to control, direct chamber cooling of the centrifuge bowl is the system most widely used. The very stable chlorinated fluorocarbon refrigerants which damage the environment are being replaced by less polluting refrigerants based on more easily degraded compounds. These refrigeration systems are standard for almost all low and high-speed centrifuges. The heat generated by the rotor can be greatly reduced if the centrifuge has a vacuum pump to reduce the air pressure in the bowl; this greatly reduces the strain on the refrigeration system. In the case of ultracentrifuges, the rotors of which run in a high vacuum, the cooling capacity required is low enough to allow Peltier-effect thermoelectric refrigeration to be used. Thermoelectric elements allow the manufacture of very compact machines. Temperature control is very important for ultracentrifuges where often one is trying to separate relatively small macromolecules on non-viscous gradients. Large variations in temperature can lead to convection currents in the liquid in the tube and this will adversely affect the quality of the separation obtained. The most accurate temperature control is obtained by using radiometer temperature sensors which can control the temperature to less than ± 0.5 °C while thermistor-based sensors are less accurate, capable of controlling the temperature only to within about ± 2 °C.

3.3 Vacuum systems

The higher the circumferential speeds, the higher is the rotor surface friction with the surrounding air. For high-speed centrifuges either there is no vacuum system (usually up to 21 000 r.p.m.), or a simple mechanical pump capable of generating a vacuum of 200–300 Pa is used for speeds of up to 30 000 r.p.m. To ensure exact temperature control for ultracentrifuges while spinning either very high-performance rotors or bulky rotors with rough surfaces at high speeds the simple vacuum system with a mechanical fore pump must be supported by a diffusion pump (giving a 20-fold reduction in pressure). Some instruments have a threshold of 40 Pa to prevent rotors from spinning faster than 50 000 r.p.m. until a better vacuum is reached; ideally, a vacuum of less than 10 Pa should be achieved.

3.4 Rotor recognition systems

In many simple types of centrifuge the maximum speed of the rotor is limited by the weight and windage of the rotor as well as the power of the motor. However, as a result of national regulations and as manufacturers seek to increase the sophistication of their machines, a number of different rotor identification systems have been devised. While simple overspeed systems employ optical sensors to read the number of sectors on a disc on the bottom of rotors, the more sophisticated rotor identification systems use arrays of magnets that vary in their positions and alignment of poles to determine both the maximum speed and the type of rotor. Another sophisticated system, based upon ultrasonic scanning of the rotor shape, has been introduced and this can also detect whether the rotor is correctly installed in the centrifuge. These sophisticated systems can not only identify the maximum speed of the rotor but can identify also the type of rotor so that it is possible to convert between speed and the centrifugal force generated at that speed using software built in to the centrifuge.

3.5 Software support systems

The original use of microprocessors was to make centrifuges more reliable, runs more reproducible, and servicing easier. However, software is being used to make even small benchtop centrifuges multilingual in their display of centrifugation parameters, giving operators guidance and warnings as well as performing conversions from speed into centrifugal force at given rotor radius positions. For ultracentrifuges, rotor and instrument log systems have been developed for built-in printers. Manufacturers have also introduced on-line networks for automatic logging of all connected instruments and a register of available rotors as well as on-line data processing and simulations. Automatic positive rotor recognition is available for low- and high-speed centrifuges as well as for some ultracentrifuges (see Section 3.4). Off-line PC-based data processing and simulation programs that run on a wide range of personal computers are also available.

4. Centrifuge rotors

4.1 Introduction

Originally centrifuge rotors were designed in a fairly empirical manner in which, having chosen the number and capacity of tubes, a round piece of metal of the appropriate size and alloy which was able to withstand the required speed

Figure 2. A computer-optimized titanium alloy ultracentrifuge rotor.

without exploding was used. However, as a result of computer-aided design methods, modern ultracentrifuge rotors often have an angular appearance (*Figure 2*). There are two reasons for machining rotors:

(a) removal of outside metal which is not directly required for holding the tubes; this is because any metal which is not required actually weakens the rotor (this is because as the rotor spins the centrifugal force acts not only on the samples but also on the rotor material)

(b) the reduction of the kinetic energy of fragments in the event of a rotor failure

The manufacturing process for a typical aluminium or titanium rotor involves forging metal into the approximate shape required followed by one or more machining steps to refine the shape.

The efficiency of rotors can be expressed in terms of their k-factors. As described in Section 6 of Chapter 1, the k-factor is determined by the speed of the rotor and the minimum and maximum radii of the sample; note that these will be different from the maximum and minimum radii of the rotors depending on whether adapters or part-filled tubes are used. The k-factors can be calculated from the equation:

$$k = 2.53 \times 10^{11} \, (\ln r_{max} - \ln r_{min})/Q^2$$

where Q is the speed of the rotor in revolutions/min and r_{min} and r_{max} are minimum and maximum radii (in centimetres) of the liquid in the tube, respectively. The k-factor indicates the length of time required to pellet a particle with a sedimentation coefficient of 1S; that is the smaller the k-factor the more efficient is the rotor. The relationship between sedimentation coefficient, s, sedimentation time, t (h) and the k-factor, k, is given by:

$$t = k/s$$

The k-factor is dependent on the speed of the rotor and all the manufacturers give the k-factors at maximum speed, but this may not always be possible. To find the k-factor at a lower speed use the equation:

$$k_{new} = k(Q_{max}/Q_{new})^2$$

Since the relationship is dependent on the square of the speed, a small decrease in speed can increase the k-factor significantly.

The k-factor assumes that the particles are sedimenting through a medium of a similar density and viscosity to water. In the event that this is not so, the k-factor may give an inaccurate estimate of the sedimentation rate of particles. As described in Section 6 of Chapter 1, k^*-factor is an alternative measure of rotor efficiency which defines the rate at which a particle sediments through a 5–20% (w/w) sucrose gradient at 5 °C. This may also give a more accurate estimation of the sedimentation of particles in isotonic sucrose solutions.

4.2 Materials used for centrifuge rotors

The stress on rotors is proportional to the square of the speed (see Section 3.1 of Chapter 1) and hence the stress on ultracentrifuge rotors is several orders of magnitude higher than that on low-speed rotors. Low-speed rotors can be made from bronze, steel, or even plastics such as polypropylene. However, most modern low-speed rotors are made from aluminium and have computer-optimized designs which combine lightness with the strength needed for higher performance. High-speed rotors are usually made from aluminium alloys although some swing-out rotors do have titanium buckets. In order to obtain the high performance required of ultracentrifuge rotors, a range of different titanium and aluminium alloys are used; the actual choice of alloy depends on the type of rotor. Advantage has also been taken of the favourable high strength-to-weight ratio of composite materials to produce rotors which are much lighter than their metal equivalents for both high-speed and ultracentrifuges.

Centrifuge rotors made from aluminium alloy must be treated with care because they are particularly susceptible to corrosion. The corrosion of aluminium rotors is greatest in acids, alkalis, and strong salt solutions. However, even dilute solutions of salts or buffers can become concentrated as a result of evaporation and so cause significant disruption to the aluminium oxide film which protects the otherwise very reactive metal. Aluminium rotors are usually anodized but this does not give ultracentrifuge rotors a high degree of protection against corrosion. The reason for this is that even dilute salt solutions can cause corrosion pitting or the protective anodizing may be scratched as a result of careless cleaning procedures (e.g. using a wire brush with a sharp metal end). It has also been suggested that the oxide film of the anodizing of ultracentrifuge rotors may crack as the rotor expands at high speed. If the anodizing film is damaged in any way then during centrifugation any liquid present in the pocket of the rotor will be forced into the crystalline structure of the metal leading to stress corrosion. Stress corrosion is the main cause of failure of aluminium rotors. Aluminium rotors also suffer from metal fatigue arising from repeated changes in shape caused by each centrifugation run. Note that stress corrosion of aluminium rotors for low-speed use only is much less of a problem since the stress on the rotor is less than the force required for crack propagation through the metal. Hence cracks do not tend to spread through the rotor even if they are extensively weakened by corrosion.

Titanium rotors are much more resistant to corrosion and indeed corrosion is unlikely to be a problem using any of the solutions normally used for the separation of biological material. Rotors made of titanium alloy are not affected by most acids and alkalis nor by most salt solutions. Since titanium is so much less reactive than the aluminium alloys used for rotors, stress corrosion does not occur, and titanium rotors also appear to be much more resistant to metal fatigue. High-performance titanium rotors can withstand many more stress cycles at their full rated speed than aluminium rotors. Derating titanium rotors from new to 90% of their maximum speed reduces the rotor stress by 20%, so increasing the life expectancy of the rotor. Titanium rotors derated from new to 80% of their maximum speed from new should last indefinitely. In the authors' experience it is very rare for a titanium rotor to fail when used within the speed range recommended by manufacturers. However, even so, repeated cycles of acceleration and deceleration up to maximum speed will eventually weaken the rotor so that derating or replacement of the rotor becomes advisable. Rotor guarantees vary from one manufacturer to another and so it is important to check with each manufacturer the exact terms of their rotor guarantee. Several manufacturers offer a rotor inspection service and on the basis of this may extend the length of the guarantee period.

One of the major innovations in rotors has been the introduction of composite materials for the construction of centrifuge rotors. Carbon composite material is made from spun or woven carbon fibres embedded in an epoxy resin matrix. The strength of carbon fibres is very directional and for maximum strength the orientation of the fibres in a composite rotor should take into account the orientation of the maximum directional forces. Some carbon fibre rotors are made from short (0.5–4 mm), randomly-orientated fibres, but for maximum strength rotors require more complex hoop windings or compressed interwoven fibre sections which are expensive to produce. While carbon fibres themselves are almost inert, the epoxy resin which binds the fibres together can make the rotors susceptible to permeation by different solvents and solutions, hence the degree of resistance will depend on the nature of the epoxy resin used and the method of manufacture. Appendix 1 gives a guide to the chemical resistance of composite materials. Composite rotors are not weakened by any of the solutions normally encountered in biological work and the high strength and rigidity of carbon fibres make the rotors resistant to fatigue. However, almost all composite rotors have metal tube holders (*Figure 3*) and so these must be taken into consideration when discussing the resistance of the rotors to corrosion and stress. Fortunately, the tube holders of most composite rotors are removable and can be replaced if they become corroded or damaged. Some rotors are claimed to be autoclavable as a result of their manufacturing process which involves curing the composite material under high pressure and temperature cycles. Autoclaving is not recommended by other manufacturers. Composite rotors are coated with epoxy paint but holes for the tube holders are machined through the composite. The extent to which the different carbon fibre rotors may absorb small amounts of water into the composite depends on the method of manufacture; if absorption does occur then any spillages of radioactive material might seriously and permanently contaminate the rotor. There are real advantages in making composite rotors in that they are 60% lighter than equivalent titanium rotors and 40% lighter than aluminium rotors. Hence, at least for the larger rotors, it is possible to reduce the amount of time needed for acceleration and deceleration and hence the total centrifugation time required. The lower weight of these rotors also reduces the stress on the drive system and makes the rotors easier to handle. Given the diversity of advice for the use and limitations of these rotors, the only general rule can be to follow the manufacturers' instructions on their use.

4.3 Classes of centrifuge rotors

Almost all preparative centrifuge rotors can be classified into one of four different types, namely swing-out (swinging bucket), fixed-angle, vertical, and

Figure 3. Carbon fibre composite rotors with their aluminium tube holders.

Table 2. Types of rotors and their applications

Type of rotor	Type of separation		
	Pelleting	**Rate-zonal**	**Isopycnic**
Fixed-angle	excellent[a]	poor	good
Vertical	poor	good	good
Swing-out	inefficient	good	adequate
Zonal	poor	excellent	adequate

[a]NVT rotors are not recommended for pelleting samples.

zonal rotors. Zonal rotors and their uses are described in detail in Chapter 8 and so will not be considered further in this chapter. Each class of rotor has markedly different characteristics in terms of its design and also, as shown in *Table 2*, their usefulness for different types of applications.

4.3.1 Swing-out rotors

In the case of these rotors, the sample tubes are loaded into individual buckets which hang vertically while the rotor is at rest. When the rotor begins to rotate the buckets swing out to a horizontal position (*Figure 4*); this process is complete by the time the rotor reaches 500 r.p.m. Hence, in these rotors the centrifugal force is exerted along the axis of the tube. However, because the centrifugal force generated is a radial force, some particles are sedimented against the walls giving rise to wall effects, but usually these do not disrupt the sample zones as they move down the gradient in this type of rotor; this is discussed in greater detail in Section 3.2.1 of Chapter 1 and Section 2.1.2 of Chapter 3.

In the case of most low- and high-speed swing-out rotors, individual buckets simply slot onto the central part of the rotor; often these rotors have a windshield to minimize the wind resistance of rotors thus maximizing the top speed and minimizing the amount heat generated. In some cases each bucket must be placed in its correct position on the rotor to ensure the correct balance of the rotor. Tubes and bottles for these rotors must be carefully balanced and placed symmetrically in the rotor.

In the case of ultracentrifuge rotors the necessity for the rotor to withstand very high centrifugal forces means that much more sophisticated designs and stronger materials are required. The original pivot pin method for attaching buckets to the rotor has dropped out of favour since it is only possible to attach a maximum of three buckets. The two commonest systems for attaching the buckets to ultracentrifuge rotors are shown in *Figure 5*. The commonest method is to hook on the buckets, and the hook may be on the bucket itself (*Figure 5a*) or on the centre of the rotor (*Figure 5b*). The security of this system depends on the detailed design of the rotors and as such varies from one rotor to another, even from the same manufacturer. The second method of attaching buckets is the ball-and-socket design (*Figure 5c*). The ease of use of this design of rotor depends

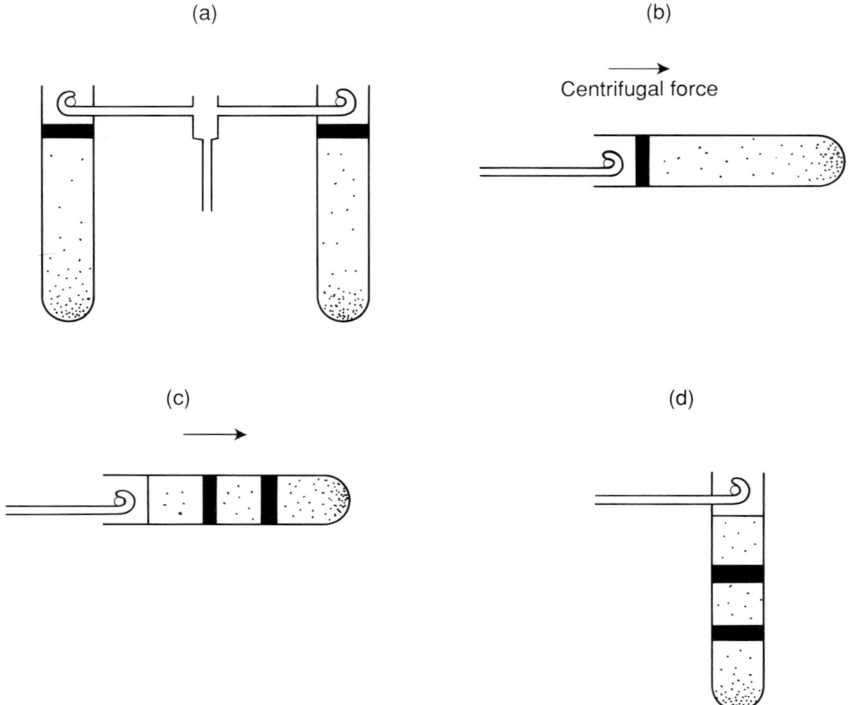

Figure 4. Separations in swing-out rotors. The tubes are filled, loaded into each bucket and attached to central body (yoke) of the rotor; at rest the buckets of the rotor hang vertically (a). As the rotor begins to move the buckets move out so that they are perpendicular to the axis of rotation (b). During centrifugation the particles sediment down the tube (c). When the rotor comes to a stop (d) the buckets return to a vertical position; there is no reorientation of the liquid in the tubes.

on the design, particularly in terms of the depth of the socket machined into the rotor. In some cases the sockets are so shallow as to result in easy, accidental detachment of the bucket. With this design it is important to keep the faces of both the top ball face of the bucket and the socket face of the rotor clean in order to ensure smooth movement of the bucket during acceleration and deceleration. To prevent uneven movements of the buckets and cold welding of titanium buckets with the titanium yoke, the rims of the buckets should be protected with a lubricant coating. Before each run a thin film of molybdenum disulphide grease (Molykote, Spinkote, Tegmol) should be applied. The other type of design is the top-loading bucket. This is an inherently safe design, but the geometry involved restricts the number of buckets to four. In fact, six-place rotors are the most useful type of swing-out rotors, allowing you to centrifuge two, three, four or six samples at the same time. However, it is important to note that, in order to keep the rotor stabilized, then *all* six buckets must be

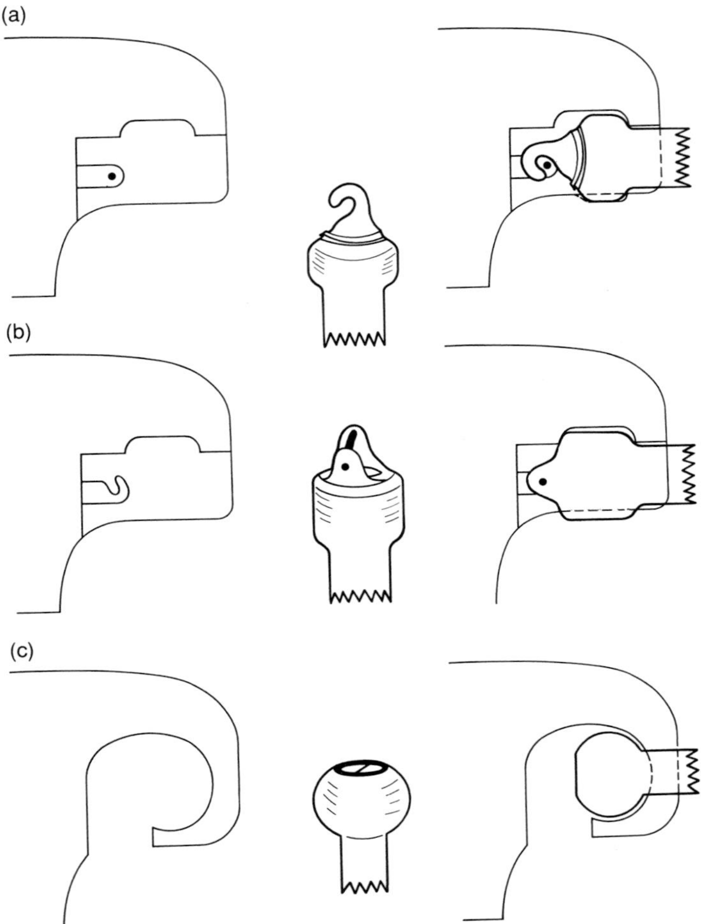

Figure 5. Methods of attaching buckets to ultracentrifuge swing-out rotors. The commonest method of attaching buckets is to hook the bucket on to the centre (yoke) of the rotor. The hook may be on the bucket (a) or the yoke (b) of the rotor. The other method of attaching the bucket is using the ball-and-socket design (c).

attached during the run. As described in Chapter 4, swing-out rotors are frequently used for the rate-zonal separation of particles on sucrose gradients. They can also be used for isopycnic fractionations although they have a much lower sample capacity than gradients of a similar volume in fixed-angle or vertical rotors, take longer to reach equilibrium, and generally (for similar tube volumes) must be run at lower speeds compared with fixed-angle and vertical rotors. However, when separating large particles such as cells and cell organelles by isopycnic centrifugation, often a swing-out rotor will give a cleaner separation because the bands are discrete and smearing along the wall as a result of wall effects is minimal. In addition, there is no reorientation of the liquid in the tube which can disrupt bands. Swing-out rotors have relatively higher *k*-factors than fixed-angle rotors of similar volume and so they are not efficient for pelleting. However, the attraction of obtaining a small pellet at the centre of the bottom of the tube often leads to their use for pelleting, particularly when the amounts of material to be pelleted are at the microgram level or less.

4.3.2 Fixed-angle rotors

In these rotors, as the name suggests, the tubes are at a fixed angle and when the rotor begins to turn the solution in the tubes reorientates (*Figure 6*). The angle of the tube in the rotor can vary from more than 50° to less than 10°.

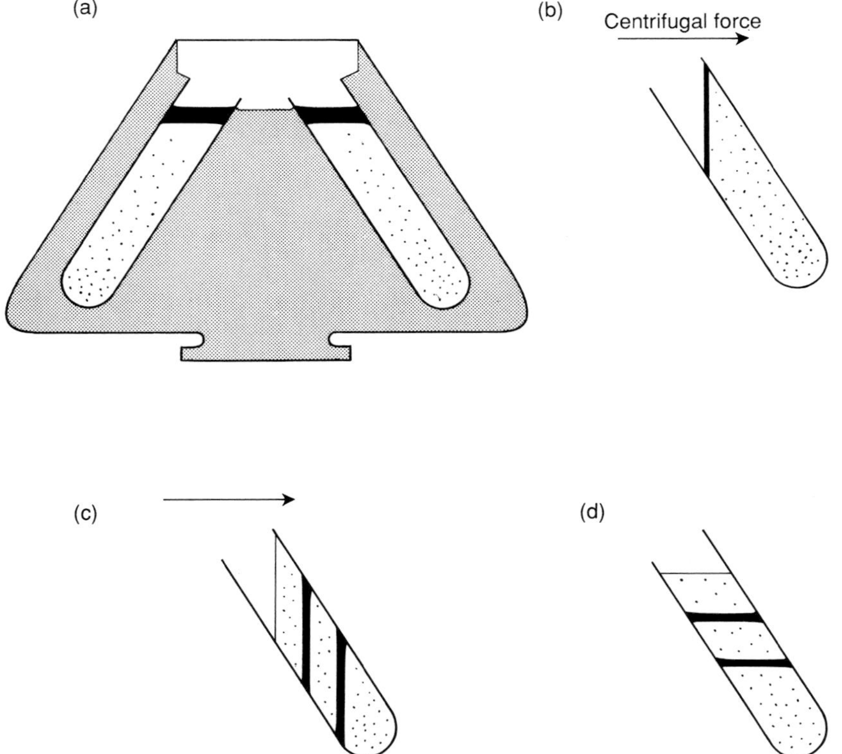

Figure 6. Separations in fixed-angle rotors. The tubes are filled and placed in the rotor pockets (a); tubes for ultracentrifuge tubes are filled completely and sealed, unsealed tubes must not be overfilled. As the rotor begins to move the liquid in the tube reorientates during acceleration (b) and the particles sediment down the tube (c). As the rotor comes to a stop (d) the liquid in the tubes reorientates as gravity becomes stronger than the centrifugal force.

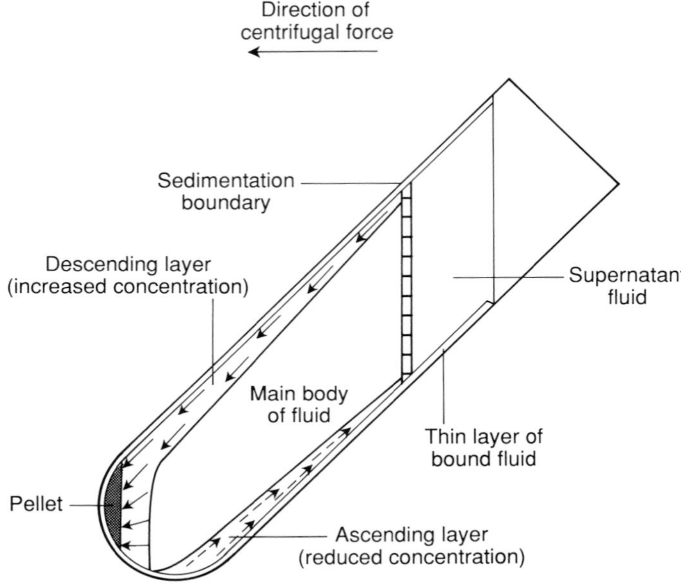

Figure 7. A diagramatic representation of pelleting in a fixed-angle rotor.

Shallow-angle rotors tend to have lower *k*-factors because of the short effective pathlength. They are also favoured for isopycnic separations because the large amount of reorientation that takes place increases the capacity of gradients (also see Section 4.2.1 of Chapter 4). Beckman Instruments Inc has designed a NVT (near vertical tube) rotor specifically to take advantage of these two characteristics for the rapid isolation of plasmid DNA on CsCl–ethidium bromide gradients (see Section 4.3.1 of Chapter 4). Fixed-angle rotors can be designed to withstand very high centrifugal forces ($>600\,000g$) enabling rotors to be made with *k*-factors of less than 30. These rotors are very efficient for pelleting because the particles are sedimented across the tube, hit the wall of the tube, and slide down to form a pellet at the bottom of the tube (*Figure 7*). NVT fixed-angle rotors are not suitable for differential pelleting because the tube angle is too shallow. Wall effects can be increased by the addition of small glass beads into the tubes and this enhances the pelleting efficiency of rotors (4), although in practice this is rarely done. The same wall effects which contribute to the efficient pelleting of particles can also disrupt sample zones as they pass down the gradient. Hence, as a general rule, fixed-angle rotors should not be used for rate-zonal separations (5) unless the particles are less than 10S. In contrast, for isopycnic separations sedimentation of particles on to the tube wall tends to have less effect on the quality of separations since they slide down the tube wall until

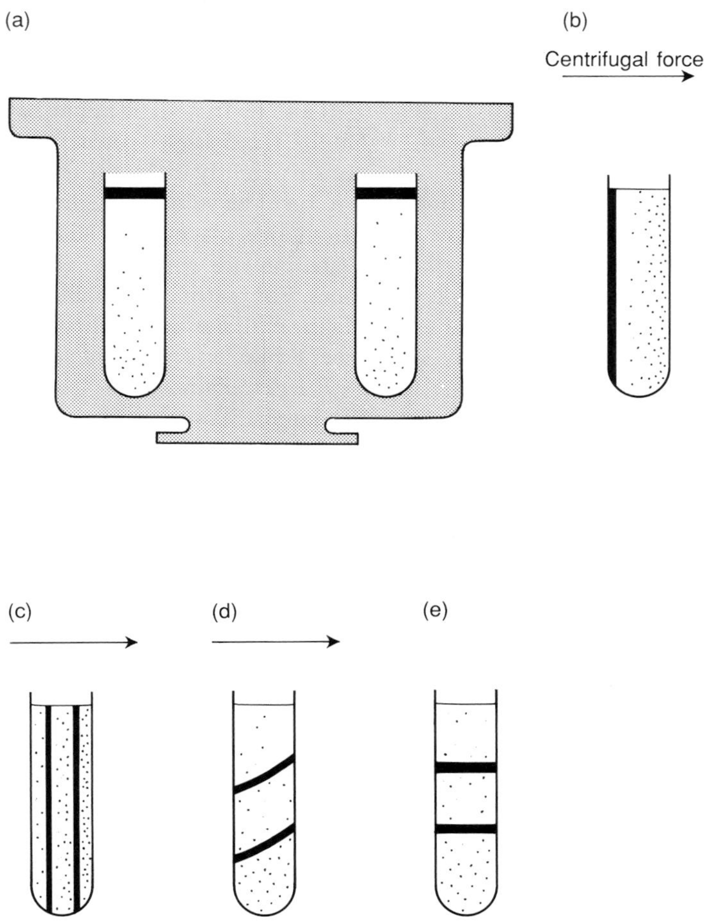

Figure 8. Separations in vertical rotors. The tubes are filled completely, sealed, and loaded in to the pockets of the rotor (a). As the rotor begins to move the liquid in the rotor reorientates through 90° (b), and particles start to sediment across the tube (c). As the rotor decelerates below 1000 r.p.m. the bands reorientate (d), and as the rotor comes to a stop this process is completed (e).

they reach their isopycnic positions. However, if the particles are very large then they may stick to the tube wall; this can be a real problem with cells or large organelles such as nuclei or mitochondria. The reorientation of the solution in the rotor increases the capacity of the rotor for isopycnic separations because the band area, which determines capacity, is so much greater after the reorientation which occurs during centrifugation (6). This effect is maximal in vertical rotors.

4.3.3 Vertical rotors

These rotors first introduced during the 1970s are manufactured for most types of high-speed and ultracentrifuge and are widely used. In this rotor the tubes are held in a vertical position (*Figure 8*) and at first sight might be considered simply as an extremely shallow angle fixed-angle rotor. However, the characteristics of this type of rotor are sufficiently different to merit separate consideration. When the rotor begins to turn the solution begins to reorientate through 90° (*Figure 8*). This reorientation takes place below 800 r.p.m. As long as the acceleration and deceleration profiles are controlled then there is no disruption of the gradient or particle zones. For reasons which are not completely understood, the resolution of gradients in small diameter tubes is spoiled if the rotor is decelerated too slowly. As a general rule, it is recommended to use rapid acceleration and braking for tubes of 1-cm diameter or less and minimal acceleration and deceleration for tubes larger than this. The short pathlength of vertical rotors, effectively the diameter of the tube, endows vertical rotors with low *k*-factors. The commonest use of vertical rotors is isopycnic centrifugation. However, the lack of wall effects, at least over the first part of the gradient (*Figure 9*), also makes them ideal for rate-zonal separations (3). It is possible to compare the theoretical resolution of vertical and swing-out rotors by using computer simulation methods (7). The design of vertical rotors allows them to generate high centrifugal forces with the minimum centrifugal force being not much less than the maximum; this feature allows vertical rotors of high-speed centrifuges to be used for rate-zonal separations which would

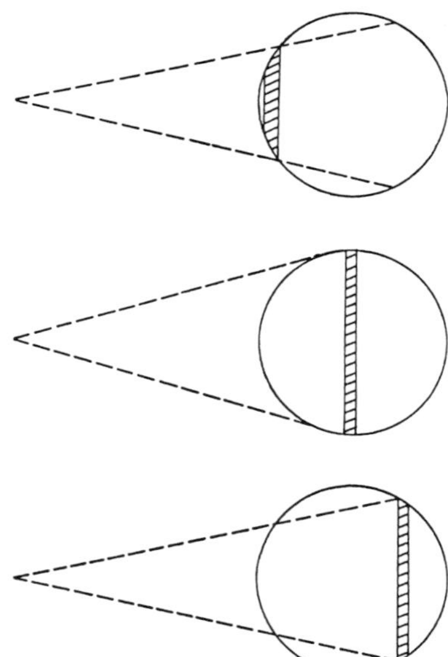

Figure 9. Sedimentation of a sample zone across a tube in a vertical rotor during centrifugation. [Reproduced from (3) with permission.]

otherwise require an ultracentrifuge and swing-out rotor. Vertical rotors are also excellent for isopycnic separations since the reorientation which occurs in the gradient maximizes both the resolution and capacity of the gradient. However, as a result of reorientation, the area of the sample band during centrifugation is larger than in other types of rotor and so, unless the sample has a sedimentation coefficient greater than 20S, sample bands will be significantly broader than equivalent bands in swing-out or fixed-angle rotors as a result of diffusion during centrifugation (8). Another problem is that any material which pellets or floats to the top of the gradient in a vertical rotor is distributed along the whole length of the tube and this can subsequently contaminate the supernatant during reorientation of the liquid in the tube at the end of the run.

4.4 Interchangeability of rotors

The drives of some types of centrifuges have become *de facto* standards; for example, all modern ultracentrifuges have the same dimensions of drive spindle, and an overspeed system which is compatible with that originally introduced by Beckman Instruments Inc. The result of this is that most ultracentrifuge rotors can be run on more than one make of machine. However, while some manufacturers support this practice, and indeed will guarantee their rotors in other makes of machine, other manufacturers specifically prohibit this practice and it voids the warranty on the rotor and centrifuge. There is some logic in this latter approach in that the dynamics of the drive systems of different manufacturers are different and rotors will behave slightly differently on each drive system depending on their design. However, during development, drive systems are often tested with a wide range of rotors and so interchangeability should not be a problem. In the case of micro-ultracentrifuges, no standard has been established and rotors of different manufacturers are usually incompatible. In the case of high-speed centrifuges, some manufacturers have made their rotors compatible with the Sorvall cone-drive system but, overall, there is only a limited degree of interchangeability of high-speed rotors. Low-speed centrifuges have a wide variety of drive shafts; often each instrument is different with incompatibilities between different models from one manufacturer. The only safe approach is to assume that none of the rotors or their components is interchangeable. When buying a low-speed centrifuge be sure to choose a drive which does not feature a keyed shaft or narrow cone drive since rotors readily become stuck on these types of shaft, making a change of rotor at best a chore and at worst impossible. The comments in this section refer to the current situation, and as rotor identification systems become more frequent and more sophisticated, the amount of interchangeability of rotors is likely to become more restricted. The golden rule must be always to check with the manufacturer before using one type of rotor on another type of machine.

4.5 Stability of rotors

One of the main considerations in both designing and using rotors is their stability during centrifugation. Besides rotating around their geometrical axis they are also subject to other directional forces, some of which may adversely affect the quality of separations. One of these forces causes precession which arises as a result of the torque of the drive tending to tip the rotational axis of the rotor; a similar motion is seen in the slow circular motion of a spinning top. This movement occurs even with perfectly balanced rotors and it is accentuated if the centrifuge is not levelled. Drive systems are damped to minimize this type of movement and, as it is a slow movement, it does not affect the quality of the separation. Another type of motion is synchronous whirl which arises from the fact that it is impossible to balance the rotors and tubes perfectly, and so the centre of gravity of the rotor will tend not to coincide with its geometrical axis. This means that as the rotor rotates the geometrical centre will describe a circular path synchronously with rotation. This type of movement is maximal at the critical speed of the rotor which depends on the characteristics of the drive system and rotor. The critical speed, where synchronous whirl is at a maximum and the rotor is least stable, is usually at about 200–300 r.p.m. Since the movement is synchronous this also does not affect the quality of the separation during centrifugation.

A more serious problem is that of asynchronous whirl which occurs at higher speeds and can seriously affect the quality of centrifugal separations. Asynchronous whirl is caused by mechanical imperfections or imbalance of the rotor or drive assembly. In this case, during acceleration and deceleration, as the rotor passes through its critical speed, the amplitude of whirl of the rotor increases significantly so that it is no longer synchronous, and this will cause the mixing of gradients as well as band broadening of the samples. Rotors and drive systems are always checked by the manufacturer for asynchronous whirl. However, instability of a rotor resulting in asynchronous whirl can, and does, occur for a number of other reasons. One frequent reason is that the centrifuge has not been levelled correctly. Another source of asynchronous whirl is the incorrect balancing of tubes. The magnitude of the effect of a small imbalance depends on the weight, design, and rotational speed of the rotor; at high speeds it can be very large indeed. Correct balancing is particularly important when centrifuging density gradients, especially when using ultracentrifuge swing-out rotors. Particular care should be taken in selecting tubes and bottles since the weight of apparently identical tubes and bottles can vary significantly from one batch to another. For precise density gradient work it is recommended that the gradients and tubes are balanced to within 0.1 g. Ultracentrifuge rotors are spun in a high vacuum so that any small leakage (for example, of the bucket of a swing-out rotor caused by a missing or damaged rubber O-ring) can cause a tube to lose several millilitres in just a few hours. If the leakage is severe then the rotor can become so unstable as to bend the driveshaft. In extreme cases the rotor may come off of the spindle, leading to catastrophic failure. Tubes must be balanced not only in terms of their actual weight but also in terms of their centre of gravity. This is particularly important in density gradient separations; tubes should not only be balanced by weight but should also contain identical gradients. Another situation which can arise is that sedimentation of samples with large amounts of suspended solids often leads to unevenly-sized pellets and so even though the bottles or tubes were originally balanced correctly the rotor is out of balance after pelleting has occurred. In the case of ultracentrifuge swing-out rotors, the caps and buckets are an integral part of the rotor and so each bucket must be attached in the correct orientation and position on the rotor to ensure the radial symmetry of the rotor thus ensuring maximum stability. In the case of ultracentrifuge rotors, the tubes and their contents must be balanced by themselves and not in the rotor buckets.

A second common cause of asynchronous whirl is running a centrifuge with a bent drive shaft. As stated previously, unbalanced rotors can bend the drive shaft. However, more usually damage to the spindle is caused when rotors are installed in or removed from the centrifuge without due care and attention. Rotors should be lifted off vertically and with the correct extractor key when one is provided, particularly when rotors are stuck on the spindle. Rotors should never be removed by wrenching or twisting the rotor at an angle to the spindle. If it is suspected that the drive shaft is bent it can be checked by placing the rotor on the spindle and spinning it slowly by hand; if the spindle is bent then the centre of the rotor will describe a small discernible circle. If the spindle is bent then it must be replaced before the centrifuge is used again. A bent spindle not only ruins the quality of centrifugal separations but also it can cause serious and expensive damage to the centrifuge. A bent spindle may also be noticed by abnormal noise and vibrations or it may cause excessive oil consumption. Systems for a built-in dynamic drive/rotor field test (DRFT) are available for some types of ultracentrifuge so that users can test both the drive system and rotors without any need to call in a service engineer. The method involves a two-step procedure in which lateral displacement of the drive with a known test rotor is compared with the results previously obtained. If the spindle is bent, it would clearly be shown by a change in the lateral displacement pattern. The behaviour of other rotors can be recorded in an expanded low-speed range and up to top speed. Plastic or dynamic deformations, for example, a defective bucket seat or an aluminium lid which has been damaged by it falling on to a hard surface, will cause alterations in the imbalance frequency. An incorrectly positioned lid as a result of a wrong or worn rubber O-ring can also be detected. Changes in the vibrational patterns of rotors are a warning sign of possible imminent failure and the rotor should not be used again until after it has been checked by the manufacturer.

Finally, it should be noted that any object that touches a spinning rotor is likely to induce asynchronous whirl, and so any attempt to restrain low-speed synchronous whirl is likely to lead to asynchronous whirl as the rotor is destabilized, and this will tend to spoil the quality of the separation achieved.

Similarly, attempts to slow down a spinning rotor by hand will also have a deleterious effect on the quality of the separation.

Rotors are also subjected to additional forces produced by excessive acceleration and deceleration during reorientation of the liquid sample; this effect is only seen below 800 r.p.m. and they are termed 'Coriolis forces'. These forces can cause severe mixing, band broadening, and boundary-zone destruction. Uneven acceleration or deceleration will also have the same effect at low (< 800 r.p.m.) speeds. Spindle stabilization by bushing or damper bearings should ensure rotor stability throughout the run and during the rotor's critical revolutions. Any rotor with a tube diameter of more than 1 cm should be run at low settings for acceleration and deceleration during reorientation of the samples. This general rule should be followed for all types of rotors, except, as mentioned in Section 4.3.3, in small-diameter tubes in vertical and shallow fixed-angle rotors where Coriolis forces are much less significant and better results will be obtained using rapid acceleration and braking.

4.6 Care of centrifuge rotors

The massive nature of most centrifuge rotors gives the impression of great strength and it is difficult to imagine circumstances which could result in the failure of a rotor. In fact, some rotors can be damaged by dropping them on to a solid surface. The repeated autoclaving of rotors, particularly those manufactured from more than one material, at 120 °C or more, can also weaken rotors. Unfortunately, often such weakening only shows up during centrifugation, usually in the form of catastrophic rotor failure.

However, as described in Section 4.2, the major problem with rotors, especially aluminium rotors, is that of corrosion in low-speed rotors and of stress corrosion in ultracentrifuge rotors. In the case of ultracentrifuge rotors, the user is strongly advised to invest in titanium rotors which are very resistant to corrosion since this will ensure greater safety and long-term savings on rotors. In the case of aluminium rotors and aluminium tube holders of composite rotors, great care must be taken to avoid any corrosion; this applies particularly to the high-performance rotors where stress corrosion can be significant. Care of aluminium rotors must start from when they are new; before they are used they should be treated with a rotor de-watering spray, either that provided by the manufacturer or a WD40 type of spray, and this treatment should be repeated regularly. After use, almost inevitably some moisture remains in the rotor pockets, and this is why aluminium rotors should always be rinsed out with distilled water after use and stored upside-down in a dry environment. The temptation to use coldrooms or refrigerators for the long-term storage of rotors should be resisted since such enclosures usually have a very humid atmosphere which can accelerate corrosion of the rotor. A typical procedure for cleaning rotors is given in *Protocol 1*.

Protocol 1. Procedure for cleaning centrifuge rotors

1. Be aware of the nature of the sample used in the rotor; if it has been used for infectious material it may be necessary to disinfect or autoclave the rotor first. To do this follow the manufacturer's instructions. *Table 3* summarizes the properties of the different types of disinfectants. If the sample was radioactive be sure to wear the necessary protective clothing.
2. Rinse out the rotor first with large amounts of tap water and then distilled water on a rubber mat in the sink. Do NOT leave the rotor to soak in water for more than a few minutes.
3. Using a warm solution of neutral detergent (detergents used for clothes or neutral laboratory detergents are usually suitable, do not use highly alkaline detergents) brush out each of the rotor pockets or buckets using a soft-bristled brush which has a soft tip. Every care must be taken to avoid any scratching of the metal during this operation. A number of manufacturers provide special brushes for cleaning rotors.
4. Rinse away the detergent solution firstly with tap water and then with distilled water and leave to drain.
5. After wiping away as much moisture as possible, leave the components to air dry; do not seal up the buckets or rotors at this time.
6. Check that each tube pocket is clean, if not repeat Steps 3–5. At the same time check the rotor for any signs of surface pitting or surface damage that could lead to corrosion.
7. Treat the rotor with a de-watering spray such as WD40 and store the rotor at room temperature.

Table 3. The effectiveness of disinfectants against different types of infectious agents

Type of disinfectant	Bacteria		Viruses		Fungi
	vegetative	spores	lipid	non-lipid	
5% Hypochlorite	+ + +	+ + +	+ + +	+ + +	+
Phenolics	+ + +	—	+ + +	+	+ + +
90% Ethanol	+ + +	—	+ + +	+	—
2% Glutaraldehyde or 37% Formaldehyde	+ + +	+ + +	+ + +	+ + +	+ + +

Titanium alloys are essentially resistant to corrosion. However, almost all titanium rotors also have some aluminium or ferrous components which are susceptible to corrosion. As an example, ultracentrifuge rotors often have an aluminium overspeed disc on the bottom that restricts the maximum speed of the rotor, and since this is a fail-safe device any damage caused by careless handling will prevent the rotor from being used. Thus the same rules and washing procedures described previously for aluminium rotors are also good practice for titanium rotors.

The same codes of practice used for metal rotors also apply to composite rotors. The reason for this is that the tube holders of the rotor are usually made from aluminium or titanium and it is very important that users are aware of the type of insert and treat them accordingly. As stated in Section 4.2, the nature of carbon fibre composite material varies both between manufacturers and in terms of the method of manufacture; because of this it is difficult to lay down general rules for the care of these rotors. It is the responsibility of the user to contact the manufacturer to ensure that they do not mistreat these rotors. If it is proposed to work with infectious material, for example, make sure that the manufacturer of the rotor can recommend a suitable decontamination method before you do the experiment.

Damage to the outside of aluminium rotors is not a serious source of corrosion. Similarly, knocks and scratches to the epoxy paint of titanium rotors do not affect their function in any way. However, a good general rule is to avoid abrasive treatment of any centrifuge rotor. Chips and scratches are often a sign of careless use and, as those responsible may go on to damage other, more sensitive components, it is usually worthwhile tracking down the culprits responsible for the damage.

Note that all rotors, irrespective of the material used for their construction, are made to a specific design density, traditionally this is 1.2 g/ml. However, some fixed-angle and vertical rotors have been designed for a density of 1.7 g/ml (the density of CsCl solutions used for DNA separations) or even 4.0 g/ml (the density of CsCl crystals). When solutions denser than the design density are used then the maximum speed of the rotor (Q_{max}) must be reduced according to the equation:

$$\text{Derated speed} = Q_{max}\sqrt{\text{RD}/\text{SD}}$$

where RD is the design density of the rotor and SD is the sample density. The same rule may apply also if adapters made of Delrin with a density of 1.4 g/ml are used in a rotor. If the use of adapters is frequent (see Section 5.3) it is recommended to buy a rotor with a design density of 1.4 g/ml or higher or to derate the rotor permanently by changing the overspeed disc (see Section 3.4) to the required lower speed. Rotors run with sample densities lower than the design density tend to have a longer life expectancy. Always check the design density of rotors because it can vary from 1.2–4.0 g/ml. In addition, when running self-forming gradients, it is important to avoid precipitation of the gradient solute, the high density of which can often result in rotor failure. To avoid such catastrophes use either the information provided by the manufacturer, usually in the form of derating charts, or a computer program to ensure the rotor is not run too fast. Similar programs are given in Appendices B and C of Chapter 1. Some types of ultracentrifuge have software which calculates the optimum speed to obtain a separation in the minimum length of time.

5. Centrifuge tubes, bottles, and adapters

5.1 Materials used for tubes and bottles

A wide variety of materials have been used for the manufacture of centrifuge tubes and bottles. Given the diversity of the sample solutions used for centrifugal separations, it is important to be aware of the limitations of each of the different types of material.

5.1.1 Glass

The very inert nature of glass makes it an almost ideal material for centrifuge tubes and bottles. Glasses can vary a great deal in their composition and this is reflected in the very different strengths of glasses. Typical soda glasses are unable to withstand centrifugal forces much in excess of 3000*g*, while borosilicate glasses such as that used for Corex glass are very much stronger and, depending on the bottle or tube and the exact formulation of the glass, can stand stresses in excess of 10 000*g*. The way the tube is supported is also very important; usually it is best to centrifuge glass tubes resting on a clean rubber pad or in a rubber adapter, so avoiding the tube resting on sharp particulate material or direct contact between glass and metal. The advice of the authors is to test the tubes or bottles before use and to avoid those with any evident damage to their surface. One of the problems encountered when using glass tubes is that small amounts of material can be lost by adsorption to the glass surface. This can be very serious if you are working with microgram amounts of material but it can be prevented if the tube is siliconized by treatment with Repelcote or Siliclad (Hopkin and Williams and Clay Adams, respectively).

5.1.2 Polycarbonate

This plastic has the advantage of being completely transparent like glass, it is very strong, and can be autoclaved. However, it is particularly sensitive to many solvents, especially those with hydroxyl groups such as ethanol and phenol. In addition, it is attacked by all alkaline solutions (this includes many laboratory detergents) even when they are quite dilute. It is also sensitive to some rotor polishes and the nuclease inhibitor diethyl pyrocarbonate. Polycarbonate is a very brittle plastic and tubes will often show stress cracking even after only one or two runs in an ultracentrifuge. Always check the condition of polycarbonate tubes before use. A number of other transparent plastics, most with a similar appearance to polycarbonate, are sold under a variety of proprietary names; note that the properties of such plastics are usually different from polycarbonate and so it is important to check the manufacturers' instructions before using them (see Section 5.1.6).

5.1.3 Polysulphone

This plastic has a slight yellowish tinge and has many of the desirable properties of polycarbonate in terms of its total transparency, strength, and that it may be autoclaved. In addition, polysulphone is resistant to alkaline solutions and, notably, also to ethanol. However, it is attacked by a number of common organic compounds, including phenol.

5.1.4 Polypropylene and polyallomer

These two plastics have similar properties, although polyallomer has marginally better chemical resistance characteristics and is more transparent. Changes in the formulation of polyallomer have made it almost transparent, although less so than polycarbonate and polysulphone tubes. Both polypropylene and polyallomer tubes can be autoclaved at 120 °C for 30 min. These plastics are softer than polycarbonate and thin-walled tubes can be punctured readily for the removal of bands or for unloading gradients.

5.1.5 Cellulose esters

The original tube material based on cellulose esters was cellulose nitrate which showed many of the characteristics of an ideal tube material in that it is hydrophilic, transparent, and easy to pierce. However, cellulose nitrate ages on storage and it also is highly inflammable when heated. It is for these reasons that it has been almost completely replaced by other cellulose derivatives such as cellulose butyrate acetate and cellulose propionate. Tubes made from these compounds exhibit many of the desirable properties of cellulose nitrate without the hazards of inflammability and explosion. None of these compounds can be autoclaved. These tubes cannot be used with caesium trifluoroacetate solutions and they are also attacked by strong acids and bases as well as a range of organic solvents.

5.1.6 Other materials

As stated in Section 5.1.2, there are a number of tube materials with proprietary names such as Ultra-clear and Polyclear whose chemical compositions are not immediately obvious and which, although resembling polycarbonate in appearance, in fact have very different properties; in the case of these two plastics they are both polyethyleneterephthalate (PET) and they have similar chemical resistance properties to polycarbonate, but neither can be autoclaved. Always check the manufacturer's information before using any type of tube that you

Table 4. Properties of centrifuge tube materials

Tube material	Appearance	Max. temp. °C	Max. RCF $\times 10^3$	Acid and alkali	Organic solvents	Sterilize[a]
Glass	transparent	300	10	+ +	+ + +	1,2,3,4
Polycarbonate	transparent	120	500	+ −	+	1,3,4
Polysulphone	transparent	120	400	+ +	+	1,2,3,4
Polyallomer	translucent	80	600+	+ +	+ +	1,3,4
Polypropylene	translucent	120	600	+ +	+ +	1,3,4
Cellulose esters	transparent	60	400	− −	+	2,3,4
Ultra-clear[b]	transparent	60	600+	+ +	+	2,3,4
Polyclear[b]	transparent	60	600+	+ +	+	2,3,4

[a]Sterilization: 1 = autoclaving; 2 = UV light; 3 = ethylene oxide; 4 = 37% formaldehyde or 2% glutaraldehyde solutions
[b]Polyethyleneterephthalate (PET)

are not familiar with. A number of other less common materials such as stainless steel are also available; in the case of this material, the high density of the metal makes it important to derate rotors appropriately (see Section 4.6).

Table 4 summarizes the properties of some of the commonest types of tube material. A full description of the compatibility of the different materials used for tubes and bottles is given in Appendix 1 of this book.

5.2 Sealing methods for tubes and bottles

Traditionally the most tedious aspect of centrifugation has been sealing the tubes or bottles used for centrifugation. The sealing of tubes is essential for the containment of biohazardous or radioactive material and, for tubes exposed to centrifugal forces of about 30 000*g* or more, it is usually necessary to fill thin-walled tubes completely in order to prevent their collapse.

Tubes for ultracentrifuge swing-out rotors are usually not capped but Quick-Seal and Ultracrimp tubes are available for some swing-out rotors. If the tubes are not sealed then, unless you are using thick-walled tubes, it is good practice to fill the tubes to within 2–4 mm of the top of the tube (but do not overfill the tubes) to ensure that they do not collapse during centrifugation. In the event that you are centrifuging a non-hazardous sample in a fixed-angle or a swing-out rotor, then it is possible to use partially-filled, thick-walled tubes without caps; these are available for almost all types of rotor, and the rigid nature of polycarbonate tubes makes them ideal for this type of use. However, it may not be possible to run ultracentrifuge fixed-angle rotors at their maximum speed unless the tubes are completely filled and capped even if you are using thick-walled tubes.

Some larger tubes and many bottles are provided with screw-on caps and, deceptively, these appear water-tight when filled with sample and handled on the

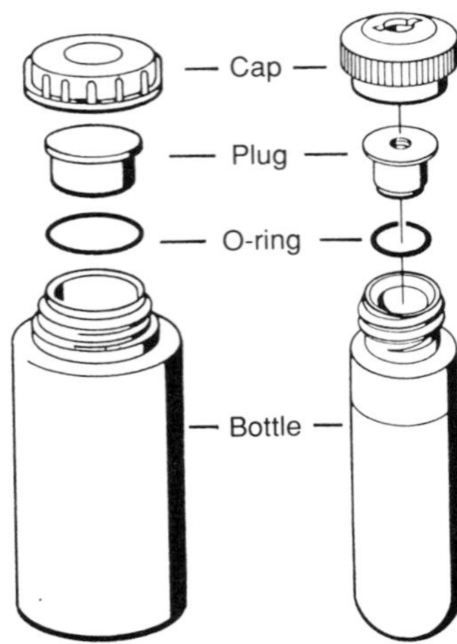

Figure 10. Illustration of O-ring seals for centrifuge tubes and bottles.

bench. However, when centrifuged these caps distort and the centrifugal force can drive liquid past the cap. To ensure a leakproof seal it is usually necessary to use an O-ring compression system for all wide-necked tubes and bottles (*Figure 10*) even for high-speed fixed-angle and vertical rotors. One other ingenious solution to the problem is to use a bottle with an offset neck (Dupont–Sorvall Dry-spin bottles), so that when the liquid reorientates during centrifugation it no longer comes into contact with the cap. Originally, caps for sealing ultracentrifuge tubes were complicated, multicomponent devices relying on the rubber O-ring compression system to seal the tubes. Such caps proved to be unreliable in the hands of inexperienced people and tedious to use for the expert; if possible avoid using this type of cap. In order to obviate the need for such complex caps, a number of different approaches have been tried including the use of screw-on caps which seal under centrifugal force; these require threaded, thick-walled tubes. There are also self-sealing caps from Kontron that use thin-walled tubes. However, the most popular type of tube now is the narrow-necked tube. The original narrow-necked tubes were the heat-sealed tubes (Quick-Seal tubes) introduced by Beckman Instruments Inc. If using this system the novice is advised to spend a little time practising tube sealing, since some skill is required in ensuring that the liquid level is at the correct height and that the top of the tube is heated the correct amount so that it seals properly without weakening the tube. Beckman Instruments Inc have also introduced narrow-necked OptiSeal tubes for NVT and vertical rotors that can be sealed manually. Another sealing system is the Ultracrimp tube introduced by DuPont–Sorvall; in this case the neck is slightly wider and is fitted with a stopper which is crimped in position using a metal cap. The original stoppers were black neoprene rubber but these have been replaced by white polypropylene stoppers

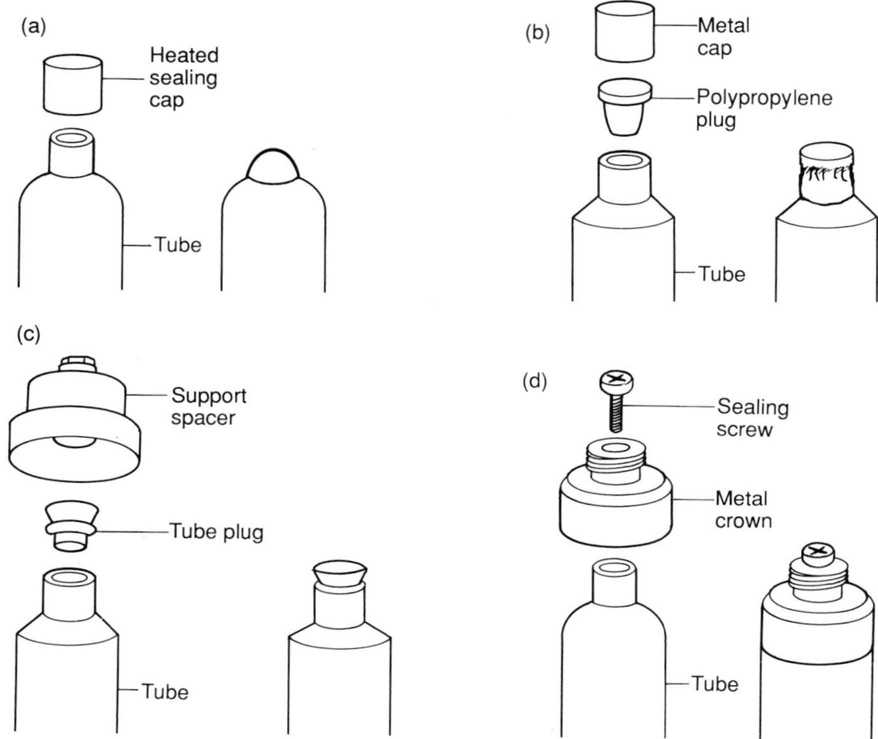

Figure 11. Types of sealing for narrow-necked tubes. (a) Quick-Seal tubes (Beckman Instruments Inc), (b) Ultracrimp tubes (Dupont–Sorvall), (c) OptiSeal tubes (Beckman Instruments Inc), and (d) Reseal tubes (Seton Scientific).

which are better able to resist the very high hydrostatic pressures which can be generated during ultracentrifugation of dense solutions in fixed-angle and vertical rotors (see Section 3.2.2 of Chapter 1). Besides these systems sponsored by their respective manufacturers, other tube-sealing systems have been devised, including one from Nalgene which has the advantage of featuring a much wider neck than the other narrow-necked tubes. Most of the previously mentioned narrow-necked tubes are permanently sealed and sample recovery requires breaking the seal, so that the tube cannot be reused. Reversible seals for narrow-necked tubes have been developed by Seton Scientific. In this case a collar around the neck serves as the cap support during the run and a simple stainless screw can be screwed into the plastic neck so forming a liquid-tight seal; this type of seal also fits the Quick-Seal tubes of Beckman Instruments Inc. *Figure 11* shows some of the different types of tube closure systems for narrow-necked tubes.

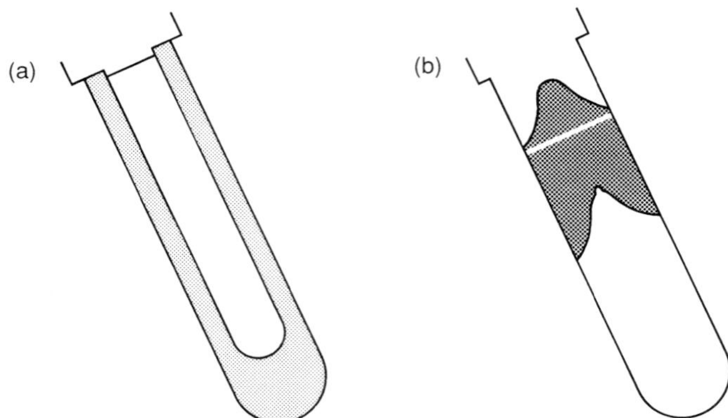

Figure 12. Adapters for rotors. Comparison of (a) conventional adapters with (b) *g*-max adapters of Beckman Instruments Inc. Note that in the *g*-max system the maximum radius and hence centrifugal force is unchanged.

5.3 Adapters for centrifuge rotors

Adapters can be very useful for reducing the effective size of the rotor pocket so that it is possible to run smaller tubes than is usual for that type of rotor. Adapters are often made from Delrin for ultracentrifuges and from rubber or Nylon for low-speed and some high-speed machines. It is important when using adapters to compensate for the change in the centrifugal force exerted on the sample as a result of changes in the position of the tube from the centre of rotation. Note that it may be necessary to derate rotors from their normal maximum speed because of the density of the adapter. Beckman Instruments Inc have a system of adapters (G-max adapters) that maximize the centrifugal force by using short wide tubes with a large spacer on top rather than a narrow thin tube (*Figure 12*) and rotors do not require derating with these adapters.

6. Centrifuge safety

There are regulations in force governing the construction of centrifuges in many countries of the world. While these vary from one country to another, they all have a common aim of ensuring the safety of the user. Almost all centrifuges have lid locks to prevent access to the rotor whilst it is rotating. It is important that the user does not attempt to circumvent this safety feature in an effort to bring the rotor to a halt faster by slowing it down by hand. Such impatience can not only result in physical injury but also is likely to have deleterious effects on the quality of the separation arising from excessive swirling in the centrifuge tubes or bottles. The rule of not touching the rotor is just as important for the older centrifuges that do not have a lid lock, but usually more willpower on the part of the user is required.

All centrifuges are designed to ensure that in the event of any catastrophic failure of the rotor or drive system all of the fragments are retained within the casing of the centrifuge. However, such events are expensive in that usually the centrifuge is damaged beyond repair and often care in use of the centrifuge can help to avoid such failures. Special care should be taken with balancing the tubes in the rotor and, in the case of swing-out rotors, ensuring that all of the rotor buckets are attached to the rotor. Tubes must be balanced as accurately as possible, at least to within 1% of each other, and the balanced tubes placed symmetrically in the rotor to ensure the quality of separation. Failure to balance the tubes correctly will lead to increased wear on the drive system and, in extreme cases, lead to failure of the drive spindle. Forcing rotors on or off of the spindle is likely to bend the spindle and this can also lead to catastrophic failures. If at all possible use a centrifuge with an out-of-balance detector since this will help to protect against such accidents.

Perhaps the greatest source of danger is the lack of cleanliness. Liquid spilt into rotors and not removed can weaken a rotor as a result of increased corrosion. The frequency of such spillages is often the result of the inexpert or naive sealing of tubes or bottles; plastic screw-on caps which appear to be water-tight normally distort under centrifugal force leading to leakages into the rotor. Also take care to ensure that the outside of the tube is dry before putting it into the rotor. The hazards arising from spillages are even greater if the sample is radioactive or bio-hazardous (e.g. infectious micro-organisms). A significant problem with ^{32}P-phosphate is that this tends to bind very tightly to metal surfaces making it very difficult to clean it from the rotor. The most important rule to follow is that all spillages in rotors must be cleaned up immediately using the procedure described in *Protocol 1* and the efficiency of the decontamination method should be checked before further use; whenever samples are hazardous the use of sealed buckets is very strongly recommended and in some cases may be mandatory based on national legislation.

In the event of the escape of samples out of the rotor and into the bowl of the centrifuge other major hazards become apparent. In the case of air-cooled centrifuges the samples are cooled by passing large volumes of air through the bowl and so any leakage of the sample can lead to contamination of the whole laboratory area around the machine. As a general rule hazardous samples should *never* be processed in air-cooled centrifuges and always centrifuged in sealed buckets if at all possible. In the case of machines with a vacuum system, samples leaking from the rotor will be sucked into the vacuum pump(s) requiring complicated and expensive decontamination procedures. Even spillages of samples into the bowl can be a potential hazard to subsequent users unless they are cleaned up immediately.

Acknowledgements

The writing of this chapter has been greatly facilitated by contributions from all major centrifuge manufacturers; special thanks are due to Colin Broadbent,

Shahla Sheikholeslam, and Aldo Taccani, who read and improved on the original drafts of this chapter and were able to correct both our misconceptions and the English grammar.

References

1. Graham, J. M., Ford, T. C., and Rickwood, D. (1990). *Anal. Biochem.* **187**, 318.
2. Abeyasekera, G. and Rickwood, D. (1991). *Biotechniques* **10**, 460.
3. Rickwood, D. (1982). *Anal. Biochem.* **122**, 33.
4. Polson, G. (1984). *Prep. Biochem.* **14**, 223.
5. Castaneda, M., Sanchez, R., and Santiago, R. (1971). *Anal. Biochem.* **44**, 381.
6. Flamm, W. G., Birnstiel M. L, and Walker, P. M. B. (1972). In *Subcellular components: preparation and fractionation* (ed. G. D. Birnie), p. 279.
7. Rickwood, D. and Young, B. D. (1981) *J. Biochem. Biophys. Methods* **4**, 163.
8. Young, B. D. and Rickwood, D. (1981). *J. Biochem. Biophys. Methods* **5**, 95.

3

Conditions for density gradient separations

MILOSLAV DOBROTA and RICHARD HINTON

1. Introduction

While the theory of centrifugation may at first appear daunting, the actual techniques involved in density gradient centrifugation are in practice quite simple and can be mastered easily by a beginner. This chapter is written primarily for those new to the techniques of density gradient centrifugation and for researchers who find that there is no published method which gives the separation which they require and must develop their own. It is, however, rarely necessary to start completely from scratch as some useful pointers to approaches can be obtained from the numerous successful applications which have been developed already. As the authors know by experience, given the different properties of rotors, gradient media, and gradient design, one can spend one's life trying to optimize separations rather than facing the real problem, the study of the properties of the separated particles. The aim of this chapter is to help people with the practical problems of designing gradient separation methods so that they can get on with real science. Hence the approach described is essentially empirical. However, theoretical considerations inevitably impose limitations on what can be achieved and it is those limits which will be addressed first.

2. Theoretical considerations

The theory of centrifugation has been discussed in detail in Chapter 1 of this book so this chapter will concentrate on issues which are concerned with gradient centrifugation.

Separations of particles in density gradients may be due to differences in size between the particles, in which case one refers to the technique as 'rate-zonal centrifugation' because the sample starts off as a narrow zone at the top of the gradient (*Figure 1a*), or to differences in density between particles, in which case one refers to the technique as 'isopycnic centrifugation' (or equilibrium banding); in this latter case particles can be either layered on top of the gradient (*Figure 1b*) or mixed within the gradient itself (*Figure 1c*) depending on the nature

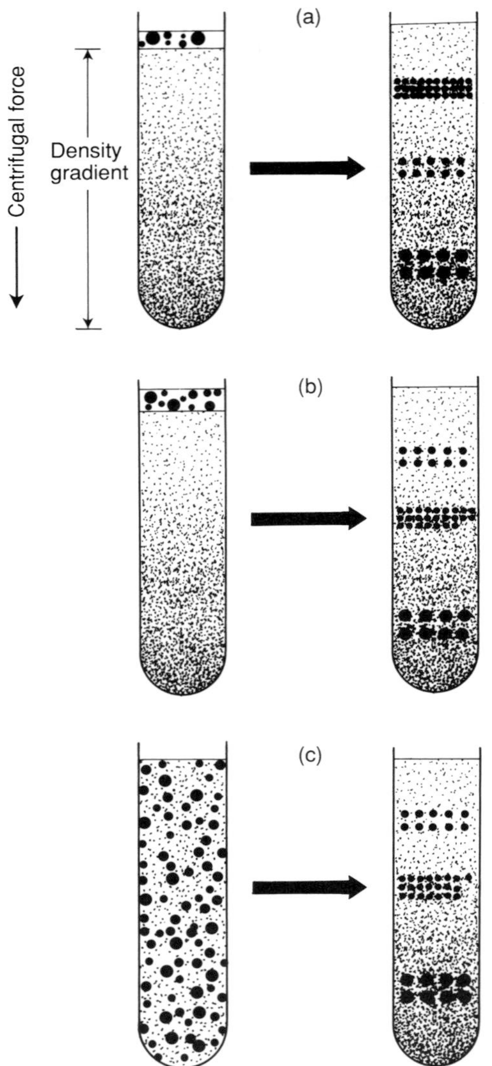

Figure 1. An illustration of the principles of (a) rate-zonal, and (b) and (c) isopycnic gradient separations. Note that in isopycnic centrifugation the samples can be layered on top of the gradient (b), or mixed throughout the solution (c).

of the particle to be separated. As will be discussed later, gradient systems may be designed to exploit differences in size and density simultaneously, but these applications remain exceptional. Furthermore, both the design of the gradient and the nature of the particle may influence the separation. In the case of separations based on particle size, experience shows that problems most frequently arise because of defects in the design of the density gradient, whereas problems when separating particles on the basis of their density are generally due to the often neglected fact that the apparent density of biological particles is dependent on the characteristics of the density gradient medium.

2.1 Theoretical considerations in rate-zonal centrifugation

2.1.1 General considerations

As discussed in Chapter 1, the rate of sedimentation of a particle in a centrifuge tube is determined by:

- the size of the particle
- the difference in density between the particle and the medium in which it is sedimenting
- the centrifugal field
- the viscous drag, which will in turn depend on the particle's surface area and shape as well as on the viscosity of the medium

The ordinary user, following established procedures, does not need to worry about these theoretical considerations but they must be remembered not only when developing new applications but also when adapting a procedure developed using one type of rotor for use with a different rotor.

2.1.2 Influence of the walls of the tube

Except in zonal rotors (see Chapter 8) sedimenting particles will interact with the walls of the centrifuge tube and it might be imagined that this could affect the separation adversely. In practice this is not always the case and experiments have shown that even with fixed-angle rotors the sedimentation of particles such as small proteins ($< 10S$) can be the same as in swing-out or zonal rotors. The reason for this apparent paradox is that the random Brownian motion of very small particles is high compared with their rate of sedimentation and their collisions with the wall are completely elastic. For larger particles wall effects can be very disruptive on gradient separations. A quantitative analysis of wall effects is given in Section 3.2.1 of Chapter 1.

2.1.3 Band broadening during centrifugation

This effect is seen mainly in rate-zonal centrifugation where there is a natural tendency for the sample bands to broaden during centrifugation; the extent of this depends on the area of the band during centrifugation and the molecular weight of the sample material. Particles with molecular weights of less than 5×10^5 diffuse sufficiently to give noticeably broader bands in vertical rotors than in swing-out rotors, even though the centrifugation time is much longer in the latter type of rotor, and this effect is due to the much greater band area in vertical rotors (1). Experimental data showing this effect is shown in *Figure 2*.

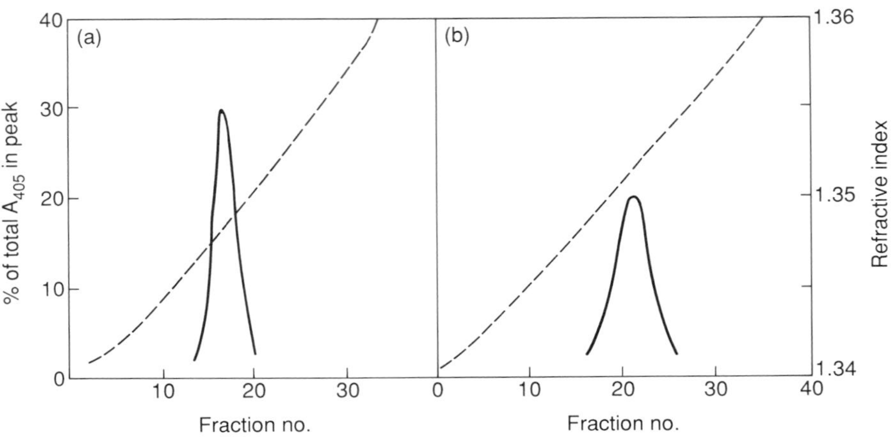

Figure 2. Comparison of the bandwidth of samples in swing-out and vertical rotors. Catalase with a molecular weight of 240 000 was separated on a 38 ml 5–20% sucrose gradient in either (a) a swing-out rotor by centrifugation for 16.5 h at 80 000*g*, or (b) in a vertical rotor by centrifugation for 3.4 h at 175 000*g*; in both cases at 20 °C (from reference 1).

In isopycnic separations the width of the bands depends on the amount of sample loaded on to the gradient (see Section 5.3.1 of Chapter 1). However, theory predicts and experiments confirm that smaller particles do form broader bands than larger particles but the hydrodynamic forces exerted on the particles as they move away from their isopycnic position are sufficient to ensure that band broadening is limited.

Band broadening with very concentrated samples can also arise because of differences between the diffusion rates of the density gradient solute and of particles and macromolecules in the sample. In theory this problem can affect both the sample zone and zones of sedimenting particles (see Section 5.2 of Chapter 1); in practice the sample zone is the only real problem. The basis of this effect is that the presence of macromolecules and particles in the sample means that the sample must contain a lower concentration of the density gradient solute than the lightest part of the gradient. Hence with a low-molecular weight solute there will be rapid diffusion of the solute from the light end of the gradient into the sample zone. The diffusion rate of the macromolecules and particles in the sample zone will, however, be less than that of the density gradient solute. Thus the overall density of the sample zone rises and may exceed that of the light end of the gradient, resulting in broadening of the sample zone. When there is a considerable difference between the concentration of the density gradient solute in the sample zone and at the start of the gradient and the tube is allowed to stand for some period this may lead to the rather spectacular phenomenon of sedimentation in droplets (*Figure 3*). The cure for this problem is to reduce the overall concentration of macromolecules and particles in the sample zone.

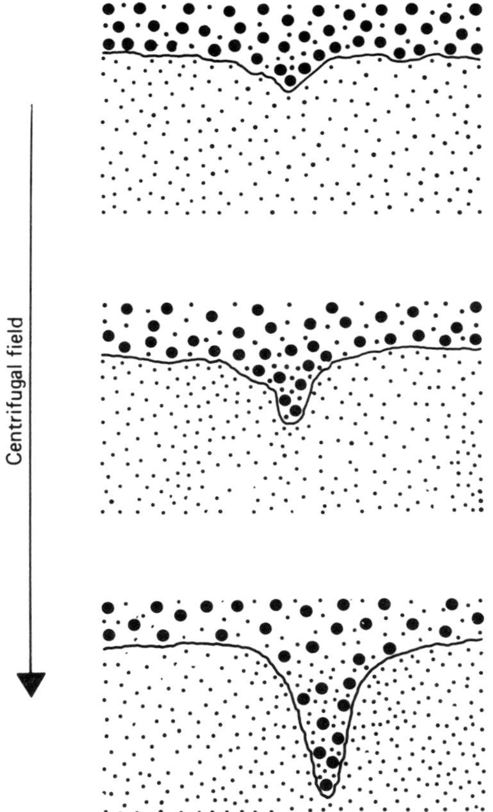

Figure 3. Sedimentation in droplets. If a sample zone containing a high concentration of a non-diffusible solute is layered over a gradient made from a diffusible solute the density of the lower part of the sample zone increases. This effect is most marked in the transient ripples developing, for example, from vibration. This will result in a part of the sample sinking in to the top part of the gradient. Over long periods, particularly if left standing on the bench, visible streamers develop as shown in the drawing (derived from reference 57).

Of equal practical importance for rate-zonal separations is the fact that gradients have a finite capacity which is determined principally by the slope of the gradient and the area of the sample band; the steeper the gradient the greater is the sample capacity. The presence of sedimenting particles inevitably increases the local density of the gradient solution. If one imagines a zone of particles of density ρ_p which occupy a partial volume, $\bar{v}_p$, of a zone sedimenting through a liquid of density ρ_l, then it may be calculated that the overall density of that fluid, ρ_t, will be:

$$\rho_t = \rho_p \bar{v}_p + \rho_l(1 - \bar{v}_p)$$

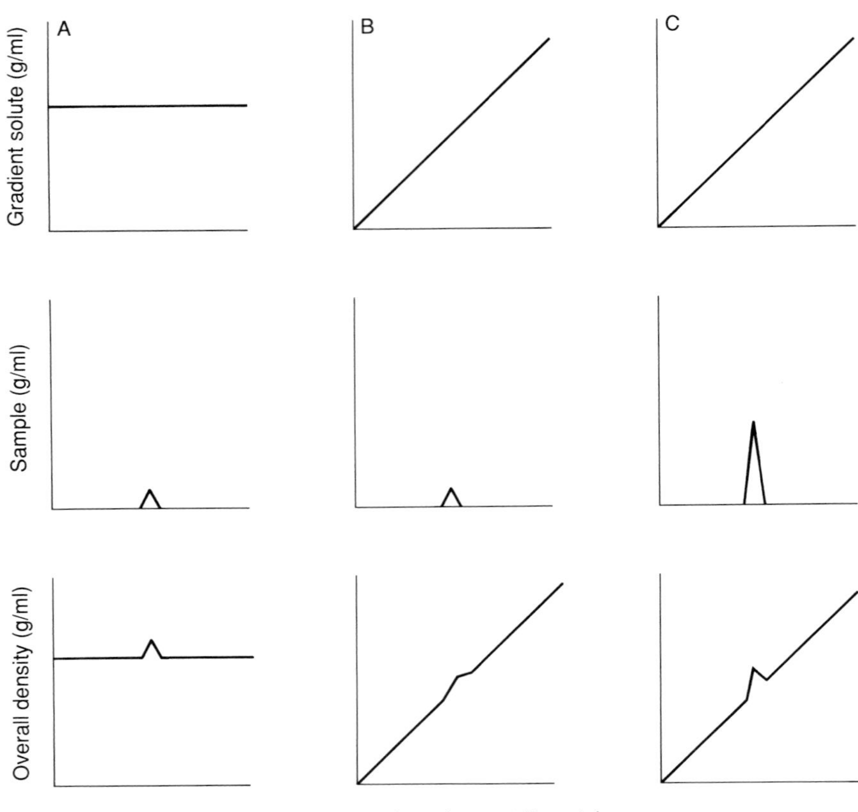

Figure 4. Capacity of gradients. In each section the top panel shows the concentration of gradient solute, the middle panel the concentration of sample, and the bottom panel the overall density. Panel A shows that a band of particles sedimenting through a uniform solution inevitably results in instability, the overall density in the band being greater than the density of the liquid below it. Panel B shows that, in the presence of a density gradient a band can be stable, the rise in concentration of the gradient solute balancing the fall in concentration of the particles. However, a given gradient can only support a certain concentration of particles and panel C shows that as particle concentrations rise, instability returns (derived from reference 57).

which rearranges to:

$$\rho_t = \rho_l + \bar{v}_p(\rho_p - \rho_l).$$

As ρ_p must be greater than ρ_l if the particles are to sediment then the overall density ρ_t must also be greater than the density of the liquid. Accordingly, if one attempted to sediment a zone of particles through a uniform liquid then one would fail, because the overall density of the liquid containing the sedimenting particles would be greater than that of the liquid below it and

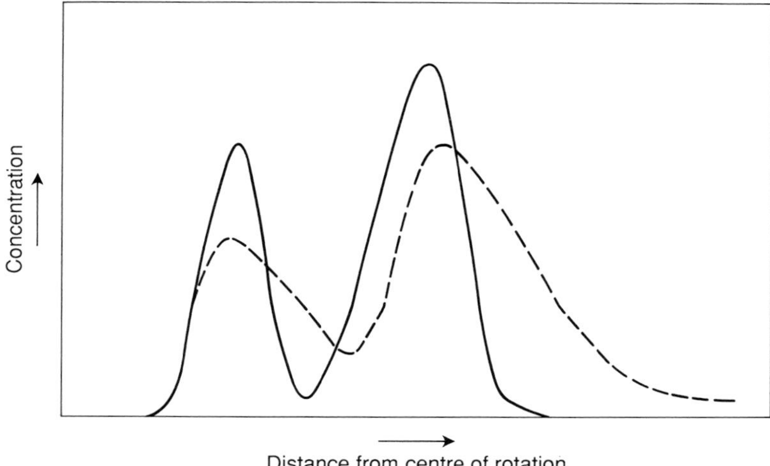

Figure 5. Effects of overloading on the position and shape of bands. Normally bands are symmetrical (—), but in overloaded gradients (---) the leading edge of the band tumbles forward giving asymmetric peaks whose centres are displaced downwards.

particles and fluid would tumble together to the bottom of the tube (*Figure 4A*). However, the presence of a density gradient can stabilize a zone of sedimenting particles. The particles still increase the overall density of the liquid but now, as shown in *Figure 4B*, the increase in concentration of the density gradient solute more than compensates for the fall in concentration of the particles in the leading edge of the zone. Accordingly, the column of liquid is stable and the particles sediment in an orderly fashion down the gradient. The steeper the slope of the gradient the greater is its capacity. It should, however, be noted that gradients have a finite capacity as shown in *Figure 4C*. If this capacity is exceeded then the bands broaden by the sample particles and liquid tumbling down the gradient together, and on fractionation of the gradient characteristic asymmetric bands are found (*Figure 5*). Besides band broadening, the samples also sediment faster than would be expected from their sedimentation coefficients, leading to erroneous estimates of their sizes; this topic is discussed further in Section 9.3 of Chapter 5.

Generally, there is no simple way of predicting the sample capacity of a gradient from mathematical theory. However, a reasonable guide is that the concentration of the sample (including all of the solutes present) should be less than 10% of the concentration at the top of the gradient; thus for a 5–20% gradient the total concentration of solutes in the sample should not exceed 5 mg/ml. Usually an empirical approach is adopted in determining sample loads; it is always better to underestimate the capacity of the gradient rather than run the risk of overloading the gradient. Some experimental procedures for

Table 1. Banding densities of macromolecules and nucleoproteins in different types of gradient medium

Type of sample	Buoyant densities in gradient media (g/ml)[a,b]					
	CsCl	Cs_2SO_4	NaI	CsTFA[c]	Metrizamide	Nycodenz
Native DNA	1.70	1.43	1.52	1.62	1.12	1.13
Denatured DNA	1.73	1.45	1.55	1.70	1.14	1.17
RNA	>1.9	1.64	1.65	1.80	1.17	1.18
Protein	1.33	1.3	–	1.45	1.26	1.27
Chromatin	1.4	–	–	–	1.19	1.18
Ribosomes	1.55	–	–	–	1.32	1.32
mRNP	1.4	–	–	–	1.2	1.2

[a] Only an approximate density can be given because the observed density depends on the exact nature of the sample and the composition of the gradient solution.
[b] Sucrose is not useful for banding any of these samples and so is not included.
[c] CsTFA is caesium trifluoroacetate, a registered trademark of Pharmacia Biosystems AB.

determining whether gradients are overloaded are given in Section 9.3 of Chapter 5.

2.2 Theoretical considerations in isopycnic centrifugation

As discussed in Section 3.1 of Chapter 1, the theory of isopycnic centrifugation is simple. If there is a difference between the density of a particle and the medium in which it is suspended then the particle will float or sink depending on whether the particle is denser or lighter than the surrounding medium; if there is no difference in density then the particles will not move in either direction and will show only the random movements of Brownian motion. Hence there would seem to be few theoretical considerations to concern the researcher. However, this is not the case, because the density of particles is not independent of the medium in which they are suspended and so an incorrect choice of gradient medium can prejudice the quality of the separation that can be obtained.

When discussing the influence of density gradient solutes on particle density two cases should be considered. First, that of particles which lack a surrounding membrane such as ribosomes, and second, that of membrane-bounded particles such as mitochondria. The first case is relatively simple provided that one remembers that the density of a small particle or macromolecule is determined not just by the intrinsic density of the macromolecule itself but also by the amount of bound water and bound metal ions. In dilute aqueous media proteins and nucleic acids are highly hydrated. However, the solutes used to form density gradients must also have a high affinity for water. Hence, as particles enter more concentrated solutions, competition develops for water molecules resulting in a loss of water from the particle and a concomitant increase in the observed density of the particles. Furthermore, there will be variations in the extent to

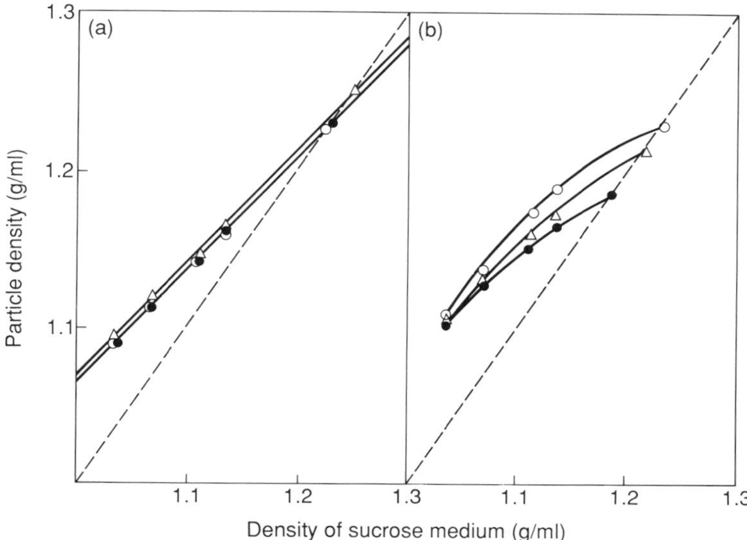

Figure 6. Effect of the permeability of particles to gradient media on their buoyant density. Variations in the density of (a) peroxisomes as indicated by the marker enzymes catalase (●), D-amino acid oxidase (○) and urate oxidase (△); and (b) mitochondria and lysosomes as indicated by the marker enzymes: acid deoxyribonuclease (○), acid phosphatase (△), and (●) cytochrome oxidase (from reference 73).

which ions in the density gradient can bind to a macromolecule or small particle. Metal ions such as caesium, mercury, silver, and even magnesium will readily bind to acidic groups of a macromolecule and modify its density. The result is a remarkable variation of the density of small particles and macromolecules between different media (*Table 1*); for a more comprehensive list see *Table 2* of Chapter 4.

The factors affecting the density of membrane-bound particles in density gradient media have been discussed by de Duve *et al.* (2). The lipoprotein membranes which surround cell organelles are freely permeable to water and impermeable to macromolecules but vary quite markedly in their permeability to small molecules such as sucrose. Furthermore, there will be marked differences in the concentrations and nature of the macromolecules within organelles. However, this physical complexity can be reduced to three theoretical 'compartments', the first being the space occupied by the membranes and macromolecules which is not affected by solute, a second is the aqueous space which is accessible to the density gradient solute, and the third is the aqueous space which is not accessible to the gradient solute but which is subject to the osmotic effects of the density gradient solute. This results in the density of the organelle changing with alterations in the concentration of the gradient solute (*Figure 6*).

Table 2. Banding densities of animal cell organelles in different types of gradient medium

Type of organelle[a]	Buoyant densities in gradient media (g/ml)[b]				
	Sucrose	Metrizamide	Nycodenz	Ficoll	Percoll
Nuclei	>1.3	1.22	1.23	—	1.09
Mitochondria	1.19	1.16	1.17	1.14	1.10
Lysosomes	1.21	1.13	1.15	—	1.06
Peroxisomes	1.23	1.22	1.22	—	1.06
Plasma membranes	1.13–1.18	1.14–1.26	1.11–1.19	1.05	1.03

[a] Densities given are for organelles of liver cells; organelles from other animal tissues may vary to some degree, and larger variations of buoyant density may be found in the organelles of plants and fungi.
[b] Only an approximate density can be given because the observed density depends on the exact nature of the sample and the composition of the gradient solution.

Hence, in the case of cell organelles, differences in membrane permeability and the osmotic effects exerted by different density gradient solutes result in marked differences in organelle banding density (*Table 2*). As can be seen, the banding densities of organelles are lower in colloidal media such as Percoll which are non-penetrant and do not exert a significant osmotic pressure, and higher in solutes such as sucrose which penetrate some membranes and which exert a significant osmotic pressure. Organelles band at even higher densities in glycerol gradients because glycerol penetrates membranes. More importantly, variations in the permeability of membranes mean that the relative banding densities of different organelles do not remain constant. Thus smooth endoplasmic reticulum fragments and plasma membranes have the same banding density in sucrose but different banding densities in Ficoll. Hence separations which are impossible in one medium become possible in another and, as discussed in Section 6.6, sequential banding first in one medium then in a second may make possible separations which could not be achieved in any single medium.

3. Gradient materials

3.1 Classification of gradient solutes

It should be emphasized that, as discussed in Section 2.2, subcellular structures and macromolecules behave differently in different gradient media. Thus equilibrium banding density of mitochondria will be considerably greater in sucrose which is partially permeant and osmotically active than in Ficoll which is non-permeant and almost completely non-osmotic. The same applies to nucleic acids which in caesium chloride solutions have a low level of hydration and so will band at a very high density, whilst in Nycodenz solutions they are highly hydrated and so band at a much lower density. The characteristics of the commonly-available gradient media are listed in *Table 3* together with their applications. It should be remembered that the presence of any gradient medium in the fractionated samples may affect the subsequent assays being carried out; for example, the interference or inhibition of some subcellular marker enzymes by sucrose is well documented. Indeed, as described in Section 7.8.4, all of the low-molecular-weight media used for the separation of large subcellular structures are likely to cause some interference in enzyme, protein, and other assays.

The density range of any water-soluble gradient solute can be slightly increased by dissolving the medium in D_2O (density = 1.11 g/ml) rather than water. D_2O also has the added advantage of being less viscous than water, thus permitting reduced centrifugation times. It is, however, rather expensive and the advantages gained are often only marginal.

3.1.1 Inorganic salts

The most commonly employed salts for density gradient work are those of the alkali metals because they are very soluble and thus achieve high densities whilst having low viscosities. However, the solutions are of high ionic strength and therefore readily break protein–protein and protein–nucleic acid ionic bonds. Gradients of caesium chloride are widely used and have been employed mainly for banding DNA, especially for the isolation of plasmid DNA. Other salts which have been used as gradient solutes include potassium tartrate, sodium bromide and iodide, caesium sulphate, formate, acetate and trifluoroacetate, and rubidium chloride.

It should be remembered that caesium and rubidium salts are extremely expensive. In the case of caesium chloride, the following protocol can be used to recycle caesium chloride after use.

Protocol 1. Recovery of caesium chloride after use

1. Starting with about a litre of CsCl gradient solution, extract it six times with an equal volume of acetone.
2. Filter off the crystals and suck dry.
3. Separate the aqueous filtrate from the acetone and heat the remaining aqueous solution *in a fume hood* with stirring to reduce the volume to about 150 ml.
4. Allow the solution to cool and filter off a second crop of crystals.
5. Combine the first and second crop of crystals and heat them in a silica dish to 600 °C to burn off all the organic material.
6. Dissolve the CsCl in a minimal amount (4–500 ml) of distilled water and filter through Whatman No 54 filter paper to remove the carbon.

Table 3. Physical properties and applications of density gradient materials

Material	Mol. wt	Max density (g/ml)	Ionic strength	Viscosity[a]	Osmolarity	Typical applications	Refs
Inorganic salts							
Caesium chloride	169	1.9	high	+	high	Nucleic acids and nucleoproteins	22
Caesium sulphate	362	2.0	high	+	high		22
CsTFA[b]	246	2.6	high	+	high	Nucleic acids	59
Rubidium chloride	121	1.5	high	+	high	Proteins	77
Sodium bromide	103	1.5	high	+	high	Lipoproteins	60, 61
Sodium iodide	150	1.9	high	+	high	Nucleic acids	62, 77
Osmotically active compounds							
Sucrose	342	1.3	non-ionic	+ +	mod.	Very many	57, 62 64, 77
Sorbitol	182	1.26	non-ionic	+ +	mod.	Cells and membranes	65
Glycerol	92	1.26	non-ionic	+ + +	high	Membranes	66
						Proteins	67
						Nuclei	68
Iodinated compounds							
Urografin	614	1.45	mod.	+	mod.	Cells	69
Metrizamide	789	1.46	non-ionic	+ +	low	Many	4, 5,
Nycodenz	821	1.45	non-ionic	+ +	low	Many	7
Polymeric compounds							
Ficoll	400 000	1.23	non-ionic	+ + +	low	Membranes and cells	9
Dextran	50–500 000	1.05	non-ionic	+ + +	low	Membranes and cells	71
Bovine serum albumin	69 000	1.12	low	+ + +	low	Cells	9, 72
Colloidal silica							
Ludox	—	1.2	low	+	low	Cells	10
Percoll	—	1.23	low	+	low	Cells and organelles	12

[a] It is difficult to locate exact figures for the viscosities of many density solutes. + indicates a solution almost as mobile as water, + + a solution with a viscosity similar to a sucrose solution of the same concentration, and + + + a solution similar in viscosity to Ficoll at the same concentration.
[b] CsTFA caesium trifluoroacetate; trademark of Pharmacia Biosystems AB.

Protocol 1. *Continued*

7. Concentrate the filtrate by heating in an evaporating dish and when a skin of crystals starts to form on the surface allow the solution to cool.
8. Filter off the crystals and concentrate the filtrate further by repeating Step 7.
9. Filter off a second crop of crystals, discard the filtrate, combine the two crops of crystals, and dry them at 130 °C.

It should also be noted that almost all salts are very corrosive to aluminum rotors and thus require great care in handling and avoiding spills. Fortunately most modern ultracentrifuge rotors are made of titanium which is resistant to the corrosive action of all types of salts. The properties of the inorganic salts and other gradient materials are listed in *Table 3*.

3.1.2 Sucrose

This readily-available sugar is by far the most popular gradient medium. Being readily soluble in water it forms solutions which cover the range 1.00–1.32 g/ml. It is thus suitable for subcellular fractionation of almost all the membrane-bound constituents of cells. Sucrose solutions have a high osmotic pressure and can remove water from membrane-bound particles causing damage, for example, to the oxidative phosphorylation system of mitochondria. Sucrose is not suitable for the fractionation of living cells as even isotonic solutions are frequently deleterious or toxic. The density, even of the saturated solution, is too low for the isopycnic separation of macromolecules or viruses. The high viscosity of concentrated sucrose at 4 °C can present handling difficulties, such as in pumping through narrow bore tubing, and its sticky nature makes it difficult to clear up when spillages occur.

Commercially-available sucrose sold for domestic purposes is generally not suitable for gradient use because it contains impurities such as protein and enzymes of plant origin, as well as metal ions and other UV-absorbing materials. Nucleases are a particular problem and it should be noted that these are lower in beet than in cane sugar. Similar impurities may also be present in standard laboratory sucrose. Such impurities can be removed easily by treatment with activated charcoal (3). However, it is best to purchase the specially purified sucrose for density gradient work that is widely available, but even this should be checked before use since even high-purity sucrose may show significant UV absorption.

3.1.3 Glycerol

Glycerol is reported to protect proteins against denaturation. However, glycerol, like sucrose, appears to inhibit some enzyme activities reversibly (55). Glycerol penetrates all biological membranes, and this property means that membrane-bound organelles band at higher densities than in sucrose. The reason why glycerol has not been more widely used is probably because the high viscosity

Structure	Mol. wt	Counter ions	Trivial names	Chemical names
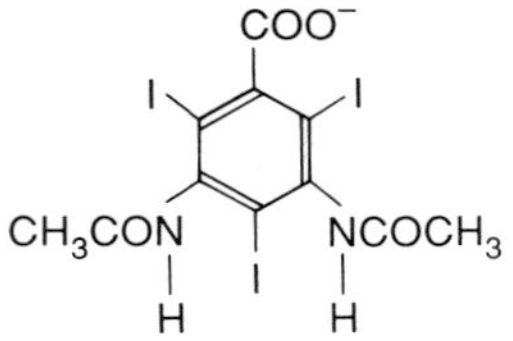 COO^-; I, I; CH_3CON (H); $NCOCH_3$ (H); I	614	MGN^+ Na^+ Na^+;MGN^+	Diatrizoate Renografin Hypaque Urografin	3,5-Diacetamido-2,4,6-triiodobenzoic acid
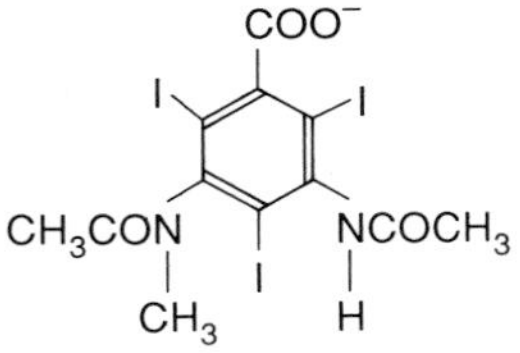 COO^-; I, I; CH_3CON (CH_3); $NCOCH_3$ (H); I	628	Na^+	Metrizoate Isopaque Ronpacon Triosil	3-Acetamido-5 (*N*-methylacetamido)-2,4,6-triiodobenzoic acid
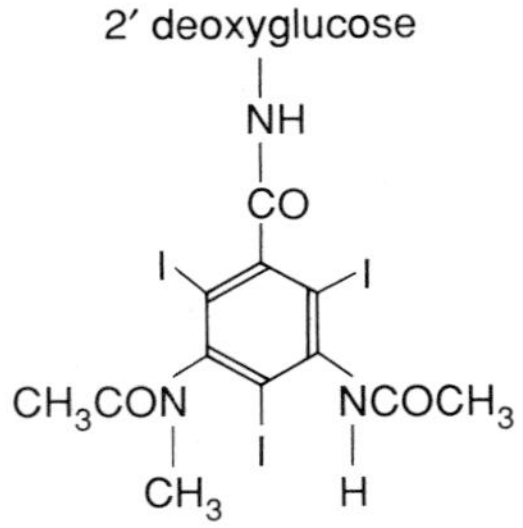 2′ deoxyglucose–NH–CO; I, I; CH_3CON (CH_3); $NCOCH_3$ (H); I	789	None	Metrizamide	2-(3-acetamido-5-*N* methylacetamido-2,4,6, triiodo-(benzamido)-2-deoxy-D-glucose
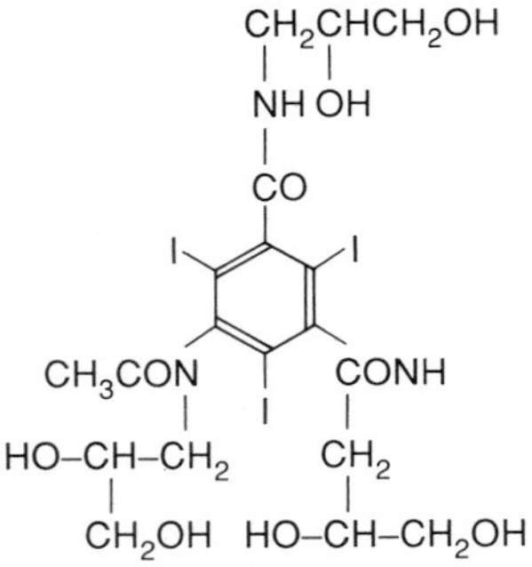 CH_2CHCH_2OH (NH, OH)–CO; I, I; CH_3CON (HO–CH–CH_2, CH_2OH); CONH (CH_2, HO–CH–CH_2OH); I	821	None	Nycodenz	*N,N′*-bis(2,3 dihydroxypropyl)-5-*N*-(2,3 dihydroxypropyl) acetamido-2,4,6-tri-iodo-isophthalamide

MGN^+=*N*-methylglucamine

Figure 7. Chemical structures of iodinated compounds used as density gradient media (derived from reference 7).

reduces the effective density range and its physico-chemical characteristics are less well documented as compared with sucrose. However, in certain areas, such as enzymology and hormone receptor research, glycerol gradients are quite widely used for rate-zonal separations.

3.1.4 Sorbitol and Mannitol

Both of these sugars have a slight advantage over sucrose in that they penetrate biological membranes more slowly. Mannitol, however, only has a very narrow density range because it starts to crystallize at concentrations as low as 15% (w/v). Sorbitol, despite, apparently, being almost as good as sucrose, has not been widely used as a gradient medium probably because it is significantly more expensive than sucrose.

3.1.5 Iodinated compounds

Although possessing advantageous properties such as low toxicity and high density, the early ionic X-ray contrast agents such as Urografin and Renografin (compounds based on tri-iodobenzoic acid; see *Figure 7*) were not widely used as gradient materials. This was probably because, in relation to the inexpensive sucrose, the advantages were marginal; these media were expensive, had high osmolarities, and tended to precipitate in the presence of calcium or magnesium ions as well as in acid solutions.

Two non-ionic X-ray contrast compounds that have been developed specifically as gradient media by Nycomed Pharma A/S are metrizamide and Nycodenz. Their structures are shown in *Figure 7*. Metrizamide especially has been used most successfully for a range of cell and subcellular fractionations and remains a favourite medium despite the advent of Nycodenz which was hailed as its successor. Both compounds have very low toxicity, low osmolality, high density, low viscosity (compared to sucrose), and are suitable for separating everything from macromolecules to viable cells by either rate-zonal or isopycnic centrifugation. When considering the separation of living cells it is worth remembering that Nycodenz is autoclavable. A number of reviews of the uses of metrizamide (4, 5) and Nycodenz (6, 7) have been published. Lists of references to applications are available from Nycomed Pharma A/S and are well worth consulting.

3.1.6 Polymeric compounds

Polymers of sugars, such as glycogen, Ficoll (polysucrose) and dextran (*Figure 8*), are essentially non-osmotic and, being biologically compatible, generally make very suitable gradient media for the fractionation of subcellular organelles and cells. The relatively low densities achievable and the high viscosities (75) may not be serious disadvantages because most membrane-bound structures band at quite low densities. Glycogen, however, is least suitable because it sediments at fairly low centrifugal speeds. It should also be pointed out that Ficoll at high densities appears to be toxic to living cells and this toxicity increases exponentially with concentration. This is due to interactions between water, Ficoll, and the salts used to maintain tonicity. At high concentrations a significant proportion of the available water becomes bound in the hydration sphere of the Ficoll (8). Hence the effective concentration of the salts in the solution increases and kills the cells. Similar problems are to be expected with

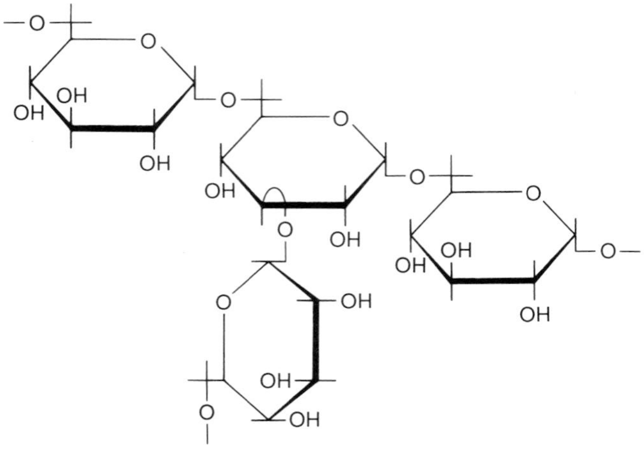

Figure 8. Chemical structure of a typical dextran derived from bacteria (from reference 77).

other high-molecular-weight solutes. The use of serum albumin has also been described (9) as a gradient material and appears to be useful for separating viable cells. However, its cost and difficulty in handling preclude its regular use.

3.1.7 Colloidal silica media

Colloidal silica sols of very small particle size (about 15 nm) were first used for banding whole cells and subcellular organelles (10). These materials which were commercially available as Ludox (du Pont) and Nalcoag (Nalco Chem Co) have densities up to 1.29 g/ml, low viscosity, virtually no osmotic effects, extremely cheap, and also able to generate 'self-forming gradients' as the particles will partially sediment at moderate speeds. Ludox, however, also has disadvantages in not being stable at slightly acidic pH, or in the presence of chloride and some divalent cations. It also contains preservatives which are toxic to cells and membrane-bound organelles. In order to overcome some of the disadvantages of Ludox, Pertoft (11, 12), who pioneered its use, helped to develop Percoll which is silica sol coated with polyvinylpyrrolidone (PVP). Percoll has all the properties of other silica sols but in addition is non-toxic, stable at physiological pH, and does not inhibit enzyme activities. It is especially suitable for fractionation of whole cells and larger subcellular organelles. However, the particulate nature of Percoll can lead to problems when separating macrophages such as Kupffer cells, which will happily phagocytose large amounts of Percoll (13). Percoll has also been reported to be taken up directly into isolated lysosomes (14). Data sheets and lists of references to applications of Percoll are available from Pharmacia-Biosystems AB.

3.1.8 Non-aqueous gradients

Although the use of organic solvents would appear to be quite unsuitable for biological materials, especially membrane-associated structures, organic gradients have been employed successfully by Siebert *et al.* (15) to fractionate very pure nuclei from a lyophilized liver homogenate. The density gradient, ranging from 1.21–1.45 g/ml, was made by mixing cyclohexane and carbon tetrachloride. The use of the organic technique provided information about localization and activity of nuclear enzymes which was not apparent from aqueous approaches because of the leaching of enzymes from organelles. An interesting and potentially very useful application of organic gradients was reported by Siebert and Hannover (16) who found that dehydrated biological molecules (e.g. proteins, chromatin, RNA, DNA and glycogen) band at characteristic densities. The lyophilized suspensions, made up in cyclohexane, can be fractionated in a cyclohexane–carbon tetrachloride gradient by centrifugation for 45 min at 1500 *g*.

3.1.9 Mixed gradients

As already mentioned, it is possible to modify the density and viscosity of gradients by making the gradient in D_2O (17). There are also ranges of commercially-available media formulations for the routine isolation of blood cells based on iodinated compounds mixed with NaCl, Ficoll, or dextran to give the correct osmolarity and density needed for a specific separation; their use is described in Section 5.1 of Chapter 7. It is also possible to design gradients whose viscosity decreases with density, for example, by starting with a high concentration of Ficoll and ending with a high concentration of Nycodenz. It is also possible to use the high viscosity of Ficoll in a mixed gradient as a barrier, to slow down the rate of sedimentation of one or more population of particles whilst the density gradient retains a continuous profile.

3.2 Choice of gradient material

The most important physio-chemical properties and features of an ideal gradient material for the separation of biological materials can be listed as:

(a) It must be readily soluble in the density range for a specific application.

(b) It must be inert and not interfere with the biological activity of the separated particles.

(c) It should, ideally, not exert a significant osmotic pressure even at high concentrations.

(d) It should be reasonably inexpensive.

(e) It should form solutions of low viscosity.

(f) It should be readily sterilized.

(g) It should not absorb in the UV range of the spectrum (i.e. <400 nm).

(h) It should be relatively easy to remove from the particles after separation.

(i) It should not interfere with chemical and enzyme assays.

(j) It should be compatible with tubes and rotor materials.

Unfortunately there is no ideal gradient material which fulfils all of these criteria and the most suitable material has to be carefully considered and chosen for a specific application. The most advantageous properties of the currently available density gradient media (*Table 3*) for the separation of macromolecules, subcellular organelles, and animal cells are surveyed in the following three sections.

3.2.1 Macromolecules

The most suitable gradient media for the separation of DNA and RNA are caesium salts, which cover the high equilibrium banding densities of nucleic acids 1.4–1.7 g/ml). It is, however, also possible to fractionate DNA and RNA in iodinated media (e.g. Nycodenz) because in these media nucleic acids remain fully hydrated and thus band at much lower densities (18). Thus non-ionic media are useful for looking at the behaviour of nucleic acids and their interaction with proteins at low ionic strength (19, 20), although separations based on base composition are not possible with these media. Nucleoproteins such as ribosomes, chromatin, and some viruses can be separated in caesium gradients but usually only after fixation with formaldehyde or glutaraldehyde (21, 22).

Apart from the classic work by Martin and Ames (23) on the separation of enzymes, there has been very little application of rate-zonal centrifugation for the fractionation of proteins. This is principally because, in most cases, size estimations of proteins can be carried out faster and more accurately using chromatographic or gel electrophoresis procedures. Serum proteins have been separated by rate-zonal sedimentation in sucrose gradients (24), but this was more of a theoretical exercise in preserving narrow bands in isometric gradients than a serious practical application. Isopycnic banding has, however, been employed for the separation of different classes of protein (25). As described in Section 4.6.1 of Chapter 4, lipoproteins are also routinely analysed on the basis of density in potassium bromide gradients (26).

Non-ionic iodinated media, in particular Nycodenz, have also proved suitable for fractionation of polysaccharides such as glycogen, porcine proteoglycans, chondroitin sulphate, and hyaluronic acids (20), although ionic media such as caesium chloride are normally preferred because their high ionic strength reduces intermolecular interactions between the highly-charged proteoglycans.

3.2.2 Cell organelles

Despite its obvious disadvantages, sucrose is still the most widely used gradient material, probably because of its low cost and a 'familiarity' earned over many years of successful fractionation. However, the media developed from X-ray contrast agents (metrizamide and Nycodenz) being biologically compatible, essentially impermeant to membranes, metabolically inactive, with good density ranges and low viscosity, have many potential advantages especially in terms of the resolution of membrane fractions. The extensive applications of iodinated media for separations of nuclei, plasma membranes, lysosomes, and peroxisomes have been reviewed by Rickwood (5, 7). Ficoll, which has little effect on membranous particles, exerts almost no osmotic pressure, and permits near ideal sample capacity for rate-zonal separations, has been used for some of the more difficult separations. Ficoll also has the added advantage of not inhibiting enzymes even when present in high concentrations, but there remains the problem that Ficoll gradients are very viscous. As discussed in Section 6.6, combinations of gradient materials will frequently permit separations impossible in any single medium.

Large membrane-bound subcellular organelles have also been fractionated with the aid of Percoll, the PVP-coated colloidal silica medium. This is despite the tendency of the colloidal silica particles to bind to membranes. Percoll is especially suitable for isopycnic work because its gradients are self-forming, but it is not suitable for rate-zonal separations because the gradient changes even during short periods of centrifugation. Percoll gradients provide an easy and rapid method for the otherwise difficult separation of brush border and basolateral plasma membrane vesicles of the kidney (27).

3.2.3 Animal cells

The iodinated media Nycodenz and metrizamide in combination with osmotic balancers can be used to form isotonic gradients which are ideal for the fractionation of animal cells. One of the earliest and most convenient uses of metrizamide is that it can separate viable cells from damaged cells (28). The uses and advantages of metrizamide and Nycodenz for separating a number of different cell types have been surveyed by Bøyum *et al.* (29). The same authors also describe the features of combination media which have been developed for the routine fractionation of various blood cells. These media, which are based on iodinated media, also contain other ingredients such as NaCl, Ficoll, and dextran, in the appropriate proportions to achieve specific densities and osmolarities. Some of these specific formulations are available as proprietary reagents (see *Table 2* of Chapter 7).

Ficoll with its excellent preservation of viability and absence of osmotic pressure has also proved suitable for the fractionation of viable cells. Its earliest applications were for the isopycnic banding of viable lymphocytes (30) but its excellent biocompatibility has made it suitable for other blood cells (31), mucosal cells (32), and spleen cells (33). It should be noted, however, that problems of toxicity may arise at very high concentrations (see Section 3.1.6).

Percoll is also an excellent medium for fractionating cells (13). It is particularly good for the rapid isopycnic banding of cells. The cell suspension is simply added to an appropriate concentration of Percoll, previously made isotonic and then centrifuged in a high-speed centrifuge until the self-forming gradient and equilibrium banding of the cells has been achieved. The main disadvantage of Percoll is that it can stick to cells, and any macrophage type of cell present in a suspension will phagocytose quite large amounts of the colloidal silica particles, with obvious consequences on their phagocytic function and a modification of their banding density. In addition, multiple artifactual bands can form in the gradient under some circumstances (40).

4. Design of initial experiments

The design of initial experiments in developing a method for the fractionation of macromolecules and macromolecular complexes is reasonably simple. If the sample is likely to have a defined size (e.g. ribosomal RNA) then one of the first steps would be to examine the rate-zonal sedimentation of the sample on a sucrose gradient. If, on the other hand, the sample is likely to be polydispersed (e.g. DNA or mitochondria) then the method of choice would be isopycnic separation using ionic or non-ionic gradients as appropriate (*Table 3*). In the case of cell organelles and viable cells the situation is often more complex. Hence, this section concentrates on the approaches which should be taken when attempting to design procedures for fractionating cell organelles and cells and these are summarized in *Protocol 2*.

Protocol 2. Strategy for separating cells and cell organelles

1. Check the literature to find out if the type of separation that you wish to do has been done before; the advent of easy access to computerized scientific databases has greatly facilitated this process.
2. If there are no reports of identical separations, are there descriptions of work with similar types of cells (e.g. different type of animal, different tissue)?
3. In the case of organelles, check the optimum conditions for homogenization in terms of the method used for cell breakage and composition of the medium (divalent cations, ionic strength and osmolarity of the medium) since both aggregation and instability of cellular structures can pose serious problems. In addition, it may be necessary to add inhibitors to prevent degradation of the intracellular components of interest.
4. Check the validity of the proposed method of identification, and in the case of cell organelles be sure that the marker enzyme or immunological marker to be used is the correct one. In the case of cells check that the staining protocol or other identification procedure is valid for your type of cells.
5. Decide what criteria need to be met in terms of the functionality of the particles after isolation (e.g. viability of cells or oxidative phosphorylation activity of mitochondria).
6. Initially use simple linear gradients to determine first the densities and then the sedimentation rates of both the particles of interest and the likely contaminants.

7. Draw a S-ρ diagram and determine whether the sedimentation characteristics of the particles of interest are sufficiently different from those of the other types of particles present.
8. If the S-ρ diagram indicates that no separation is likely in that gradient medium, return to Step 6 and try a different gradient medium.

The guidelines given in *Protocol 2* should provide useful indications as to the approach to be adopted. However, this chapter will only provide an introduction to the types of techniques used for separating the different types of biological particles, later chapters in this book will provide detailed descriptions of the precise techniques which can be used.

Obviously, when working with such complex systems, the approaches will depend on the nature of the sample, the equipment available, and the subsequent use of the purified material and so there is no single 'right' way. However, it is important to maintain a systematic approach in trying to optimize a separation technique. Nowhere is the old adage that 'if you do not understand a problem you are not likely to solve it' more true than in the fractionation of cell components. Hence, before attempting any separation it is necessary to consider the following:

(a) What components are present in the sample?

(b) What are their physical properties (i.e. size and density)?

(c) How can they be identified after the separation?

Clearly, the first question depends on the nature of the sample, while the last question is considered in Section 7.8 of this chapter and in Section 4 of Chapter 6. However, determination of the physical properties of the components of the sample and their representation in an easy-to-understand form can be discussed in a general way.

In order to design a separation strategy for any sample it is helpful to draw up what is termed a S–ρ diagram (*Figure 9*). In this the size of the particle is plotted on one axis and its isopycnic banding density on the other; note that the pattern of this diagram will vary depending on the type of gradient medium being considered. In some cases presentation of data in this form immediately makes clear the approach which needs to be adopted. As an example, it is clear that polysomes (*Figure 9a*) differ only in size and must be separated by rate-zonal centrifugation while membrane fractions which are very heterogeneous in size (*Figure 9b*) are best separated by isopycnic banding. However, often the pattern is more complex like that shown in *Figure 9c*. Whatever the pattern, it is now possible to think logically about how best to carry out the required separation.

Obviously at the start of a set of fractionation experiments normally one only has approximations of the size and density of the various components of

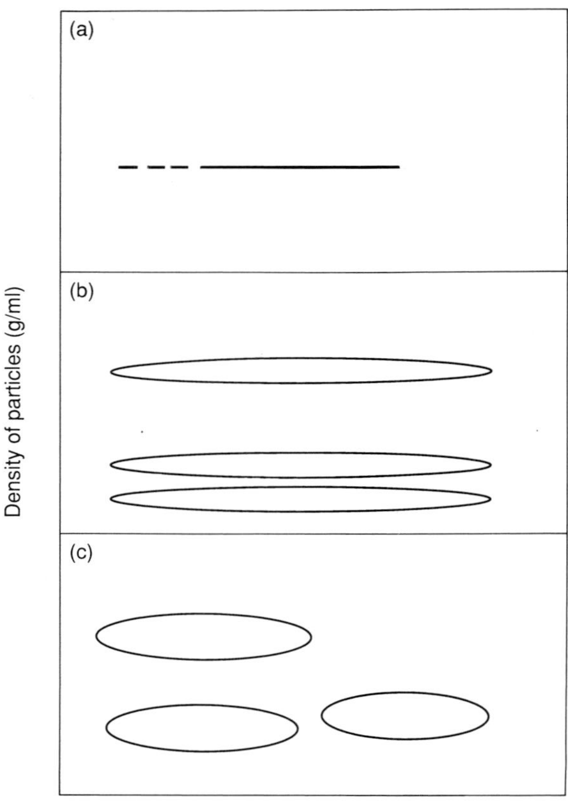

Figure 9. Examples of simple S-ρ diagrams: (a) polysomes are identical in density and can only be separated by rate-zonal centrifugation; (b) membrane fragments, very heterogeneous in size, should be fractionated by isopycnic centrifugation; (c) represents the more complex situation where it is necessary to use both rate-zonal and isopycnic centrifugation in order to obtain pure fractions.

the mixture. However, the scientific literature can usually provide guidance which will indicate the sort of separation that is likely to be successful. It is important that early experiments are designed for ease of interpretation so that they can be used to refine the S-ρ diagram. In the following discussion it is assumed that particles in the mixture differ both in size and in density. If this is not the case then obviously some of the experiments listed below can be neglected.

4.1 Choice of gradient medium and buffer for the initial studies

Factors affecting the choice of gradient solute have been discussed earlier (Section 3.2). Conventionally, sucrose is used as the solute of choice for the separation of cell organelles but, if possible, experiments should be carried out to check that exposure to the high concentrations of sucrose (isopycnic banding densities may require up to 1.8 M sucrose) do not damage the particles. As discussed in Section 7.8.4, the presence of high concentrations of sucrose and most other low molecular weight solutes in assay media results in the inhibition of many enzymes. This inhibition is fully reversible and not indicative of damage. The other frequent problem with sucrose gradients is their high viscosity which can seriously detract from the quality of the separation. If high viscosity is a problem or if the sample of interest is damaged by high concentrations of sucrose, then the use of other materials such as metrizamide or Nycodenz should be considered. There is no such 'obvious' choice for living cells but Nycodenz, metrizamide, and Percoll are probably the gradient media of choice, although the last mentioned is normally only employed for isopycnic separations.

Gradients for work with organelles and cells are usually buffered to about pH 7.4. For work with cell organelles concentrations of buffer should be kept low, for example 5 mM Tris or bicarbonate appear to be satisfactory but it is unlikely that concentrations up to 20 mM will damage cell organelles. In some cases salts may be needed to stabilize structures; for example magnesium ions may be required to stabilize nuclei or to prevent the release of ribosomes from the endoplasmic reticulum. In other cases it may be necessary to add a chelating agent such as EDTA to prevent lipid peroxidation caused by free ions of transitional elements such as iron, or membrane aggregation induced by divalent cations such as calcium and magnesium.

4.2 Preparation of the sample for initial experiments

On balance the authors recommend minimal sample preparation for the initial experiments on organelle separation and often the use of whole homogenates is to be recommended. This does, however, place very severe limits on the amounts of material that can be loaded on to the gradients and this, in turn, can lead to difficulties in detecting the separated components. If some pre-separation by differential pelleting is essential one should be very aware of the risks of organelles aggregating when pelleted and breaking on resuspension. Similar considerations apply to living cells and samples of cells should be checked for viability and aggregation by light microscopy before and after separation. It should be noted that these comments only apply to initial experiments. For routine use the authors regularly pre-purify the starting material by differential pelleting to increase the amount of material that can be loaded on to each gradient.

4.3 Choice of gradients for initial experiments with cell organelles

As stated previously, the initial experiments should be designed so that the results are easy to interpret. Start by trying simple, linear density gradients.

Table 4. A guide to centrifugation conditions for initial rate-zonal and isopycnic separations of different types of samples

Type of sample	Type of separation	
	rate-zonal[a]	isopycnic
Cells	300g for 10 min	2000g for 20 min
Nuclear fraction	1000g for 10 min	40 000g for 60 min
Mitochondrial fraction	8000g for 20 min	80 000g for 90 min
Microsomes	60 000g for 30 min	100 000g for 300 min
Polysomes	100 000g for 100 min[b]	100 000g for 20 h
Soluble fraction (proteins and RNA)	100 000g for 18 h[b]	300 000g for 48 h[c]

[a]Assumes that sucrose gradients are used except for cells where an isotonic gradient of Nycodenz or Ficoll should be used.
[b]If the approximate sedimentation coefficient is known and a 5–20% sucrose gradient at 5 °C is used then a more precise estimate of the conditions required can be deduced from the k^*-factor of the rotor (see Section 6 of Chapter 1).
[c]Fractionation in a non-viscous gradient medium such as CsCl.

In a completely unknown system it is probably best to begin with the determination of particle densities by isopycnic banding. A gradient extending from 0.5 M to 2 M sucrose should band all membrane-bounded organelles. Centrifugation times are obviously determined by the smallest organelle in the sample and by the centrifugal force generated by the rotor. Some typical combinations are shown in *Table 4*. Having determined the banding densities of the different components of the mixture one can then proceed to examine their range of sizes. The gradients employed for this purpose usually extend from a density slightly above that of the sample solution up to the banding density of the lightest organelle. Thus for a sample suspended in 0.25 M sucrose the gradient would commence at 0.4 M sucrose. In most tissues the lightest organelles are denser than 0.8 M sucrose so that rate-zonal gradients extending from 0.4 M to 0.8 M sucrose are generally appropriate, giving adequate space to separate the lightest organelles from soluble material diffusing from the sample zone while ensuring that the vast majority of organelles do not reach their banding density. The gradients should be formed over a small 'cushion' of 2 M or 2.2 M sucrose sufficient in volume to fill the hemispherical base of the tube. The viscosity and density of this solution prevent a pellet forming so ensuring a quantitative recovery when the gradient is fractionated.

4.4 Initial conditions for fractionating living cells

As stated previously, iodinated media such as Nycodenz or metrizamide and colloidal silica (Percoll) are currently the most widely used materials for the separation of living cells. The experimental approach with the two types of gradient media varies. Nycodenz and metrizamide gradients are normally preformed. Percoll gradients for fractionating living cells may be preformed by using a gradient maker, by pre-centrifugation, or formed *in situ* during the centrifugal separation of the cells. However, the last of these methods is not recommended because the conditions for forming the correct shape of gradient may not be compatible with the nature of the sample and, in addition, artifactual bands of cells may form in the gradient if the cells are present while the gradient is forming during centrifugation (40).

Lists of applications are available both from Nycomed Pharma A/S (for Nycodenz and metrizamide) and Pharmacia-Biosystems AB (for Percoll). These should always be consulted and the advice given in this section is primarily for cases where no similar separations can be located. When considering initial conditions for Nycodenz or metrizamide one should bear in mind that most living cells band at densities of between 1.03 g/ml and 1.13 g/ml in these media, while dead cells are permeable to these materials and so band at markedly higher densities (> 1.15 g/ml). Hence the gradient should extend beyond the maximum expected density of living cells to ensure a clean separation of damaged cells and viable cells. When designing the gradient it is important to remember that living cells are sensitive to changes in the tonicity of the surrounding medium, so the sample and the lighter end of the gradient must contain sufficient salts to maintain isotonic conditions. As a general rule, cells tolerate hypertonic conditions better than hypotonic conditions with cells remaining viable even when the tonicity of the medium exceeds 450 mOsm. Since both Nycodenz and metrizamide are osmotically active it is necessary to use NaCl or glucose to maintain isotonic conditions (6); recipes for such gradients are also given in Section 5.2 of Chapter 7. Typically, for initial experiments one would use a linear gradient extending from a density of 1.05 g/ml to 1.15 g/ml. Centrifugation for 20 min at 2000*g* is sufficient to band even small cells such as lymphocytes. In the case of Percoll the density of living cells ranges from about 1.04 g/ml to 1.10 g/ml. Dead cells are lighter than living cells in this medium. Hence a gradient ranging from 1.04 g/ml to 1.10 g/ml with a cushion of density 1.15 g/ml might be suitable. Centrifugation for 30 min at 1000*g* is normally sufficient to band cells in this non-viscous medium. As a general rule, and to the consternation of many biochemists, density gradient separations of cells work best at room temperature (approx. 20 °C); cells separated at 5 °C are more prone to aggregation and appear to suffer more in terms of disruption of their normal metabolic behaviour. Separation of mammalian cells at physiological temperatures (37–40 °C) is seldom undertaken and indeed few centrifuges are able to maintain this type of temperature with any degree of accuracy.

5. Preparation of gradients

5.1 An overview of methods

Density gradients may be prepared in several ways depending on the type of gradient required and personal preferences. Gradients can be divided into

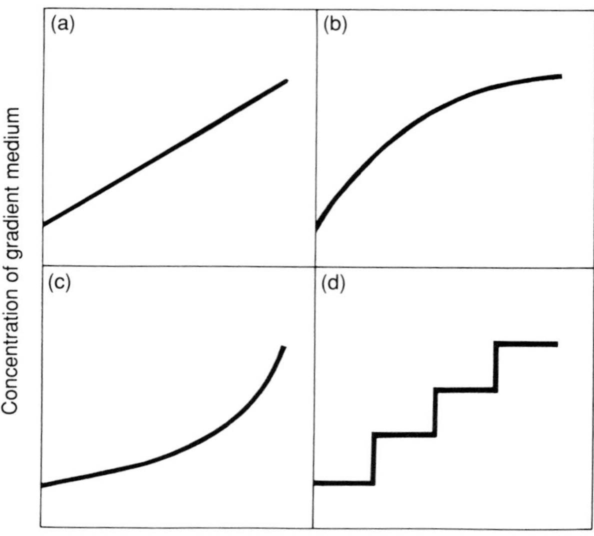

Figure 10. Examples of different types of gradient profiles. Continuous gradients may be (a) linear, (b) convex, or (c) concave, and can be used for rate-zonal or isopycnic separations; discontinuous (step) gradients (d) are used only for isopycnic separations.

continuous and discontinuous (step) gradients (*Figure 10*); continuous gradients are usually linear but non-linear, particularly exponential, gradients have been advocated for particular separations; gradients with more complex shapes are only used for centrifugal separations in zonal rotors.

Discontinuous gradients can be prepared simply by underlayering the required solutions one under the other, starting with the lightest solution first. There are a range of methods for preparing continuous gradients. Some methods such as freezing and thawing of a uniform solution in a centrifuge tube are not recommended because of the uncertain shape and concentration range of the gradients produced by this method. Gradients can be allowed to form over several hours by allowing discontinuous gradients to diffuse. Another approach, which is only applicable to readily sedimentable solutes, depends on the ability of the solute to form its own gradient during centrifugation. However, again the shape and steepness of the gradient depends on the type of medium and the centrifugal force as well as the type of rotor used. Hence it is difficult to ensure that the correct shape of the gradient is obtained. Complex gradients are best formed with the aid of specialized gradient formers which can be programmed to produce an exact profile. The choice of which method to adopt depends on the shape of the gradient required, the nature of the gradient material itself, the equipment available and, of course, on personal preferences.

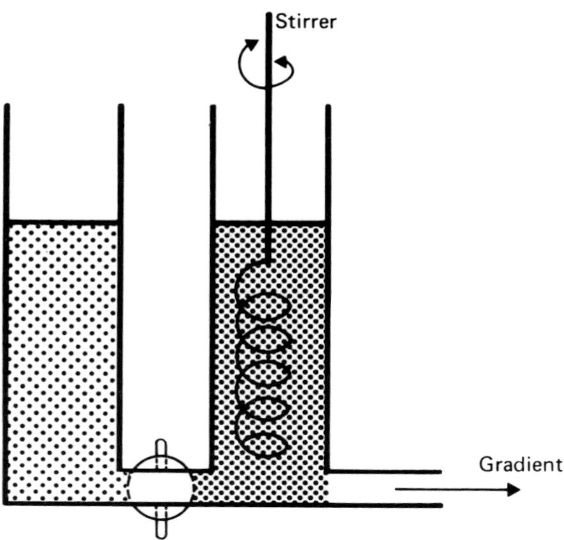

Figure 11. A simple apparatus for the preparation of linear gradients (redrawn from reference 57).

5.2 Preparation of linear gradients using diffusion methods

Gradients of readily-diffusible solutes may be prepared by diffusion. A step gradient of just two solutions layered in a centrifuge tube will form a continuous, almost linear, gradient after standing for about 24 h. The diffusion process can be speeded up by starting with a series of 3–5 steps layered into the tube. In addition, diffusion may be facilitated by laying the sealed tube containing the step gradient horizontally, thus increasing diffusion between the layers; using this method a linear gradient forms in less than an hour. An apparatus for forming multiple density gradients by accelerated diffusion under controlled and reproducible conditions is described in the next section.

5.3 Preparation of linear gradients using gradient mixers

Although the literature contains references to various gradient profiles, many of these are highly specialized and are mainly of curiosity value. By far the most common profile of gradients employed in swing-out tubes is the gradient whose concentration (thus density) increases linearly with volume. In a tube of equal cross-sectional area all the way down, these gradients are also linear with distance from the centre of rotation in swing-out rotors.

This type of gradient can be made easily using an apparatus like that illustrated in *Figure 11*. It consists of two chambers with an outlet, and a stirrer in the mixing chamber whilst the other chamber is the reservoir. To make a gradient dense end first, the denser solution is put into the mixing chamber with stopcock closed and an equal amount (volume usually but strictly it should be the same weight) of the lighter solution is put in the reservoir. For most efficient mixing it is best to use an overhead paddle stirrer. If such a stirrer is not available then a magnetic stirring bar can be used. When using a magnetic stirring bar it is best to put the same size of stirring bar in both the mixing chamber and the reservoir since this helps to avoid backflow from the mixing chamber into the reservoir. Opening the outlet from the mixing chamber, before the stopcock between the chambers, also helps to minimize the problem of backflow. After opening the stopcock between the chambers, mixing of the liquid flowing from the reservoir, as liquid flows from the mixing chamber, will result in a linear gradient streaming slowly from the outlet down the wall of the centrifuge tube to form the gradient. An even flow down the wall of the tube will ensure that the gradient of continuously decreasing concentration will be evenly layered in the tube. To make a gradient light end first, as may be necessary with non-wettable centrifuge tubes, the light solution is placed in the mixing chamber and the dense in the reservoir. In this case the gradient is delivered to the bottom of the tube using a glass capillary tube or a long stainless steel needle.

Rather than making gradients singly it is more practical to make three or six gradients simultaneously using a multichannel peristaltic pump. Obviously each pump line must deliver the same flow rate to achieve identical gradients and this is not always easy to achieve in practice. It is also important to bear in mind the total volume of gradient needed for the multiple tubes and thus the size of the gradient maker. Minor adjustments needed to balance the tubes to within the 0.1 g limit recommended by manufacturers may be made by under-layering with some dense solution using a syringe with a long needle; alternatively, low density solution can be added carefully to the top before the sample is loaded on to the gradient.

The principle of the two-cylinder gradient maker can be improved to produce more precise linear profiles by accurately controlling the flow rates in and out of the mixing chamber. Accurate delivery of the solution from the reservoir by a pump (rather than gravity) and withdrawal of the gradient from the mixing chamber at exactly twice the flow rate will yield a linear gradient (34). However, it is not clear that precisely linear gradients offer any distinct advantages in terms of obtaining better results. Proprietary gradient makers for generating single or multiple gradients for a variety of tube sizes are available from the leading ultracentrifuge manufacturers.

An elegant and rapid approach to generating up to six identical linear gradients in various sizes of centrifuge tubes employs an automatic sequence of tilting and rotating the tubes (35). Such a system is commercially available from BioComp Inc and is also marketed by Nycomed Pharma and Du Pont–Sorvall. The sequence of operation begins with layering equal volumes of the light and dense gradient solutions and then capping the centrifuge tube. The instrument then tilts the tubes and rotates them at low speed forming a smooth gradient in as little as two or three minutes (*Figure 12*). Recommended angles of tilt, speed, and duration of rotation for each type of gradient solute and each type

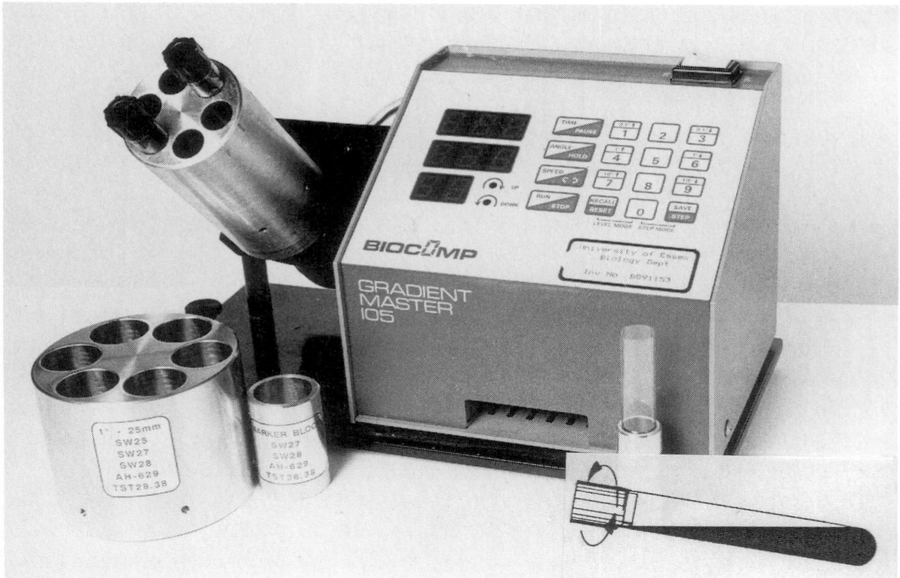

Figure 12. A commercial density gradient machine. Gradients are prepared by layering the light solution over the dense and placing the tubes in the apparatus. A smooth gradient is generated by slow rotation of the tubes for 1–2 min. The gradient shape is determined by the time of rotation and the angle of the tubes, both of which can be pre-programmed (35).

of tube are provided with the instrument. Once established the exact conditions may be stored and recalled for each subsequent experiment.

5.4 Preparation of step and discontinuous gradients

Apart from linear gradients, the other frequently used type of gradient is the step or discontinuous gradient. However, its use is primarily restricted to isopycnic separations where they are used to obtain sharper bands than would be obtained using a continuous gradient. Step or discontinuous gradients can be made simply by carefully layering, with the aid of a pipette or a syringe, a series of gradient solutions starting with the densest at the bottom. However, it is usually simpler, using a long syringe needle (a spinal needle is very suitable), to load the lightest solution first. Passing the needle through the gradient causes remarkably little disturbance to the interfaces even in narrow tubes. After layering the gradient always check that there are sharp, refractive interfaces between each of the solutions as this is a useful indication that the gradient has been made correctly. Once prepared, discontinuous gradients should be used immediately. If required, the sharp steps can be 'rounded' by leaving the gradient standing vertically for about 2 h.

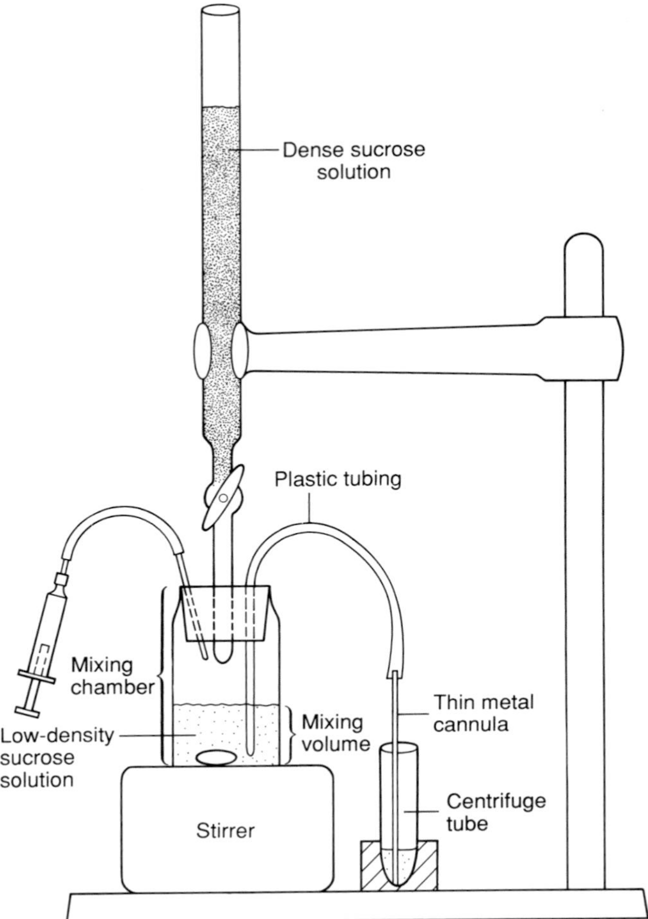

Figure 13. Apparatus for the preparation of 'exponential' density gradients. The mixing vessel is closed so that outflow is exactly balanced by inflow. This apparatus produces gradients of the shape shown in *Figure 14*.

5.5 Formation of exponential gradients

This shape of gradient is not generally used for swing-out rotors except for the occasional need for isokinetic gradients. The exponential shape is, however, ideal for zonal rotors whose non-linear radius/volume relationship is such that a gradient exponential with volume is approximately linear with radius.

Exponential gradients are formed when the volume in the mixing chamber remains constant. The simplest apparatus for making such gradients is shown in *Figure 13*. For a gradient normally formed light end first, the denser solution from the reservoir is fed into an air-tight mixing chamber which contains the

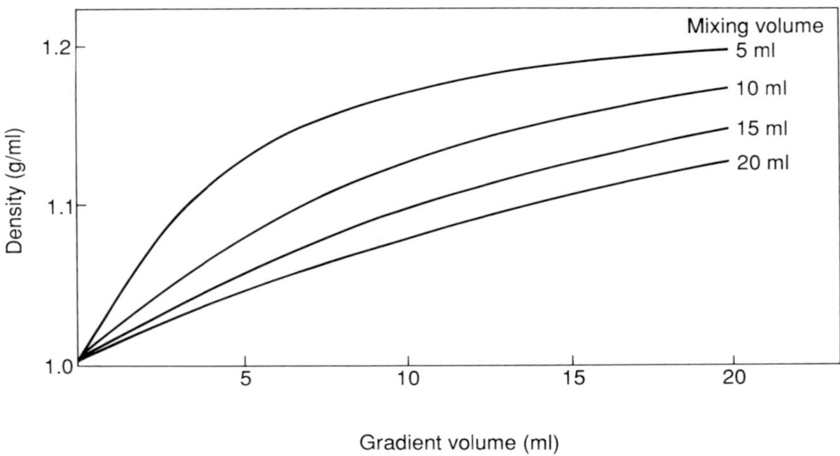

Figure 14. Exponential gradients formed by the type of apparatus illustrated in *Figure 13*. The total volume of gradient formed in each case is 20 ml. The liquid initially present in the mixing vessel is assumed to have a density of 1.0 g/ml, and that in the reservoir a density of 1.2 g/ml. The curves illustrate how the volume of liquid in the mixing vessel affects the gradient shape.

lighter solution and is vigorously stirred to ensure effective mixing. As the gradient solution from the mixing chamber is fed or pumped to the tube or rotor it is replenished at the same rate with the denser solution from the reservoir. The constant mixing chamber volume can also be maintained by a two-pump (or two identical tubes in a peristaltic pump) system which ensures that the rates of flow in and out of the mixing chamber are exactly the same. The shape of exponential profiles which can be produced by such a system depends on the volume of solution in the mixing chamber, the density difference between light and dense solutions, and the volume of the gradient produced. Examples of steep (small mixing volume) and shallow (large mixing volume) exponential profiles which can be formed are illustrated in *Figure 14*. The authors have described the use of an inexpensive and versatile gradient maker that can generate concave, linear, and convex gradients (36). It has also been reported that a syringe can be modified to make a simple, inexpensive exponential gradient maker (37). A particularly elegant, commercially available, apparatus for exponential gradients has a piston-adjustable, variable-volume mixing chamber, with a carefully shaped inlet and stirrer recess. (Bio Sep–USA).

5.6 Preparation of complex profile gradients

Complex profile gradients, such as combinations of linear or 'exponential' sections are used most frequently in zonal rotors. Complex shape gradients may be formed with programmable gradient makers which reproduce the exact shape of a template or graph. These commercially-available gradient makers were developed mainly for large volume gradients used in zonal rotors and chromatographic methods and so they are usually not suitable for forming gradients in tubes of most swing-out rotors.

5.7 Self-forming gradients

The molecules of density gradient solutes must, of course, be denser than the water in which they are dissolved and will hence themselves sediment in a centrifugal field. In the case of solutes such as sucrose this sedimentation is so slow under the normal conditions used in density gradient centrifugation that it may be ignored. In contrast, alkali metal salts, such as caesium chloride, used in the fractionation of nucleic acids are so dense, and the centrifugal forces required to band the sample so large, that significant sedimentation of the gradient solute does occur leading to the formation of useful gradients. Iodinated gradient media such as Nycodenz and metrizamide will also form useful gradients when centrifuged and self-forming gradients of these media have been used to separate a wide range of macromolecules and macromolecular complexes (see Section 4.2.2 of Chapter 4). While it usually takes several hours of ultracentrifugation to generate gradients from uniform solutions, some gradient materials, such as Percoll, consist of a colloidal suspension of silica particles with diameters of up to 15 nm. These particles sediment in quite low centrifugal fields (10–30 000 *g*) in less than half an hour to form a useful gradient; very high centrifugal fields will result in complete pelleting of the gradient material.

5.7.1 Gradients of alkali metal salts

One real advantage of working with self-forming gradients of alkali metal salts is that there has been a vast amount of work done on predicting the shape and steepness of gradients, and this is particularly true for caesium chloride gradients. Full details of the equations given here have been described elsewhere (77).

The slope of the gradient formed by alkali metal salts is given by the equation:

$$d\rho/dr = \omega^2 r/\beta^\circ = 1.1 \times 10^{-2} Q^2 r/\beta^\circ$$

where ω is the angular velocity in radians/sec, r the distance in centimetres from the axis of rotation, β° is a constant depending on the concentration and properties of the density gradient solute (see *Table 5* for values), and Q is the rotor speed in revolutions/min. From this equation others can be derived giving the density range and the maximum and minimum densities within the gradient.

The density range $(\rho_2 - \rho_1)$ is given by:

$$\rho_2 - \rho_1 = \frac{\omega^2}{2\beta^\circ}(r_2^2 - r_1^2) = \frac{1.1 \times 10^{-2} \times Q^2}{2\beta^\circ}(r_2^2 - r_1^2)$$

Table 5. Values of $\beta°$ for ionic gradient media

Gradient solute	$\beta°$ values ($\times 10^{-9}$) for solutions with initial densities (g/ml) of							
	1.1	1.2	1.3	1.4	1.5	1.6	1.7	1.8
CsCl	—	2.04	1.55	1.33	1.22	1.17	1.14	1.12
Cs_2SO_4	—	1.06	0.76	0.67	0.64	0.66	0.69	0.74
CsTCA[a]	—	—	—	1.18	—	—	—	—
CsTFA[b]	—	—	—	—	—	1.29	—	—
KBr	6.20	3.80	3.05	—	—	—	—	—
KI	4.28	2.55	1.96	1.73	—	—	—	—
NaBr	7.70	5.20	—	—	—	—	—	—
NaI	5.10	3.19	2.82	—	—	—	—	—
RbBr	—	2.15	1.56	1.34	1.22	—	—	—
RbCl	—	3.42	2.76	2.25	—	—	—	—
RbTCA[a]	—	—	—	—	1.33	1.38	—	—

[a] TCA is the trichloroacetate salt.
[b] TFA is the trifluoroacetate salt; CsTFA is a registered trademark of Pharmacia-Biosystems AB.

where ρ_1 and ρ_2 are the densities at the top and bottom of the gradient and r_1 and r_2 are the distances (in centimetres) of the meniscus and bottom of the gradient from the axis of rotation, respectively.

In order to calculate the densities of the light and dense ends of the gradient one must first calculate the isoconcentration point, that is the point where the density of the gradient is the same as that of the original solution. This distance, r_c, from the centre of rotation is given by the equation:

$$r_c = [1/3(r_1^2 + r_1 r_2 + r_2^2)]^{1/2}$$

It should be noted that these equations strictly apply only to swing-out rotors but for practical purposes they may also be used for vertical and fixed-angle rotors.

The density at any other point may then be calculated using the equation:

$$\rho_r = \rho_i - \frac{1.1 \times 10^{-2} \times Q^2}{2\beta°}(r_c^2 - r^2)$$

where ρ_r is the density at a point r cm from the centre of rotation and ρ_i is the density of the original homogeneous solution.

Finally, while considering the theory of self-forming gradients, the time of centrifugation must be considered. The centrifugation time in hours, t, needed for the gradient to reach true equilibrium is given by the equation:

$$t = k(r_2 - r_1)^2$$

where k is a constant which is inversely proportional to the diffusion coefficient of the solute. For CsCl at 20 °C, k equals 5.6. Hence doubling the length of the gradient will increase the centrifugation time required fourfold. The time taken for particles to reach their equilibrium banding density will clearly depend on their size. With a non-viscous solute such as caesium chloride this time is given by the equation:

$$t = \frac{9.83 \times 10^{13} \times \beta^{\circ}(\rho_{\mathrm{p}} - 1)}{Q^4 r_{\mathrm{p}}^2 s_{20,\mathrm{w}}}$$

where ρ_{p} is the density of the smallest particle in the mixture, and $s_{20,\mathrm{w}}$ is its sedimentation coefficient, and r_{p} is the distance of the band of particles from the centre of rotation at equilibrium.

Calculations using the equations given in this section for the length of time required for the formation of an equilibrium gradient and the banding of particles in the gradient will indicate that, in the case of a typical 12 ml gradient in a fixed-angle rotor, the centrifugation time required is about 100 h. In contrast, experimental data indicates that good separations in such gradients can be obtained in less than a third of the calculated times. This discrepancy arises because the equations given relate to equilibrium situations and it is not necessary to obtain a true equilibrium gradient in order to obtain a good separation. This topic is discussed in greater detail in Section 5.3 of Chapter 1.

Computer modelling of density gradient centrifugation can be used for predicting the isopycnic separation of macromolecules by equilibrium banding. This is useful because, first, both the positions of particles and the gradient shape change during centrifugation so that it is much more difficult to predict intuitively the effects of alterations in conditions than it is with stable gradients. Second, the long centrifugation times involved make proceeding by trial and error very tedious. Third, some prior calculations should be made in any case because if the centrifugal force or initial concentrations of the gradient material are too high then there is the risk of crystallization of the gradient solute at the base of the tube. This is extremely dangerous because the crystals will cause local stress on the rotor which can result in catastrophic rotor failure. The source codes of two programs which can be used for calculating the shape of isopycnic gradients are given in Appendices B and C of Chapter 1.

Equilibrium banding of nucleic acids and other molecules involves relatively long centrifugation times depending on the type of rotor and the centrifugal force which can be applied. Centrifugation times can be shortened in three ways. First, keep the gradient pathlength as short as possible, and rotors of choice are vertical or NVT rotors but partially-filled tubes in shallow fixed-angle rotors will often do equally well; if at all possible avoid using swing-out rotors because these have a long pathlength and low capacity. Usually the minimum time of centrifugation is determined by the time taken to form the gradient but this limitation can be circumvented by using pre-formed gradients. Simple two or three step gradients can be used for this purpose. These will rapidly diffuse during centrifugation to give a continuous gradient. The shape of the final equilibrium gradient is determined by the average density of all the liquid in the tube and the speed of centrifugation, not by the initial distribution. If, on the other hand, the minimum time of centrifugation is determined by the time taken for particles to reach their equilibrium densities, then the relaxation method of Anet and Strayer (38) can be used to shorten the centrifugation time required. In this case, centrifugation is initially at the maximum allowable speed of the rotor. Note that as the average density of the gradient is likely to exceed the design density of the rotor this will have to be calculated (see Section 4.6 of Chapter 2); failure to derate could result in disintegration of the rotor. Centrifugation at high speed will result in rapid banding of the particles at the centre of a steep gradient. The speed is then decreased for the last 1–2 h of the run to that needed to give a shallower gradient spanning the density range of the particles, causing the particle bands to move apart and so improving resolution (*Figure 15*). By using a short pathlength rotor in a micro-ultracentrifuge plasmid DNA banding can be achieved by centrifugation for 2–4 h (see Section 4.3.1 of Chapter 4).

The main problems arise from the very low viscosity of the solutions used since this makes gradients very susceptible to disturbance during deceleration and during removal from the rotor. It should be noted that at high centrifugal speeds gradients are very stable. Most mixing occurs at low speeds (< 1000 r.p.m.) and it is here that minimum braking is necessary in most cases. For the same reason, the apparatus used in gradient fractionation should be carefully designed to avoid density inversions (see Section 7.6). A minor problem, which sometimes occurs with larger particles, is pelleting of the sample before the gradient has time to form. This may be avoided simply by layering the sample over a solution denser than any of the particles.

5.7.2 Iodinated gradient media

Gradients of Nycodenz or metrizamide self-form quite quickly in vertical or fixed-angle rotors. The same parameters of gradient pathlength, centrifugal force, and the length of centrifugation that determine the shape of alkali metal salt gradients are also responsible for determining the shape of these gradients. However, in the case of these gradient media, temperature is also important since small increases in the temperature can have a marked effect in decreasing the viscosity of the gradient and hence increasing the rate at which the gradient forms.

The rates of diffusion of these compounds are much lower than for the salts of alkali metals and so a cushion of dense medium placed at the bottom of the gradient will modify its profile (39). There have been essentially no mathematical analyses of the formation of these gradients and so guidance on the best concentration of gradient medium to use and the choice of the optimum centrifugation conditions must remain rather empirical, although some guidance

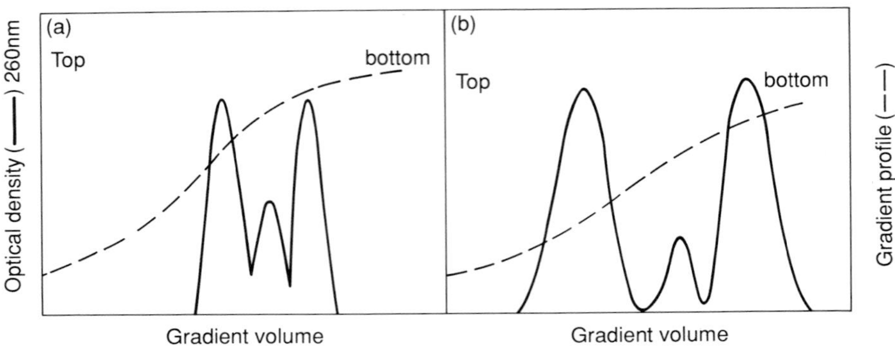

Figure 15. Use of the relaxation method to band DNA in CsCl gradients. Particles are initially banded by centrifugation at the highest possible speed; this results in a steep gradient and closely packed bands (a). The rotor is then slowed, diffusion flattens the gradient, and the bands move apart maximizing the resolution of the gradient (b).

can be obtained from the scientific literature. In fact the equilibrium shape of gradients in fixed-angle rotors tends to be rather exponential in shape and so not ideal for maximum resolution of particles in the gradient. However, because the gradients form so slowly it is possible to manipulate the shape of the gradients in terms of centrifugation conditions so that the resolution of gradients is maximized (39).

5.7.3 Percoll gradients

As in the case of iodinated gradient media, it is not possible to predict the shape of Percoll gradients other than by empirical methods. The coated silica particles have a negative charge and this stabilizes the colloidal suspension. During centrifugation the particles will sediment forming a gradient. Hence the major factor in the formation of a density gradient with Percoll is depletion of colloidal silica at the top of the tube and enrichment of the colloidal particles in the lower part of the tube. Hence, under normal conditions of use, the shape of a Percoll gradient is dependent on both the speed and the time of centrifugation (*Figure 16*). However, the ionic composition of the gradient affects the stability of the colloid and gradients will form much faster in the presence of metal ions because of their interaction with the negatively-charged silica particles. Because of the large size of the particles of Percoll, diffusion is very slow and so, once formed, gradients show little change in shape even after standing for some weeks.

The method of choice for the preparation of Percoll gradients is very dependent on the final application. Gradients may be pre-formed either using the apparatus discussed in Section 5.3, or pre-formed by centrifugation at a speed higher than that which will be required for the final separation, or formed during centrifugation. Note that if the sample is present during the time when the gradient is self-forming, then local discontinuities in the gradient can

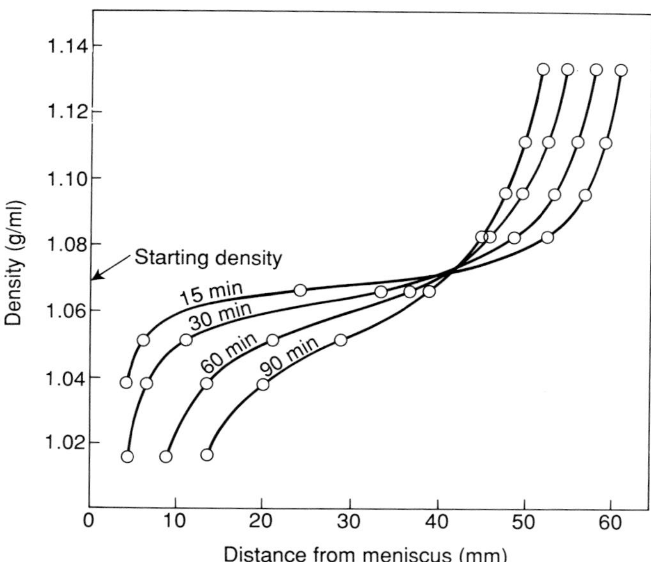

Figure 16. Development of a gradient on centrifugation of a uniform solution of Percoll. The figure shows the gradient, as indicated by the position of density marker beads (shown as circles) after centrifugation at 20 000g_{av} in a fixed-angle rotor (figure courtesy Pharmacia-Biosystems AB).

cause the formation of multiple artifactual bands of the sample within the gradient (40).

When separating large particles such as living cells, pre-formed linear gradients are probably best for rate-zonal separations and for initial experiments on isopycnic banding. Percoll does not sediment significantly under the conditions required to separate living cells and results from linear gradients are generally easier to interpret than those from sigmoidal gradients formed by Percoll during centrifugation. On the other hand, for routine separations, gradients pre-formed by centrifugation may be advantageous. A sigmoidal gradient gives good resolution in the centre of the gradient while concentrating low and high density contaminants in small volumes at the top and bottom of the gradient. When preparing the gradient the initial density should be the median density of the particles of interest. The gradient is then formed by pre-centrifugation at 10 000–30 000*g* for 20–30 min. The samples are then layered over the gradient and banded at low speed (approx. 1000*g*), so minimizing the risk of damage to the cells. These figures are given solely as examples. The shape of gradients is critically dependent on rotor geometry. At the time of writing, excellent booklets and reference lists are produced by Pharmacia-Biosystems AB and the authors would strongly recommend anyone designing a separation in a Percoll gradient for the first time to consult them.

Gradients formed during centrifugation are used mostly in the separation of smaller particles such as cell organelles. Percoll has been used mostly for the separation of the larger membrane-bounded organelles but can be used for particles down to the size of bacteria. The principles of gradient design are the same as for the pre-formed gradients discussed in the previous paragraph but when separating smaller particles the pelleting of Percoll becomes more of a problem.

5.8 Preparation of iso-osmotic gradients

Iso-osmotic gradients are gradients in which the osmotic environment is constant throughout the gradient. These gradients are essential for the fractionation of cells and they have also been used to investigate the osmotic behaviour of animal membrane-bound organelles. In the case of gradient media which do not exert any osmotic effects, essentially the colloidal silica media such as Percoll, an iso-osmotic gradient, can be made simply by adding an osmotic balancer such as NaCl or glucose to the medium and forming the gradient using one of the ways recommended in Section 5.7. Note, however, that Percoll solutions cannot be sterilized by autoclaving after the osmotic balancer has been added. In the case of polymeric media such as Ficoll and bovine serum albumin, these also exhibit low osmolarity in dilute solution, and iso-osmotic gradients for rate-zonal separations can be made by adding an osmotic balancer to the gradient medium before preparing the gradient. The iodinated gradient media, especially Nycodenz which is autoclavable, are widely used for cell separations (see Chapter 7), but because these media are osmotically active it is necessary to prepare the gradient from two or more iso-osmotic solutions that are made by diluting isotonic Nycodenz (28% w/v) with either 0.75% NaCl or 4.1% glucose (33). All Nycodenz solutions can be autoclaved after preparation as required.

6. Optimization of experimental conditions

It is unlikely that the initial experiments will provide satisfactory separations so that further experiments will have to be undertaken to optimize the separation conditions. These will follow several stages. The first stage will probably be fairly simple. If carrying out a rate-zonal separation then all the material of interest may be concentrated at the top or bottom portions of the gradient. A simple adjustment of centrifugation speed (remember that the centrifugal force is proportional to the square of the speed) or time of centrifugation will often solve this problem. Likewise, in isopycnic banding the slope of the gradient can be adjusted to cover only the density range of particles of interest. It is possible that these simple alterations will solve the problem. If not then more elaborate steps must be taken. Two groups of problems may be identified. When particles are separated on the basis of size it may be that at the speeds and times needed to separate the smallest particles of interest from the sample zone, the largest particles are pelleted. More usually the problem is that neither rate-zonal sedimentation nor isopycnic banding alone gives complete separation of particle populations. These problems can be tackled in a number of different ways.

6.1 Positioning of sample in the gradient

Normally a sample is layered on top of the gradient but there are circumstances where positioning the sample under the gradient or even, for isopycnic separations, distribution throughout the gradient are desirable. Loading the sample under the gradient has a considerable theoretical attraction, since the risk of hydrodynamic instability (sedimentation in droplets) is removed and with it the possibility of contamination of particulate fractions by soluble proteins spreading down from the sample zone. The disadvantages of this approach are that the sample is exposed to very high concentrations of the gradient solute and high hydrostatic pressures (see Section 3.2.2 of Chapter 1). The advantages of placing the sample at the bottom of the gradient are greatest for samples, such as serum lipoproteins, where the density is inversely related to the size, and also for membrane fractions (see Section 3.3.1 of Chapter 6). Loading the sample under the gradient can be used for both rate-zonal and isopycnic fractionation or in combined 'S–ρ' protocols (see, for example, 41).

Distribution of the sample throughout the gradient which is usual practice for self-forming gradients may also be advantageous for the separation of samples such as living cells, where there is a high risk of aggregation. Examples of this procedure are given in Section 5.2.1 of Chapter 7.

6.2 Extending the fractionation range of a gradient

The range of particle sizes which can be separated at any one time can be markedly increased by use of non-linear gradients. As discussed in Chapter 1, the rate of particle sedimentation is proportional to the difference in density between the particle and the medium and inversely proportional to the viscous drag. In convex gradients the larger particles are slowed by the high viscosity of the denser part of the gradient while the slower sedimenting small particles are less affected. It should be emphasized that while such convex gradients may solve the problem of fractionation range they may introduce another problem because the shallow slope at the top of the gradient inevitably means that it has a low capacity and so such gradients are easily overloaded.

6.3 Analysis of the causes of poor separations

Imperfect separation of two particles may be due to technical problems in the separation or to heterogeneity among the particles themselves. If the separation is on the basis of particle density then the cause may be that the smaller particles in the denser population have not reached their banding densities. Repeating the separation using a longer centrifugation time can be used to test this possibility. If this is not the cause of the problem and there is still doubt on

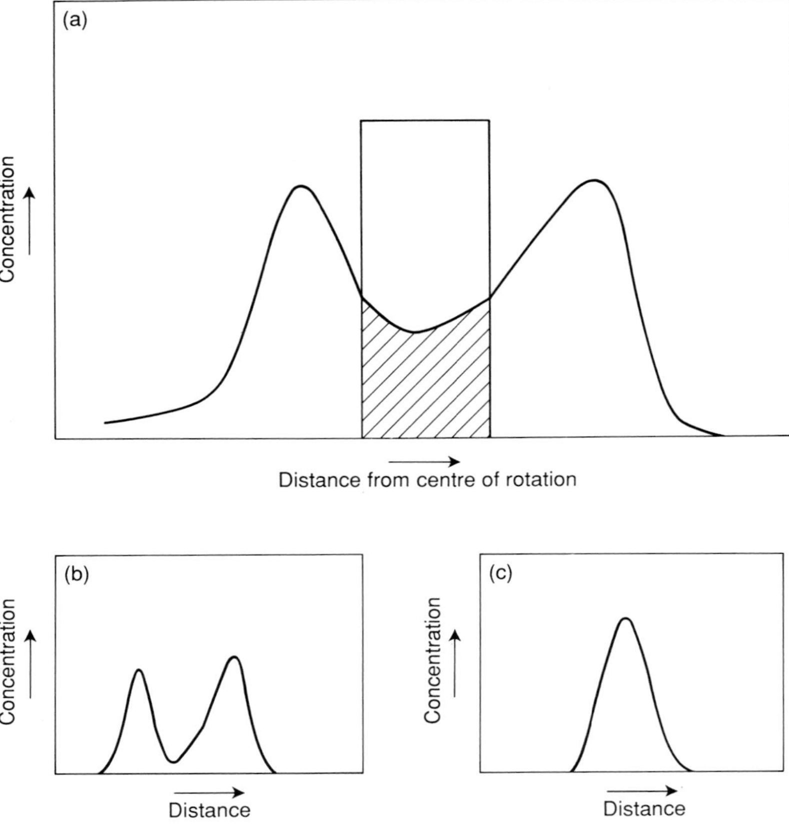

Figure 17. An illustration as to how an apparently poor separation (a) may be due to artifacts of centrifugation or to the presence of minor components. This question may be resolved by recentrifugation of material from between the two major peaks under the same conditions; (b) and (c).

on the cause of the poor separation this can be identified by a simple experiment; collect material from the overlap region, concentrate it, remove excess gradient solute, and recentrifuge. If the problem is technical, recentrifugation will give two distinct bands, but if the problem is due to heterogeneity of the particles then only a single major band will be seen (*Figure 17*).

6.4 Sharpening particle bands

Band broadening due to technical problems afflicts separations by rate-zonal sedimentation much more than isopycnic separations. Problems may arise during application of the sample to the gradient, or to broadening of the sample or sedimenting bands due to hydrostatic or hydrodynamic instability. Examination of the shape of the bands may provide a clue to the problem for, as mentioned previously in Section 2.1.3, band broadening due to hydrostatic or hydrodynamic instability results in asymmetric bands. Generally, however, no cause is apparent. In such cases the first thing is to make a rough estimate of the width of the sedimenting bands as compared to the sample band; as a general rule the sample volume should be in the range of 2–5% of the gradient volume with an initial thickness of less than 3 mm. If this calculation shows that there has indeed been band broadening rather than over-optimistic expectations then you should check your sample loading technique using a coloured sample. This should, if possible, be a 'real' sample to which a dye such as 0.1% Blue Dextran has been added, and the advantage of this is that it is then possible to check the second possibility which is that the presence of the particles has raised the overall density of the sample above that of the light end of the gradient. Only when these simple checks have been made should hydrostatic or hydrodynamic instability be suspected. To test for these effects, samples containing the normal amount of material and samples containing the minimum detectable amount of material should be run in parallel. Differences in resolution indicate gradient overloading. This may be solved by use of steeper gradients or by reducing the total amount of material in the sample; for example, by pre-purification of the sample prior to gradient fractionation.

It is also possible to obtain zone sharpening of the sample bands in rate-zonal centrifugation. This can be done in two different but complementary ways. First, it is often advantageous to load the sample in a low viscosity solution, usually this can be achieved simply by omitting sucrose. In this situation the sample particles at the bottom of the sample layer begin to sediment into the gradient but they encounter a sharp increase in viscosity at the top of the gradient causing the particles to slow down, and this then allows particles from the upper part of the sample to catch up with the particles at the top of the gradient and they then sediment together as a band which is narrower than the original sample band. Zone sharpening will also occur in the gradient since the front of the band is always moving into denser, more viscous solution, and so the particles in the less viscous solution move faster and keep the band narrow. An example of this type of zone sharpening is shown in *Figure 18*. As a general rule, steeper, more viscous gradients (5–30% or 10–40% w/w sucrose) will give sharper bands than shallower (5–10% or 5–20% w/w sucrose) gradients. However, there are limits to this approach in that for rate-zonal separations the density of the gradient must always be less than that of the particles. Higher viscosity gradients also require higher centrifugal fields as otherwise resolution may be lost as a consequence of the need to increase the time of centrifugation significantly, which in turn allows greater diffusion of the sample bands.

6.5 Problems arising from overlap in particle properties

If incomplete separation of particle populations is due to a genuine overlap in sedimentation rate or in density one must return to the S-ρ diagram (*Figure 9*)

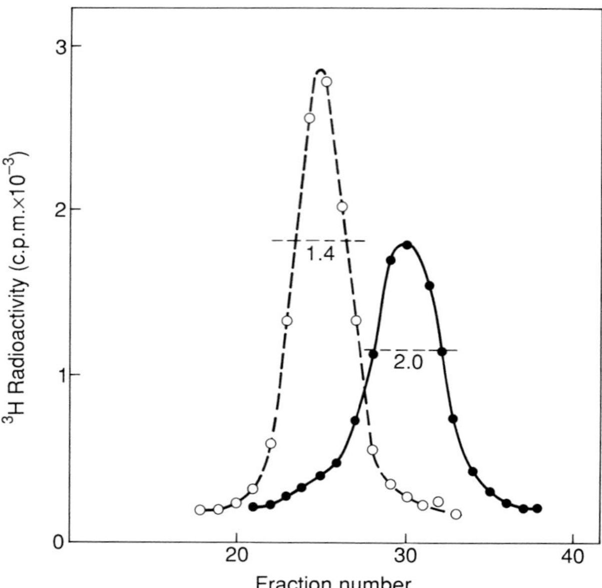

Figure 18. Gradient-induced zone sharpening. Rate-zonal separation of ^{3}H-DNA from bacteriophage lambda on either a 5–20% linear sucrose gradient (●–●) or a 5–30% linear sucrose gradient (○---○). After centrifugation the gradients were fractionated and the peaks of DNA analysed. Measurement of the peaks at 60% of their height indicates that the zone of DNA is significantly narrower in the 5–30% gradient. (Adapted from reference 42.)

to determine alternative separation conditions. Data obtained in the earlier experiments may be used to refine this. In many cases one can now see that particles whose populations have overlapping sedimentation rates will differ in their density while particles with similar densities differ markedly in size. If this is the case then a complete separation may be obtained by successive applications of rate-zonal sedimentation and isopycnic banding; it is normally convenient to perform the rate-zonal sedimentation as the first step. The fractions containing the required particles may then be taken, pooled, and either applied to an isopycnic gradient or used directly as the 'light' solution to form such a gradient. Carrying out the steps in this order avoids the need to concentrate the particles between the two separations.

6.6 Problems where populations of particles overlap in size and density

In many cases populations of particles overlap in both size and density (e.g. in liver small mitochondria and large, light lysosomes) and in such cases a complete separation cannot be achieved, but frequently subpopulations of each organelle can be purified. In such a situation careful thought must be given to the causes of the heterogeneity. Following homogenization it is thought that nuclei, lysosomes, and probably peroxisomes are released in the same form as they are present in the living cells. The long tubular mitochondria which exist in the living cell fragment to approximately spherical structures while the various membrane systems fragment to vesicles. This fragmentation will inevitably result in a range of particle sizes and, because the ratio of membrane to contents varies with particle size, in a range of particle densities. Where this is the cause of the heterogeneity then a subpopulation of fragments is likely to be representative. On the other hand, heterogeneity in size or density may reflect real heterogeneity of organelles within a particular cell type or between the different cell types of a given tissue, and both factors appear to contribute to the remarkable heterogeneity of kidney lysosomes (43). In this case, of course, a subpopulation will definitely not be typical of the whole population. Careful choice of gradient material can be of help in minimizing overlap; remember that the S–ρ diagram will change from one gradient medium to another since particles behave very differently in terms of their observed density, depending on the nature of the density gradient solute. As discussed earlier in this chapter (Section 2.2), differences in the extent to which different solutes penetrate cell structures and differences in their osmotic effects means that not only do absolute banding densities vary with the medium, but so do relative banding densities (*Tables 1* and *2*). For macromolecules a wide range of ionic and non-ionic gradient media are available and the medium of choice will depend on the exact type of separation required. Similarly, in the case of cell organelles, if one has failed to obtain a separation in sucrose one should examine relative banding densities in a low-viscosity, low-osmolarity medium such as an iodinated medium (e.g. Nycodenz or metrizamide), or a colloidal medium such as Percoll. One may also consider the use of glycerol, which is fully permeant but, as discussed previously, the high concentrations needed to band organelles make for long centrifugation times. Frequently a single medium will not give the required separation and successive steps may be required as used for the separation of endosomes (44). The choice of media for use with single cells is more restricted but, in addition to the iodinated media and Percoll, gradients of Ficoll and bovine serum albumin may be used, although both have serious drawbacks (see Section 3.2.3).

6.7 Modification of the properties of particles

In some cases initial experiments make it clear that it will be difficult to obtain an adequate separation using any of the available density gradient solutes because there is almost complete overlap in both size and density between the particles of interest, irrespective of the gradient medium used. In this case one must seek to modify the physical properties of the particle selectively. For macromolecules the density can be modified by changing the ionic composition of the gradient (e.g. addition of silver or mercury salts to Cs_2SO_4 gradients for DNA

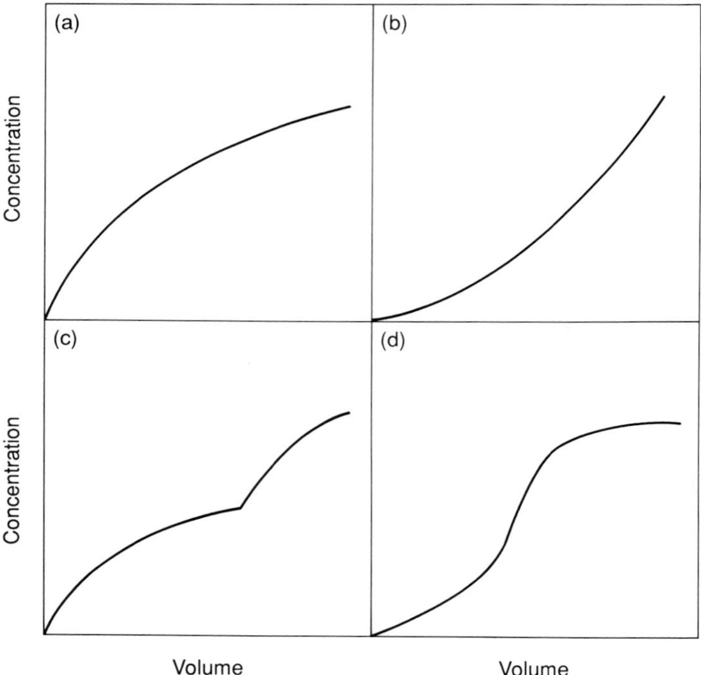

Figure 19. Various gradient profiles for zonal rotors. Exponential gradients such as that shown in (a) are especially useful in zonal rotors to produce gradients which are linear with radius. Inverse exponential gradients (b) which may be prepared with a 'double pump' apparatus (57) increase the size range of particles which may be separated on a single gradient. More complex gradients (c) can be generated by the superimposition of simple segments. Sigmoid gradients (d) can be generated by a programmable gradient maker; it is dangerous to use such gradients for rate-zonal separations as bands will sharpen in the steep central section, resulting in overloading as they enter the flatter final section.

separations) or by the use of centrifugation in the presence of specific ligands such as ethidium bromide that bind differentially to DNA with different conformations. For further details the reader should consult Section 4.3 of Chapter 4. In the case of cell organelles it may also be possible to modify the density of the particle. Lysosomes, for example, may be loaded with exceptionally light or dense material (45). Attempts have also been made to exploit the ability of some enzymes to produce dense, insoluble, products when supplied with suitable substrates (e.g. 46) but these have not been popular as might be expected, possibly because of the rather exacting conditions needed for the density-labelling procedures to be successful. Detailed descriptions and protocols of the density perturbation methods used for separating membrane fractions are given in Section 3.4 of Chapter 6. In the case of cells, modification

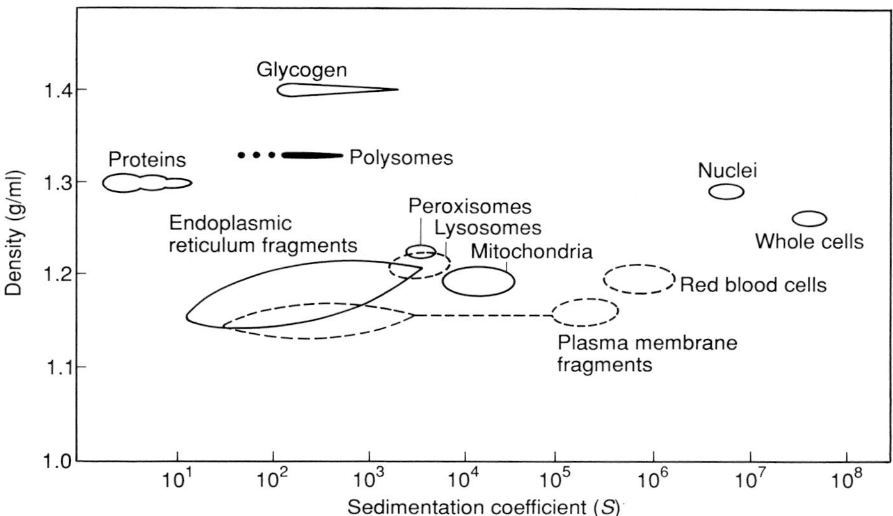

Figure 20. An S–ρ diagram for rat liver. The diagram shows the distribution of the different components present in the total liver homogenate. The densities of the denser components such as nucleoproteins are estimated values only (derived from reference 57).

of the sedimentation rate normally involves linkage of the structures of interest to large particulates; for example, via antibodies to cell surface components. Such procedures, exemplified by the use of magnetic beads in lymphocyte separations (48), are allied to other methodologies and so fall outside the scope of this chapter.

6.8 Uses of complex gradients

So far this chapter has concentrated on the use of simple linear gradients. Such gradients should always be used for initial experiments because the results are easy to interpret and frequently give sufficiently good separations. However, other gradient shapes may improve separations once the size and density distributions of particles in a mixture are properly understood. Some examples are given in the following sections.

6.8.1 Complex continuous gradients

Because of the sophisticated equipment required and the minimum volumes which can be handled by this equipment, these gradients are essentially restricted to zonal rotors. Four of the most common shapes are shown in *Figure 19*. The first gradient (a) gives a gradient which is linear with distance in zonal rotors, the second (b) would be used to increase the fractionation range in a rate-zonal separation, the increasing density and viscosity at the bottom of the gradient slowing the larger particles. The third gradient form (c) maximizes the amount of sample which can be loaded on to the gradient. This type of gradient would

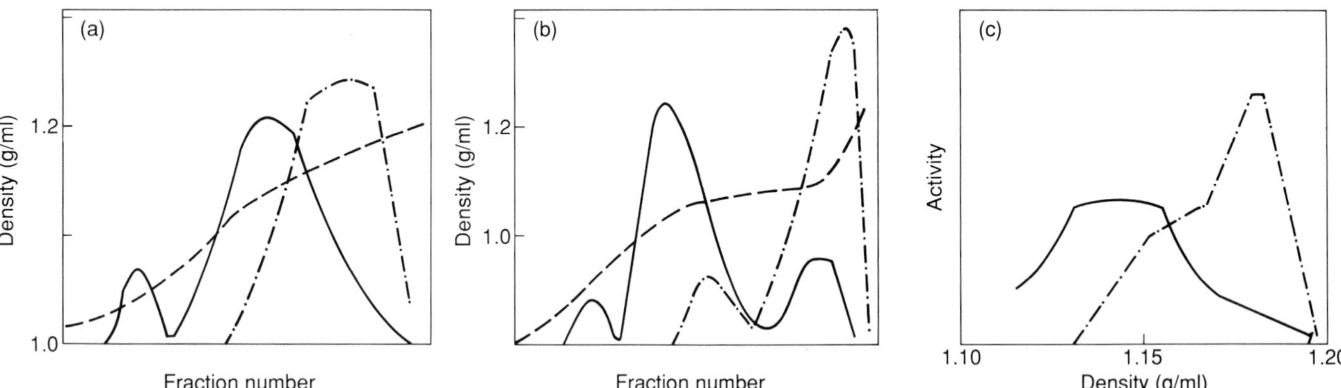

Figure 21. Artifactual 'separations' produced by flat gradient sections in isopycnic gradients. Section (a) shows the separation of plasma membrane fragments (—) and endoplasmic reticulum fragments (- · - ·) on a near-linear gradient, section (b) and (c) the same separation on a gradient with a flat central section with the activities of marker enzymes for the two organelles plotted (b) against fraction number and (c) against density. This shows that the apparent separation of the membrane fragments into two populations in (b) is an artefact of gradient shape (redrawn from reference 57).

be used when separating particles from a sample which contains large amounts of soluble proteins and where the maximum density is limited by the isopycnic banding density of some components. The fourth gradient type has an initial section in which the smaller components of a mixture are separated by sedimentation rate, and a second section in which larger particles are separated by isopycnic banding. The S–ρ diagram for a system of particles where such a gradient is useful is shown in *Figure 20.*

The examples given here show how gradients can be designed to solve particular problems. It is difficult to give general rules for gradient design except to say that:

(a) The design of gradients depends on understanding the relevant properties of the particles involved and that means, in most cases, preparing an S–ρ diagram.

(b) In rate-zonal separations, where the slope of a gradient is reduced there will be a reduction in capacity which may result in hydrostatic instability and band broadening.

(c) When separating samples by isopycnic banding, use of gradients containing both a concave and a convex section may cause problems in interpretation as a result of apparent rather than real heterogeneity (*Figure 21*).

In addition to using gradients complex in shape one can also, in principle, form gradients containing more than one solute. This will not normally cause problems if the two solutes have similar diffusion coefficients (e.g. gradients of mixed caesium salts can be used to fractionate RNA) or if the second solute does not affect the density profile, as is the case of urea added to CsCl gradients for the fractionation of nucleic acids and proteins. If there are marked differences in diffusion coefficient major problems from hydrodynamic instability may arise if there are any zones in the gradient where the concentration of the less diffusible solute is falling while that of the more diffusible solute is rising.

6.8.2 Step gradients

Step gradients can prove a very useful way in which to separate and concentrate particles simultaneously. Clearly, step gradients can only be used for isopycnic separations. The design of this type of gradient again needs knowledge of the S–ρ diagram. The concentration of the solutions forming the steps obviously requires information about the relative density of the different components of the mixture. However, knowledge of the relative sizes is also important because there is a risk that particles passing through an already-formed band may aggregate with material already present. Hence for samples where there is a direct relationship between size and density, the samples are best layered over the gradient, and where there is an inverse relationship samples are best layered under the gradient. There is one exception to this rule. Samples containing a high-molecular-weight solute such as Ficoll should never be layered over gradients of low-molecular-weight materials, otherwise sedimentation in droplets is almost certain to happen.

7. Practical considerations

7.1 Choice of centrifuge rotor

Detailed descriptions of the various types of centrifuge rotor and their range of applications have been given in Section 4.3 of Chapter 2. In terms of gradient separations, swing-out rotors are the most useful since they are the rotors of choice for all types of rate-zonal separation as well as being the preferred rotors for the isopycnic banding of cells and cell organelles. An additional advantage is that the tubes of swing-out rotors are the easiest to fractionate because they are parallel sided and fit into any of the various types of gradient unloaders. Swing-out rotors are not really suitable for isopycnic separations of macromolecules in CsCl gradients and so if a suitable fixed-angle rotor is available then it should be used for this type of separation. Fixed-angle rotors can also be used for the rate-flotation of samples (e.g. lipoproteins). While vertical rotors have the advantage of requiring much shorter run times than swing-out rotors, they are not to be recommended for the development of a fractionation procedure because often material will pellet if the gradient and run conditions have not been correctly optimized, and this pelleted material can contaminate the whole length of the gradient as the liquid in the tube reorientates at the end of the run. In addition, gradients in vertical rotor tubes are more difficult to fractionate than tubes of swing-out rotors. However, once the fractionation method has been optimized then it is worth considering whether there is any advantage to be gained from using a vertical rotor for the gradient separation step.

Once you have decided which rotors to use, transfer them to an environment which is at the same temperature as the centrifugation run so that the rotors equilibrate to the correct temperature before use; usually 1–2 h in a refrigerator or incubator is sufficient.

7.2 Choice of type of centrifuge tube

A detailed description of the different types of centrifuge tube that are available has been given in Section 5.1 of Chapter 2. Most centrifuge tubes which are suitable for gradient work are made of either polyallomer, polycarbonate, polysulphone, or PET. Polyallomer tubes are relatively inexpensive and are easily recognized because they are translucent; the actual degree of transparency depends on the source of the tube and the thickness of the wall. This plastic is of moderate strength but has very good chemical resistance and is easy to pierce. New tubes tend to be so water repellent that gradients may not flow evenly down the side of the tube but form large droplets which fall down and disturb the gradient already in the tube. To overcome this problem, new polyallomer tubes may be soaked briefly in chromic acid to etch the surface so that it becomes wettable (47) or, if facilities are available, then glow discharging of PET tubes has been reported to have a similar effect (49). These treatments may, however, increase the adsorption of proteins and nucleic acids on to the tube wall and this can be a major problem if you are separating sub-microgram amounts of sample. Rinsing tubes out with a dilute solution of protein or SDS solution may help to solve this problem but it may not always be feasible. Polycarbonate and polysulphone tubes are transparent, very strong, readily wettable, and thus on balance the most suitable for gradient separations. However, polycarbonate tubes are brittle and thus unsuitable for gradient unloading methods that involve piercing tubes.

Generally, centrifuge tubes which are used in swing-out rotors tend to be fairly thin-walled. This is because the centrifugal force is along the axis of the tube and, provided that the tube full to within 3 mm of the top, is dry on the outside, and fits well in the bucket it is most unlikely to deform or collapse even at high centrifugal forces. The most common failure occurs when these criteria are not met and the tube may then concertina if the centrifugal field exceeds about 30 000*g*. Thick-walled tubes that can be run partially full are available for most rotors and these can be used for short-column gradients (50). Heat sealed (Quick-Seal) and Ultracrimp tubes are also available for swing-out rotors, but because of the difficulty of unloading gradients in this type of tube they are not recommended unless the containment of the gradient and its sample are of paramount importance (e.g. if it is highly infectious or radioactive).

The tubes designed for fixed-angle rotors are generally stronger. Thinner tubes are intended to be completely filled and sealed so that the liquid inside hydraulically supports the tube, thus preventing it from collapsing. Thick-walled tubes made of polycarbonate are essential if they have to be used partially filled. A particular use of such tubes is with short column gradients, which permit rapid and effective equilibrium spins to be achieved in fixed-angle rotors. Of the variety of tubes available for fixed-angle rotors it is important to note that generally only straight-walled tubes are suitable for gradient fractionation by upward displacement because of the difficulties of getting an airtight seal at the top (see Section 7.4). However, straight-walled tubes usually require the use of the rather complicated and rather inefficient multi-component cap with its rubber O-ring. This rather difficult system contrasts with the ease and efficiency of use of the narrow-necked tubes, which have resulted in this type of tube becoming the tube of choice for both vertical and fixed-angle rotors. In order to simplify the unloading of gradients from narrow-necked tubes Seton Scientific sells a tube cutter which cuts off the top cone section of the tube leaving a smooth top, allowing the gradient to be unloaded using any of the standard methods. If narrow-necked tubes are to be used for gradient fractionations then this device is strongly recommended. A detailed description of methods of sealing tubes is given in Section 5.2 of Chapter 2.

7.3 Preparation and composition of the sample

The success of a gradient fractionation procedure will depend on the condition of the sample just as much as on the shape and composition of the gradient. Ideally, membranous organelles and whole cells should be prepared in a medium and at a temperature which preserve their integrity and viability. This is not necessarily 4 °C since, as described in Section 4.4, cooling may damage viable cells. It is also important to ensure that the sample preparation is synchronized with gradient preparation so that once the sample is ready it is not necessary to wait while the gradient is prepared. Not only should gradients be prepared ready for use but they should also be pre-equilibrated to the correct temperature before use; this may take several hours if the gradients are simply left in a refrigerator or incubator.

The rather critical conditions required for rate-zonal sedimentation dictate that a number of criteria must be observed for a successful separation whilst, in contrast, isopycnic banding is much more forgiving of technical errors.

(a) For rate-zonal separations it is most important for the sample to have a lower overall density than the top of the gradient. The reasons for this are clearly explained in Section 2.1.3. In practice it is prudent to check the density of the sample and ensure that this is lower than the least dense gradient solution.

(b) The volume (i.e. thickness) of the sample band should be kept as small as possible for maximal resolution. It is important to realize that in rate-zonal sedimentation the sedimenting bands cannot be narrower than the sample band unless they approach equilibrium or undergo band-sharpening (see Section 6.4). In practice all bands broaden but to a lesser degree than might be expected because diffusion is balanced by the natural band-sharpening effect due to the increasing viscosity of the denser gradient solution.

Problems associated with sample band composition and density, such as the phenomenon of streaming ('droplet formation'—described in Section 2.1.3) or tumbling (overloading), may be minimized by applying the sample as an inverse gradient. In this, the leading edge of the sample band has the same density (or gradient solute concentration) as the top of the gradient but an infinitely low concentration of sample. The sample concentration then increases whilst the gradient solute decreases; hence the term inverse gradient (51). In practice such inverse gradient samples are much easier to make with large sample volumes as would be employed with zonal rotors. It is difficult to make inverse gradients with small samples (see Section 7.4) and the improvement in resolution obtained is usually marginal. Careful preparation and layering of samples is usually a much better recipe for success.

For isopycnic banding the volume of sample may be as large as 80% of the tube volume allowing concentration of particles from very dilute samples.

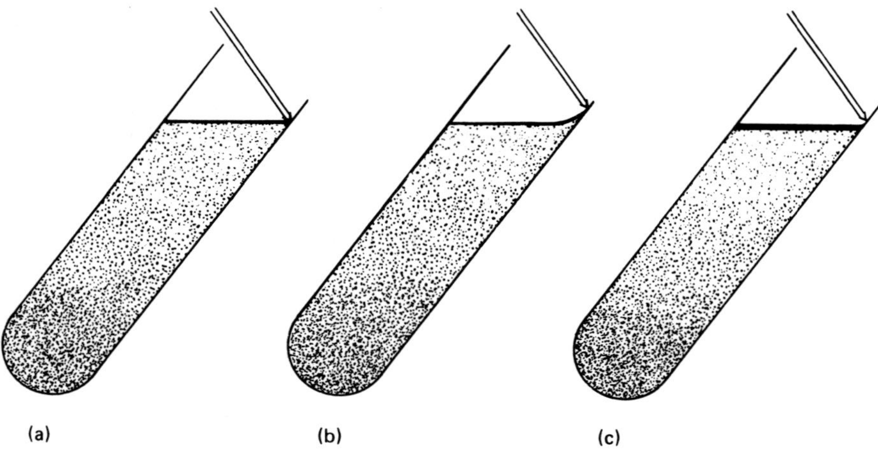

Figure 22. Layering the sample on to a gradient. The pipette or syringe is touched against the meniscus, drawn up slightly, and the sample layered slowly on to the gradient. The liquid from the pipette or syringe is directed at the wall of the tube, minimizing mixing with the top of the gradient (derived from reference 57).

Also the sample concentration does not affect band capacity and thus can permit the fractionation of highly concentrated samples. However, as the amount of material in the sample increases so does the width of the band (see Section 5.3.1 of Chapter 1).

7.4 Application of samples on to the gradient

The application or layering of the sample band on top of a gradient, after a little practice, can be performed by anyone accustomed to using a pipette (see *Figure 22*). With the tube held or located vertically (or slightly tilted, if preferred), the tip of a pipette containing the sample is touched against the meniscus at the edge of the tube. The tip is then moved very slightly upwards to draw the meniscus up the wall of the tube. Finally, the sample is allowed to run slowly out of the pipette to form a well-defined layer on top of the gradient. This is easily achieved with a manual pipette. If this is ruled out on grounds of safety a 'loose' syringe (glass syringes are usually best) or a bulb-controlled pipette are suitable alternatives. It is recommended that novices should practice by layering a dye sample on to a 'light' gradient solution.

As stated in the previous section, there should be advantages in loading samples in the form of an inverse gradient and a rather crude inverse wedge sample can be applied in the following way. After drawing up the sample into a pipette this is then immersed into some of the least dense gradient solution and an equal volume of this is allowed to flow up into the pipette. The mixing of the gradient solution with the sample will form a crude inverse gradient which may now be layered on to the gradient in the tube. However, as stated previously, it is only worth resorting to such intricacies if it is clearly beneficial in terms of resolution of the sample.

For rate-flotation techniques (commonly used for lipoproteins) the sample should be suspended in the densest gradient solution and applied under the gradient; this can be done most conveniently using a syringe fitted with a long needle. It is essential in this case not to allow air bubbles to enter the gradient as they may disturb the gradient and carry with them traces of sample thus contaminating the gradient.

The application of samples for isopycnic banding is less critical provided that the gradient is not disturbed. The technique of equilibrium flotation may, in some cases, offer distinct advantages over sedimentation, particularly for the separation of membrane fractions (see Section 3.3.1 of Chapter 6).

In some cases it is advantageous to overlay the sample with a low-density solution, usually with sample buffer. This is recommended particularly for rate-zonal separations in vertical rotors in order to avoid possible disturbance to the sample layer during reorientation of the gradient that occurs at low speed. In addition, in zonal rotors it is imperative to use an overlay to move the sample away from the central core in order to form a narrower band, and also to push the sample into a region of sufficient centrifugal force to permit sedimentation. This is discussed in greater detail in Chapter 8.

7.5 Conditions during centrifugation

As well as accurate speed control and duration of centrifugation the control of temperature of the sample and gradient are most important for reliable and reproducible fractionations. Accurate temperature indication is indeed vital for calculating sedimentation coefficients as the viscosity of the gradient must be precisely known (see Section 9.2 of Chapter 5). If possible, use an ultracentrifuge with temperature control which is accurate to within 0.5 °C or less. To confirm the temperature of the run it is useful to spin a gradient without sample at the same time as the sample and check its temperature with an accurate thermocouple probe. The choice of centrifugation conditions depends on the nature of the sample.

Whilst smooth acceleration and deceleration are important for all gradient work, these parameters must be carefully controlled when gradients reorient in fixed-angle rotors or vertical rotors. Fortunately, most of the modern drive systems have developed to the point when all high-speed centrifuges and ultracentrifuges are equipped with programmed acceleration modes, different acceleration, and deceleration rates which minimize the disturbance of gradients, even when working with non-viscous salt gradients. As there are still quite a few older machines in use it is worth bearing in mind that for all swing-out rotor gradient work the rotor should be accelerated as slowly as possible up to 1000 r.p.m. and it should be brought to rest with

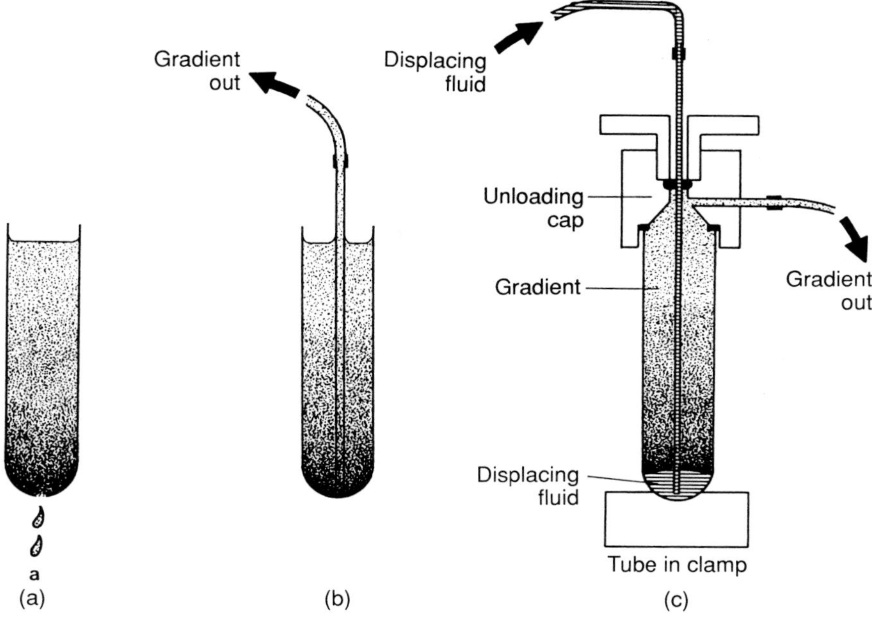

Figure 23. Methods for collecting density gradients after centrifugation: (a) the bottom of the tube is pierced and fractions collected as the gradient drips out; (b) the gradient is pumped out via a narrow tube inserted through the gradient to the tube bottom; or (c) collection by upward displacement, where the gradient is displaced upwards through an unloading cone by delivering a dense liquid to the bottom of the centrifuge tube via a narrow cannula.

the brake switched off below about 1000 r.p.m. Most modern ultracentrifuges can be programmed to do this automatically.

7.6 Recovery of fractions from the gradient

Having achieved a successful fractionation the problem now is how best to collect the separated bands as discrete fractions. Before describing the methods of how to fractionate gradients it is worth pointing out that a single discrete band (e.g. plasmid DNA or mitochondria) may be easily and rapidly removed using a Pasteur pipette with a bent tip or a syringe, provided that it is the only band which is needed and it is readily visible and identifiable.

In order to collect the whole gradient in a series of fractions three different methods may be employed (*Figure 23*). The oldest and perhaps simplest (although expensive on tubes) is to pierce the bottom of the tube carefully, and collect the gradient as it drips out. It is, however, cumbersome, even using a commercial piercing apparatus. It has been suggested that the gradient may be sucked out via a pump from the bottom of the tube via a narrow capillary tube

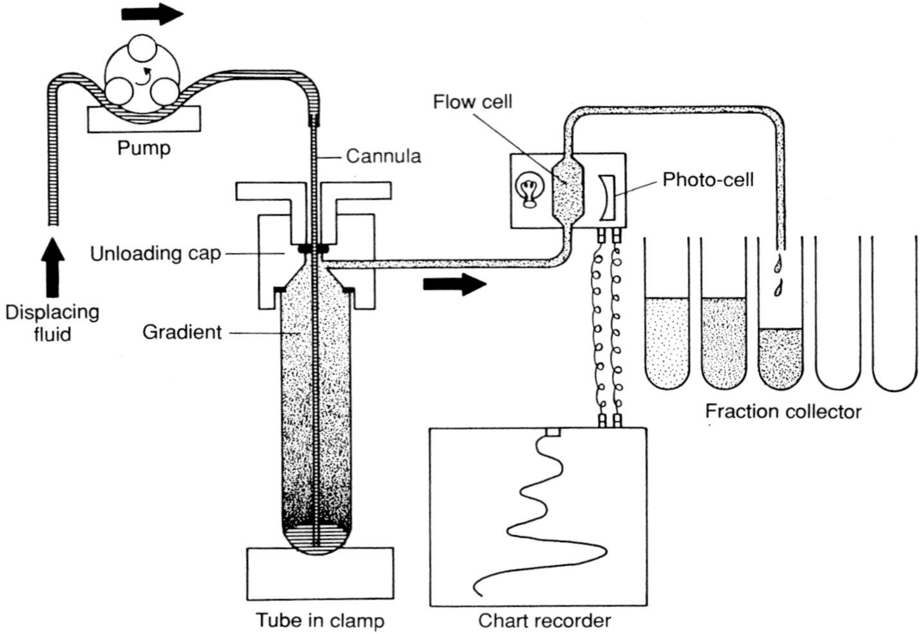

Figure 24. Continuous monitoring. The gradient is displaced upwards toward an unloading cap by introduction of a dense displacing fluid at the tube bottom, using a peristaltic pump or, preferably, a motorized syringe. The displaced gradient is monitored optically by passing it through a spectrophotometer flow cell before collection as a series of fractions.

which has been carefully lowered through the gradient. However, this method cannot be recommended because it has a serious disadvantage in that density inversion of the solution occurs and the fractionated components must pass through the pump, so losing resolution of the gradient and possibly degrading the samples. The third and probably the best method involves the pumping of a dense liquid to the bottom of the tube and the subsequent upward displacement of the gradient via a tapered cap. The most suitable pump for this type of displacement is a very even flow-rate-perfusor type syringe pump if it is available, otherwise a normal peristaltic pump can be used. Alternatively, for low viscosity displacement fluids, it is possible to use a burette. The apparatus for holding, sealing the tube, and inserting a cannula for the displacing liquid is commercially available from a number of manufacturers. For successful recovery of gradients by this method it is important to seal the tube at the top and to ensure that the cannula for the displacing liquid is also well sealed. Make sure that there is no skin of low-density material floating on top of the gradient; if there is then carefully remove it using a paper tissue or a stainless steel spatula. The cannula must be completely filled with the dense displacing liquid (no air bubbles) before it is inserted slowly, but smoothly in one movement, to the bottom of the tube. The flow rate chosen for displacement must be sufficiently slow to allow bands which may drag on the walls of the tube and through the displacing cap to pass without any loss of resolution. Tube runs should have a narrow bore, be kept short, and directed upwards to avoid mixing by density inversion which is a particular problem with non-viscous gradients such as CsCl gradients. A complete system for unloading the gradient, and monitoring and collecting fractions is illustrated in *Figure 24*.

It is also worth noting that displacing solutions should either be of the same material as the gradient solute or immiscible with it in order to avoid rapid diffusion and mixing of the gradient and displacing liquid. Cheap materials like sucrose can be used for sucrose gradients and in this case the displacing solution can be discarded; the problem here is the high viscosity of concentrated sucrose solutions. More expensive materials like Nycodenz can be re-used a number of times provided that their density is checked and 'topped up'. However, probably the best displacing liquids are dense fluorocarbons which are immiscible with water and organic solvents, inert, low viscosity, and fairly cheap. A typical material is FC43 (3M Company) which is also available from Nycomed Pharma A/S as 'Maxidens'. This type of displacement material can be reused many times, but care must be taken with continuous-flow UV monitors not to contaminate the flow cell as the fluorocarbon is difficult to wash out.

The recovery of fractions from zonal rotors is also best achieved by 'upward' displacement. The dense displacing solution is pumped to the edge line, through the rotating seal, and via the channels which lead to the periphery of the rotor (equivalent to the bottom of the tube). The gradient comes out light-end first out of the centre line of the seal assembly, ready to be monitored (see Section 7.7) and collected as fractions. The procedure for unloading zonal rotors is discussed in greater detail in Chapter 8.

7.7 Determining the density gradient profile

Although the profile of the gradient will be known approximately from the intended shape of the generated gradient, it is generally good practice to check the exact profile by measuring the density of the individual fractions. The gradient profile can then be plotted along with other parameters analysed in the fractions. Accurate measurement of the density of fractions is also essential for the calculation of sedimentation coefficients. This is discussed in detail in Chapter 5.

A range of different techniques such as pycnometry and refractive index conversions for each solute can be used to determine the density of the individual gradient fractions. However, this task has been simplified by the advent of direct-reading, digital, portable (even temperature reading) densitometers. A range of such instruments offering a wide range of accuracy and features is available from Paar Scientific. The portable model DMA 35, which reads the density to three decimal places, is ideal for most gradient work. However, a disadvantage of these instruments is that at least 0.5 ml of sample is required for the measurement and so this method is not useful for small (< 14 ml) gradients. As the measurement is non-destructive and the sample can be recovered it is most convenient with low-volume fractions to take the density measurement first before carrying out any other analyses.

If a density meter is not available or if the fractions are smaller than 0.5 ml then refractive index measurements are a convenient means of determining density by subsequent conversion from tables available for that particular gradient solute (see Appendix 3). The most suitable instrument for this is the Abbé type refractometer which has good accuracy and only requires a small amount of sample, typically 20–30 μl are sufficient. Note that when taking refractive indices it is important to make allowance for temperature and any other dissolved solutes before applying the equations given in Appendix 3. In the case of mixed gradients, (that is, where there are two or more gradient solutes), then it is necessary to make a calibration curve of density against refractive index. Even so, refractive index can only be seen as a guide for mixed gradients and accurate measurements should be made using a densitometer or pycnometry.

Whilst there are continuous-flow type instruments available for monitoring both density and refractive index these offer no particular advantage as they add further complications into the displacing system and are generally much more expensive.

Determination of density from refractive indices can only be applied to true solutions and so this method cannot be used with colloidal gradient media. Density marker beads are available from Pharmacia-Biosystems AB in ten different densities, each colour coded for easy identification. They are especially useful for following the progress of self-generating gradients and were indeed developed in conjunction with Percoll. Rather than include the beads in with a biological sample it is best to load the beads on to a separate gradient and use this as a reference. Obviously for clear identification they are best used with completely transparent polycarbonate tubes. It must be remembered that the banding densities of these cross-linked dextran beads are specified for Percoll gradients and that their density will differ if they are used with other types of gradient materials. After centrifugation, measure the position of each band of beads along the length of the gradient and plot the result (see *Figure 16*).

7.8 Detection of separated components

It is difficult to lay down general rules for the detection of material following fractionation of a density gradient as these will clearly depend on the exact nature of the sample. However, certain general principles can be laid down and common problems identified.

7.8.1 Visual inspection of the gradient

During the initial development of a separation method it is strongly recommended that transparent centrifuge tubes are used so that gradients can be inspected and, if necessary, drawings or photographs made before fractionation. Indirect illumination is the most useful method for spotting the position of light-scattering bands but, in some cases, examining band colour using direct illumination is useful. It is also sensible to collect fractions in clear tubes and examine their colour by standing the tube rack on a white surface. While these quick checks are only of limited value in locating the separated components they are by far the easiest methods for detecting aggregation or precipitation of material in the gradient, and can be useful for analysing the position of mitochondrial material which is brownish or detecting DNA in ethidium bromide–CsCl gradients.

7.8.2 Monitoring the distribution of material during fractionation of gradients

During method development it is always useful to monitor the distribution of material during fractionation of the gradient. Either column monitors, as used in chromatographic separations, or spectrophotometers fitted with flow cells can be used. In either case a recurrent problem is misting of the flow cell in hot and humid weather. Placing a beaker or sachet of dessicant in the sample chamber of the detector some while before fractionating the gradients usually solves this problem. Another common problem is direction of the flow through the flow cell. This should always be arranged so that light liquid lies above denser liquid in the flow cell itself, otherwise unacceptable mixing will occur. The direction of flow through the cell thus depends on whether the gradient is being fractionated by dripping from the bottom or by upwards displacement (*Figure 23a* or *c*). Upwards displacement generally results in fewer problems, for in this case small air bubbles will tend to be cleared from the cell by the flowing liquid. A final practical problem is to ensure that the flow cell is washed well between fractionations, and is filled either with water in the case of fractionation by upwards displacement or with dense gradient solution in the case of downward displacement. Failure to observe these precautions will result in turbidity in the cell and false readings.

Monitoring the absorbance of a displaced gradient thus presents few technical problems. Choice of the wavelength at which to monitor can be somewhat more tricky. For materials such as nucleic acids or nucleoprotein particles there is clearly no problem. These materials have well-characterized absorption maxima at 260 nm and the absorbance trace will provide an accurate measure of the concentration of particles. The situation is different when fractionating cells and membrane-bounded organelles. These scatter as well as absorb light and, at the concentrations normally found, in separations are opaque in the ultraviolet. In the authors' experience, at the concentrations normally employed for preparative fractionations, a wavelength of about 540 nm gives satisfactory results. Because light scattering is inversely proportional to the fourth power of the wavelength, sensitivity can be adjusted up or down by using lower or higher wavelengths respectively. It should be noted that light scattering has a complex relationship with particle size so that monitoring, in this case, indicates the position of bands but not their concentration. Monitoring of gradients during method development is generally strongly recommended. It is less essential during routine separations but monitoring may well be useful in providing an early indication of a technical problem.

In principle, particular cells or cell organelles could be labelled by reaction with specific antibodies linked directly or indirectly to fluorescent dyes. The position of these particles could then be identified using a flow fluorimeter. While the output from a density gradient could, in principle, be monitored for radioactivity, it is in practice much better measured 'off line' after fraction collection.

7.8.3 Identification of cells and organelles in gradient fractions by light and electron microscopy

Light microscopy has very limited value in identifying cells or cell organelles separated on density gradients. Apart from nuclei and the large sheets of plasma membrane, stabilized by junctional complexes, cell organelles are too small to be recognized under the light microscope. In the case of cells these are difficult to recognize unequivocally by simple staining procedures except in exceptional cases. This situation is changing with the development of increasing varieties of monoclonal antibodies specific for particular cell types.

The majority of cells and the larger cell organelles can be identified by electron microscopy and, for full validation of a separation of such particles, electron microscopic examination of the isolated fractions should be carried out at least once. However, this technique is far too laborious for the routine characterization of fractions, with the possible exception of virus particles which may be identified following negative staining. Careful thought must be given to sample preparation prior to electron microscopy of fractions, especially those separated by isopycnic banding. It is normal to collect material prior to electron microscopy by dilution and centrifugation. If the fraction contains particles of unequal size the pellet will be stratified, with the largest particles being concentrated at the base. Alternatively, particles may be isolated from the gradient fraction by filtration (52). Whichever method is chosen the reader is strongly advised to consult one of the methods manuals on electron microscopy and to appreciate the amount of work that such analyses involve.

7.8.4 Biochemical identification of separated cells and cell organelles

Again it is only possible to discuss general principles. Fractions may require monitoring for protein, lipid, and nucleic acid content, for the presence of radioactive tracers, and for enzymes used to mark the presence of particular organelles. The presence of low-molecular-weight gradient materials in the assay mixture interferes in the Lowry reaction for protein (53) and probably with other methods as well. Such materials also interfere with enzyme assays (8, 54, 55) and with the counting of radioactive macromolecules (56). The last is especially insidious in that the efficiency of counting of low-molecular-weight internal standards is not affected. Gradient materials containing sugars may interfere with methods for the assay of nucleic acid which depend on measurement of ribose or deoxyribose, while those containing aromatic rings (e.g. Nycodenz and metrizamide) may interfere in methods for measuring nucleic acid which depend on a final measurement of the extinction at 260 nm. This is by no means an exhaustive list. The moral is clear; in the long run time is saved if assays are checked for interference by gradient solutes before any actual measurements on samples are undertaken.

There is little point in a general discussion on the assay of chemical components of separated fractions but it is necessary to consider the question of 'markers' for subcellular organelles. As has been stated earlier most cell organelles cannot be recognized by light microscopy, and electron microscopy is both tedious and unable to distinguish between vesicles derived from the different systems of internal membranes. Hence there is a need for 'markers', chemical components (normally enzymes) which are specifically localized to particular compartments. Appendix 4 of this book lists typical assays for cell components and frequently-used marker enzymes. The choice of these marker assays depends on the tissue and the complexities are beyond the scope of the present article. Given here are examples of possible pitfalls.

(a) Membrane flow in the endocytic and exocytic pathways inevitably results in 5–10% mis-sorting. Hence there will always be some transfer of 'markers' from one compartment to another even though retrieval pathways will keep concentrations low.

(b) Enzyme assays measure an activity, not an enzyme. The classic example here is glucose-6-phosphatase activity. A specific enzyme catalysing this reaction is found in the endoplasmic reticulum of hepatocytes and provides an excellent marker. This enzyme is also found in the kidney but is confined to the proximal convoluted tubule. The low levels of glucose-6-phosphatase activity found in other tissues are due to non-specific phosphatases which have a 'random' location.

(c) Certain enzymes are inactive as first synthesized and are activated during transfer to their final location. The site of activation varies. Thus acid phosphatase is active in the trans-Golgi elements of some tissues but not others.

There can thus be no simple guide to the choice of markers for particular cell types and for cell organelles. Choices must be rooted in knowledge of the tissue under study. Markers used in cell fractionation studies should be linked back to cell components which have been localized to particular organelles by morphological methods; this is discussed further in Section 4 of Chapter 6.

Generating one's own markers, by labelling of components of the exocytic pathway by carefully-timed injection of radiolabelled precursors, and of components of the endocytic pathway by provision of suitable, labelled ligands for the various receptors, has proved its value over the years. Nevertheless, selection and use of markers remains difficult and, unfortunately, is not an area where published methods, other than on the best characterized tissues or from the best laboratories, can be relied on blindly. The results obtained from the fractionation of cells or cell organelles must be collated with morphological, histochemical, and functional studies on the tissue under investigation if gross errors are to be avoided.

Acknowledgements

The authors are very grateful to the editor of this book, Dr David Rickwood, for many constructive suggestions, and to Miss R. Caldwell for typing the manuscript.

References

1. Young, B. D. and Rickwood, D. (1981). *J. Biochem. Biophys. Methods* **5**, 95.
2. De Duve, C. J., Berthet, J. and Beaufay, H. (1959). *Prog. Biophys. Biophys. Chem.* **9**, 325.
3. Steele, W. J. and Busch, H. (1967). In *Methods in cancer research* (ed. H. Busch), Vol 3, p. 61. Academic Press, New York.
4. Rickwood, D., Birnie, G. D., and Hell, A. (1973). *FEBS Letters 33*, 221.
5. Rickwood, D. (ed.) (1976). *Biological separations in iodinated density gradient media.* IRL Press Ltd, Oxford.
6. Ford, T. and Rickwood, D. (1982). *Anal. Biochem.* **124**, 293.
7. Rickwood, D. (ed.) (1983). *Iodinated density gradient media: a practical approach.* IRL Press, Oxford.
8. Bach, M. K. and Brashler, J. R. (1970). *Exp. Cell Res.* **61**, 387.
9. Harwood, R. (1974). *Int. Rev. Cytol.* **38**, 369.
10. Pertoft, H. and Laurent, T. C. (1969). In *Modern separation methods of macromolecules and particles* (ed. T. Garritson) p. 71. John Wiley, New York.
11. Pertoft, H., Laurent, T. C., Laas, T., and Kagedal, L. (1978). *Anal. Biochem.* **88**, 271.
12. Pertoft, H., Rubin, K., Kjellen, L., Laurent, T. C., and Klingeborn, B. (1977). *Exp. Cell Res.* **110**, 449.
13. Henell, F. and Glauman, H. (1982). In *Sinusoidal liver cells* (ed. D. L. Knook and E. Wisse) p. 353. Elsevier Bio Medical Press, Amsterdam.
14. Marzella, L., Ahlberg, J., and Glaumann, H. (1980). *Exp. Cell Res.* **129**, 460.
15. Siebert, G., Furlong, N. B., Romen, W., Schlatterer, B., and Jaus, H. (1973). In *Methodological developments in biochemistry*, Vol. 4, Subcellular Studies (ed. E. Reid), p. 13. Longman, London.
16. Siebert, G. and Hannover, R (1977). In *Membranous elements and movement of molecules* (ed. E. Reid), p. 189. Ellis Horwood, Chichester.

17. Kempf, J, Egly, J. L., Stricker, Ch., Schmitt, M., and Mandel, P. (1972). *FEBS Lett.* **26**, 130.
18. Birnie, G. D., McPhail, E., and Rickwood, D. (1974). *Nucleic Acid Res.* **1**, 919.
19. Rickwood, D. and McGillivray, A. J. (1977). *Exptl. Cell Res.* **104**, 287.
20. Ford, T. and Rickwood, D. (1983). In *Iodinated density gradient media: a practical approach* (ed. D. Rickwood), p. 23. IRL Press, Oxford.
21. Jackson, V. (1978). *Cell* **15**, 945.
22. Hancock, R. (1970). *J. Mol. Biol.* **48**, 357.
23. Martin, R. G. and Ames, B. N. (1961). *J. Biol. Chem.* **236**, 1372.
24. Spragg, S. P., Morrod, R. S., and Rankin, C. T. Jnr (1969). *Sep. Sci.* **4**, 467.
25. Hutterman, A. and Guntermann, U. (1975). *Anal. Biochem.* **64**, 360.
26. Kelley, J. L. and Kruski, A. W. (1986). *Methods in Enzymol.* **128**, 170.
27. Boumendil-Podevin, E. F. and Podevin, R. A. (1983). *Biochim. Biophys. Acta* **735**, 86.
28. Seglen, P. O. (1976). In Biological separations In *Iodinated density gradient media: a practical approach* (ed. D. Rickwood), p. 107. IRL Press, Oxford.
29. Bøyum, A., Berg, T., and Blomhoff, R. (1983). In *Iodinated density gradient media: a practical approach* (ed. D. Rickwood), p. 147. IRL Press, Oxford.
30. Nanni, G., Baldini, I., and Ferro, M. (1969). *Boll. Soc. Ital. Biol. Sper.* **45**, 935.
31. Peake, P. W. (1979). *J. Insect Physiol.* **25**, 795.
32. Scott, W. N., Yoder, M. J., and Gennaro, J. F. (1978). *Proc. Soc. Exp. Biol. Med.* **1158**, 565.
33. Gorzynski, P. M., Miller, R. G., and Phillips, R. A. (1970). *Immunol.* **19**, 817.
34. Bock, R. M. and Ling, N. S. (1954). *Anal. Chem.* **26**, 1543.
35. Coombes, D. H. and Watts, R. M. (1985). *Anal. Biochem.* **148**, 254.
36. Hinton, R. H. and Dobrota, M. (1969). *Anal. Biochem.* **30**, 99.
37. Dreyer K., Opalka, B., and Schulte-Holthausen, H. (1991). *Nucleic Acid Res.* **19**, 1348.
38. Anet, R. and Strayer, D. R. (1969). *Biochem. Biophys. Res. Comm.* **34**, 328.
39. Birnie, G. D. and Rickwood, D. (1976). In *Biological separations in iodinated density gradient media* (ed. D. Rickwood), p. 193. IRL Press Ltd, Oxford.
40. West, C. M. and McMahon, D. (1976). *Anal. Biochem.* **76**, 589.
41. Hinton, R. H. and Mullock, B. M. (1978). *Clin. Chim. Acta* **82**, 31.
42. Fritsch, A. (1975). Beckman Instruments Int s.a., Geneva.
43. Andersen, K. J., Haga, H. J., and Dobrota, M. (1987). *Kidney International* **31**, 886.
44. Mullock, B. M., Luzio, J. P., and Hinton, R. H. (1983). *Biochem. J.*, **214**, 823.
45. Dean, R. T. (1977). In *Lysosomes, a laboratory handbook* (ed. J. T. Dingle), (2nd edn) p. 1. Elsevier, Amsterdam.
46. Leskes, A., Siekevitz, P., and Palade, G. E. (1971). *J. Cell Biol.* **49**, 288.
47. Wallace, H. (1969). *Anal. Biochem.* **32**, 334.
48. Gaudermach, G., Leivestad, T., Ugelstad, J., and Thorsby, E. (1986). *J. Immunol. Meth.* **90**, 178.
49. Zollinger, M., Marmet-Bratley, M. O., and Karska-Wysocki, B. (1984). *Anal. Biochem.* **142**, 88.
50. Voelker, P. and Furst, A. (1989). *Beckman Applications sheet DS-756*, Beckman Instruments Inc, Palo Alto, CA.
51. Meuwissen, J. A. T. P. (1973). In *Methodological development in biochemistry* (ed. E. Reid), p. 29. Longman, London.

52. Baudhuin, H., Evrard, P., and Berthet, J. (1967). *J. Cell Biol.* **32**, 181.
53. Gerhardt, B. and Beevers, H. (1968). *Anal. Biochem.* **24**, 337.
54. Hinton, R. H., Burge, M. L. E., and Hartman, G. C. (1969). *Anal. Biochem.* **29**, 248.
55. Hartman, G. C., Black, N., Sinclair, R., and Hinton, R. H. (1974). *Methodological developments in biochemistry* (ed. E. Reid), Vol. 4, Subcellular Studies, p. 93. Longman, London.
56. Dobrota, M. and Hinton, R. H. (1973). *Anal. Biochem.* **56**, 270.
57. Hinton, R. H. and Dobrota, M. (1976). *Density gradient centrifugation.* North Holland, Amsterdam.
58. Wattiaux, R., Wattiaux de Coninck, S., Ronveaux-Dupal, M. F., and Dubois, F. (1978). *J. Cell Biol.*, **78**, 346.
59. Carter, C., Britton, V. J., and Haff, L. (1983). *Biotechniques* **1**, 142.
60. Lindgren, F. T., Jensen, L. C., and Hatch, F. T. (1972). In *Blood lipids and lipoproteins, quantitation, composition and metabolism* (ed. G. J. Nelson), p. 181. Wiley Interscience, New York.
61. Hinton, R. H., Al-Tamer, Y., Mallinson, A., and Marks, V. (1974). *Clin. Chim. Acta* **53**, 355.
62. Birnie, G. D. (1972). (ed.) *Subcellular components, preparation and fractionation.* Butterworths, London.
63. Dobrota, M. (1971). In *Separations with zonal rotors* (ed. E. Reid), p. Z-2.1 and p. Z-6.1. Wolfson Bioanalytical Centre, University of Surrey, Guildford.
64. Price, C. A. (1982). *Centrifugation in density gradients.* Academic Press, New York.
65. Cline, G. B. and Dagg, M. K. (1973). In *Methodological developments in biochemistry* (ed. E. Reid), Vol. 3, p. 61. Longman, London.
66. Wallach, D. F. H. (1967). In *Specificity of cell surfaces* (ed. B. D. Davis and L. Warren), p. 129. Prentice Hall, Englewood Cliff, New Jersey.
67. Schreier, M. G. and Staehelin, T. (1973). *Nature New Biol.* **242**, 35.
68. Johnson, I. R. and Mathias, A. P. (1972). In *Subcellular components* (ed. G. D. Birnie), p. 53. Butterworths, London.
69. Hinton, R. H. and Mullock, B. M. (1976). In *Biological separations in iodinated density gradient media* (ed. D. Rickwood), p. 1. IRL Press, London.
70. Mathias, A. P. and Wynter, C. V. A. (1973). *FEBS Letters* **33**, 18.
71. Graham, J. M. (1972). *Biochem. J.* **130**, 1113.
72. Miller, R. G. and Phillips, R. A. (1969). *J. Cell Physiol.* **73**, 191.
73. Beaufay, H. (1966). La centrifugation en gradient de densite. Thesis, University Catholique de Louvain.
74. Handbook of Chemistry and Physics (1977). (57th Edn). Chemical Rubber Company.
75. Pretlow, T. G., Boone, C. W., Shrager, R. I., and Weiss, G. H. (1969). *Anal. Biochem.* **29**, 230.
76. Ridge, D. (1978). In *Centrifugal separations in molecular and cell biology* (ed. G. D. Birnie and D. Rickwood). Butterworths, London.
77. Birnie, G. D. and Rickwood, D. (ed.) (1978). *Centrifugal separations in molecular and cell biology.* Butterworths, London.

Appendix A: Properties of sucrose solutions

Concentration		Molarity	Density (g/ml)		Refractive index		Viscosity (mPas)	
w/w	w/v		5°	20°	5°	20°	5°	20°
0	0.00	0.000	1.0000	0.9882	1.3345	1.3330	1.515	1.000
2	2.01	0.059	1.0078	1.0060	1.3375	1.3359	1.585	1.053
4	4.06	0.118	1.0157	1.0139	1.3407	1.3388	1.674	1.112
6	6.13	0.179	1.0237	1.0218	1.3437	1.3418	1.780	1.177
8	8.24	0.241[a]	1.0318	1.0299	1.3467	1.3448	1.905	1.251
10	10.38	0.303	1.0400	1.0381	1.3497	1.3478	2.057	1.333
12	12.56	0.367	1.0483	1.0465	1.3530	1.3509	2.220	1.426
14	14.77	0.431	1.0568	1.0549	1.3562	1.3541	2.410	1.531
16	17.02	0.497	1.0653	1.0635	0.3593	1.3573	2.635	1.650
18	19.30	0.564	1.0740	1.0722	1.3625	1.3605	2.875	1.786
20	21.62	0.632	1.0829	1.0810	1.3646	1.3638	3.137	1.945
22	23.98	0.700	1.0918	1.0899	1.3692	1.3672	3.460	2.124
24	26.38	0.771	1.1009	1.0990	1.3724	1.3706	3.838	2.331
26	28.81	0.842	1.1101	1.1081	1.3757	1.3740	4.282	2.573
28	31.29	0.914	1.1195	1.1175	1.3792	1.3775	4.807	2.855
30	33.81	0.988	1.1290	1.1270	1.3826	1.3811	5.435	3.187
32	36.37	1.063	1.1386	1.1366	1.3864	1.3847	6.187	3.581
34	38.98	1.139	1.1484	1.1464	1.3900	1.3884	7.106	4.052
36	41.62	1.216	1.1583	1.1562	1.3937	1.3921	8.234	4.621
38	44.32	1.295	1.1683	1.1663	1.3975	1.3959	9.651	5.315
40	47.06	1.375	1.1785	1.1765	1.4018	1.3998	11.44	6.167
42	49.84	1.456	1.1889	1.1867	1.4055	1.4037	13.76	7.234
44	52.68	1.539	1.1994	1.1973	1.4096	1.4076	16.77	8.579
46	55.56	1.623	1.2100	1.2078	1.4139	1.4117	20.72	10.30
48	58.49	1.709	1.2208	1.2186	1.4181	1.4158	25.99	12.51
50	61.48	1.796	1.2317	1.2296	1.4225	1.4199	33.18	15.43
52	64.51	1.885	1.2428	1.2406	1.4268	1.4242	43.18	19.34
54	67.60	1.975	1.2541	1.2519	1.4313	1.4284	57.42	24.68
56	70.74	2.067	1.2655	1.2632	1.4356	1.4328	78.27	32.12
58	73.94	2.160	1.2770	1.2748	1.4400	1.4372	109.5	42.78
60	77.19	2.255	1.2887	1.2865	1.4445	1.4417	159.1	58.49

[a] Isotonic sucrose is 0.25 M (−8.54% w/v).
Source: reference 63.

Appendix B: Properties of glycerol solutions

Concentration % (w/v)	Density (g/ml)			Viscosity (mPas)		
	5°	10°	20°	5°	10°	20°
0	0.999 4	0.999 1	0.997 7	1.505 0	1.307 0	1.002 0
1	1.002 0	1.001 6	1.000 1	1.555 1	1.349 3	1.031 9
2	1.004 5	1.004 1	1.002 5	1.606 5	1.392 6	1.062 5
3	1.007 1	1.006 6	1.005 0	1.659 3	1.436 9	1.093 8
4	1.009 7	1.009 1	1.007 4	1.713 5	1.482 4	1.125 8
5	1.012 2	1.011 7	1.009 9	1.769 1	1.529 1	1.158 6
6	1.014 8	1.014 2	1.012 3	1.826 4	1.577 9	1.192 2
7	1.017 4	1.106 7	1.014 8	1.885 3	1.626 3	1.226 6
8	1.020 0	1.019 3	1.017 2	1.945 9	1.676 9	1.261 9
9	1.022 5	1.021 8	1.019 7	2.008 4	1.729 0	1.298 2
10	1.025 1	1.024 4	1.022 2	2.072 9	1.782 6	1.335 4
11	1.027 7	1.026 9	1.024 7	2.139 4	1.837 8	1.373 8
12	1.030 3	1.029 5	1.027 2	2.208 1	1.894 8	1.413 2
13	1.033 0	1.032 1	1.029 7	2.279 1	1.953 6	1.453 9
14	1.035 6	1.034 6	1.032 2	2.352 7	2.014 4	1.495 8
15	1.038 2	1.037 2	1.034 7	2.429 0	2.077 3	1.539 1
16	1.040 8	1.039 8	1.037 2	2.508 2	2.142 5	1.584 0
17	1.043 4	1.042 4	1.039 7	2.590 5	2.210 2	1.630 4
18	1.046 1	1.045 0	1.042 2	2.676 2	2.280 5	1.678 5
19	1.048 7	1.047 6	1.044 7	2.765 5	2.353 7	1.728 5
20	1.051 4	1.050 2	1.047 3	2.858 8	2.430 0	1.780 5
21	1.054 0	1.052 8	1.049 8	2.956 3	2.509 7	1.834 7
22	1.056 7	1.055 4	1.052 4	3.058 5	2.593 0	1.892 3
23	1.059 3	1.058 0	1.054 9	3.165 8	2.680 3	1.950 4
24	1.062 0	1.060 6	1.057 5	3.278 6	2.771 9	2.012 4
25	1.064 6	1.063 3	1.060 0	3.397 4	2.868 3	2.077 4
26	1.067 3	1.065 9	1.062 6	3.522 9	2.969 9	2.145 8
27	1.070 0	1.068 5	1.065 2	3.655 7	3.077 2	2.217 9
28	1.072 7	1.071 2	1.067 7	3.796 4	3.190 8	2.294 0
29	1.075 4	1.073 8	1.070 3	3.946 0	3.311 3	2.374 5
30	1.078 1	1.076 5	1.072 9	4.105 3	3.439 4	2.459 9
31	1.080 8	1.079 1	1.075 5	4.275 4	3.575 9	2.550 6
32	1.083 5	1.081 8	1.078 1	4.457 5	3.728 1	2.647 4
33	1.086 2	1.084 5	1.080 7	4.652 8	3.878 0	2.750 7
34	1.088 9	1.087 2	1.083 3	4.863 0	4.045 8	2.861 3
35	1.091 6	1.089 8	1.086 0	5.089 7	4.226 4	2.980 0
36	1.094 3	1.092 5	1.088 6	5.334 9	4.421 5	3.107 9
37	1.097 1	1.095 2	1.091 2	5.600 9	4.632 7	3.245 8
38	1.099 8	1.097 9	1.093 8	5.890 4	4.862 1	3.395 2
39	1.102 5	1.100 6	1.096 5	6.206 4	5.111 9	3.557 3
40	1.105 3	1.103 3	1.099 1	6.552 4	5.385 0	3.733 8
41	1.108 0	1.106 0	1.101 8	6.932 5	5.684 4	3.926 7
42	1.110 8	1.108 8	1.104 5	7.351 6	6.013 8	4.137 9
43	1.113 5	1.111 5	1.107 1	7.815 3	6.377 4	4.370 2
44	1.116 3	1.114 2	1.109 8	8.330 3	6.780 3	4.626 6
45	1.119 1	1.117 0	1.112 5	8.904 5	7.228 6	4.910 4
46	1.121 9	1.119 7	1.115 1	9.547 4	7.729 2	5.225 9
47	1.124 6	1.122 4	1.117 8	10.270	8.290 7	5.578 1
48	1.127 4	1.125 2	1.120 5	11.087	8.923 3	5.972 8
49	1.130 2	1.127 9	1.123 2	12.014	9.639 3	6.417 1
50	1.133 0	1.130 7	1.125 9	13.071	10.454	6.919 5

Source: reference 76.

Appendix C: Properties of caesium chloride solutions

Concentration		Molarity	Density	Ref. index
w/w	w/v		(g/ml) 20°	20°
0.00	0.00	0.000	1.0000	1.3330
2.00	2.03	0.120	1.0153	1.3345
4.00	4.12	0.245	1.0311	1.3361
6.00	6.27	0.373	1.0475	1.3378
8.00	8.50	0.505	1.0643	1.3395
10.00	10.80	0.641	1.0818	1.3412
11.00	11.98	0.711	1.0907	1.3421
12.00	13.17	0.782	1.0997	1.3430
13.00	14.39	0.855	1.1089	1.3439
14.00	15.63	0.928	1.1183	1.3449
15.00	16.89	1.003	1.1278	1.3458
16.00	18.17	1.079	1.1375	1.3468
17.00	19.47	1.156	1.1473	1.3478
18.00	20.79	1.235	1.1573	1.3488
19.00	22.14	1.315	1.1674	1.3498
20.00	23.51	1.396	1.1777	1.3508
22.00	26.33	1.564	1.1989	1.3529
24.00	29.24	1.737	1.2207	1.3550
26.00	32.27	1.916	1.2433	1.3572
28.00	35.40	2.103	1.2666	1.3595
30.00	38.66	2.296	1.2908	1.3618
32.00	42.03	2.496	1.3158	1.3642
34.00	45.54	2.705	1.3417	1.3666
36.00	49.18	2.921	1.3685	1.3692
38.00	52.96	3.146	1.3963	1.3718
40.00	56.90	3.380	1.4251	1.3744
42.00	61.00	3.623	1.4550	1.3772
44.00	65.27	3.877	1.4861	1.3800
46.00	69.73	4.141	1.5185	1.3829
48.00	74.37	4.417	1.5522	1.3860
50.00	79.23	4.706	1.5874	1.3891
52.00	84.30	5.007	1.6421	1.3924
54.00	89.62	5.323	1.6625	1.3959
56.00	95.19	5.654	1.7029	1.3995
58.00	101.05	5.001	1.7453	1.4033
60.00	107.21	6.367	1.7900	1.4074
62.00	113.71	6.754	1.8373	1.4117
64.00	120.59	7.162	1.8875	1.4163

Source: reference 74.

Appendix D: Properties of Metrizamide and Nycodenz solutions

Metrizamide solutions (20 °C)

Concentration (% w/v)	Molarity (mol/l)	Density (g/ml)	Refractive index	Viscosity (mPas)	Osmolarity (mOsm)
0	0.000	0.9982	1.3330	1.0	0
10	0.127	1.0512	1.3483	1.3	107
20	0.253	1.1062	1.3646	1.6	180
30	0.380	1.1612	1.3809	2.3	247
40	0.507	1.2162	1.3971	3.6	320
50	0.633	1.2712	1.4133	6.0	385
60	0.760	1.3262	1.4295	11.0	440
70	0.887	1.3812	1.4456	26.0	—

Nycodenz solutions (20 °C)

Concentration (% w/v)	Molarity (mol/l)	Density (g/ml)	Refractive index	Viscosity (mPas)	Osmolarity (mOsm)
0	0.000	0.999	1.3330	1.0	0
10	0.122	1.052	1.3494	1.3	112
20	0.244	1.105	1.3659	1.4	211
30	0.365	1.159	1.3824	1.8	299
40	0.487	1.212	1.3988	3.2	388
50	0.609	1.265	1.4153	5.3	485
60	0.731	1.319	1.4318	9.5	595
70	0.853	1.372	1.4482	17.2	1045

Source: reference 7.

Appendix E: Properties of Ficoll solutions

Concentration		Temperature 4 °C		
% w/w	% w/v	Refractive index	Density (g/ml)	Viscosity (mPas)
0.00	0.00	1.3346	1.0004	1.56
4.00	4.06	1.3392	1.0145	3.40
6.00	6.13	1.3420	1.0215	5.01
8.00	8.23	1.3454	1.0290	7.37
10.00	10.37	1.3484	1.0365	10.35
12.00	12.53	1.3514	1.0441	14.27
14.00	14.73	1.3538	1.0518	20.21
16.00	16.96	1.3579	1.0597	27.42
18.00	19.21	1.3608	1.0673	38.33
20.00	21.50	1.3645	1.0752	52.31

Source: reference 75.

4

Centrifugal methods for characterizing macromolecules and their interactions

DAVID RICKWOOD

1. Introduction

Much of the early centrifugation work was centred on the behaviour of macromolecules, particularly proteins, during sedimentation. The discrete high molecular weights of proteins and nucleic acids made them ideal for extensive studies using the analytical ultracentrifuge. This emphasis reflected the importance of these molecules in the structure and function of cells. Subsequently, a great deal of analytical work has been carried out using preparative ultracentrifuges. Much of this work was focused on the sedimentation of macromolecules and the effects of changes in conformation and macromolecular interactions. Macromolecules can be separated on the basis of their sedimentation rate, determined primarily by their size and conformation (rate-zonal centrifugation), or on the basis of density in different types of gradient solution (isopycnic centrifugation). However, the development of sophisticated electrophoretic techniques (1,2) has resulted in much of the work on investigating the molecular mass of macromolecules now being carried out using one of the electrophoretic methods. In addition, the rapid sequencing methods now available for DNA and proteins (3,4) make it possible to determine the size of molecules in a very precise manner, much more accurately than any centrifugation or electrophoretic method.

Most of the current applications involve using centrifugation to characterize hydrodynamic properties of native macromolecules and to separate a wide range of biological samples on the basis of density. Exceptions to this general rule are where other techniques such as electrophoresis are inappropriate. As an example, the presence of the non-protein components of glycoproteins and lipoproteins makes them migrate in an anomalous fashion on polyacrylamide gels, leading to inaccurate estimations of size. In addition, centrifugal separations are useful in situations where the fractionation needs to be carried out at high ionic strength. The compositions of electrophoresis buffers are restricted because of the heating effects of the electric current, while in centrifugation the range of media that can be used is almost without limit in terms of ionic concentration and non-aqueous separations. Hence, rate-zonal and isopycnic separations both remain extremely important fractionation procedures in preparative centrifugation.

While Chapter 3 has discussed the general aspects of gradient separations, this chapter will describe the types of centrifugal separations used for various types of macromolecules. The quantitative aspects of rate-zonal centrifugation are described in Chapter 5.

2. Precautions for working with macromolecules

The inherently large size of macromolecules means that they are very susceptible to degradation. Hence one of the main principles in working with macromolecules is to avoid degradation of samples since even a single break in a molecule will cause it to change its properties and can often prevent any further useful analysis.

Degradation can be the result of physical treatment, or to be more precise mistreatment, or can arise from enzymic degradation. An example of the former case is the shearing of large linear DNA molecules. Native DNA is a very long, fairly stiff molecule which fragments as a result of shearing stresses such as pipetting; shearing stresses readily fragment DNA to a minimum size of about 500 bp. Most other types of macromolecule, even large RNA molecules, are much more resistant to shearing since they do not have the same rigid, rod-shaped structure.

The more usual problem with the degradation of macromolecules is that of enzymic digestion arising as a result of contamination of the sample by hydrolytic enzymes. Degradation can occur either during the isolation procedure by endogenous enzymes in the sample or as a result of contamination of the sample during its preparation. The three methods for avoiding enzymic degradation are the use of inhibitors, the use of denaturing conditions, and deproteinization of the sample. Of the three options, the most widely applicable technique is the use of inhibitors. *Table 1* is a list of some of the methods used for inhibiting nuclease activity when working with nucleic acids, and some of the protease inhibitors currently used are described in *Table 2* of Chapter 6. Particularly when using some of the more specific inhibitors (e.g. some of the protease inhibitors), it is desirable to use a combination of different inhibitors to ensure the complete absence of enzymic degradation. On the other hand, some inhibitors (e.g. diethylpyrocarbonate) may cause experimental artifacts by modifying the properties of the sample or be difficult to remove from the sample (e.g. aurintricarboxylic acid). Denaturing conditions and deproteinization of solutions are used primarily for isolating nucleic acids when it is imperative to use all possible means in order to avoid degradation of these very large molecules. In the preparation of RNA, normally all solutions are autoclaved and/or treated

Table 1. Inhibitors of nuclease activity

Inhibitors	Active conc.	Target enzymes	Mode of action
EDTA	1–10 mM	Non-specific	Chelates divalent cations required by nucleases
Sodium dodecyl sulphate (SDS)	0.1–1%	Non-specific	Binds to and denatures proteins
Diethyl pyrocarbonate (DEPC)	0.1%	Non-specific	Inactivates nucleases by covalent modification
Ribonucleoside vanadyl complex	10 mM	RNases	Binds to active site
Heparin	1 mg/ml	Basic proteins	Complexes with proteins
Ribonuclease inhibitor	1000 units/ml	RNase	Protein that binds only to RNase
Aurintricarboxylic acid	10 µM	Nucleases	Binds to and inactivates nucleases
Bentonite	3 mg/ml	Basic proteins	Adsorbs proteins
Cadmium salts	1 mM	Nuclear RNases	Binds to enzymes

with 0.1% diethylpyrocarbonate in order to inactivate all types of nucleases. In addition, the samples can be deproteinized by extracting them with phenol or phenol/chloroform mixtures prior to fractionation. A typical protocol for the isolation of RNA is given in *Protocol 1*, and alternatively it is possible to use the kind of centrifugation method described in *Protocol 4*.

Protocol 1. Deproteinization of RNA preparations

1. Homogenize the sample in nine volumes of ice-cold 20 mM Tris–HCl (pH 8.0), 1 mM EDTA.
2. Add a tenth volume of 10% SDS, then add an equal volume of phenol:chloroform:isoamyl alcohol (50:50:2 by volume) containing 0.1% 8-hydroxyquinoline and mix well to give an emulsion.
3. Centrifuge the mixture at 10 000*g* for 10 min.
4. Remove the upper aqueous layer, repeat the extraction with phenol/chloroform, and centrifuge as in Step 3.
5. Add a tenth volume of 3 M ammonium acetate to the aqueous extract, mix, and then add two volumes of ethanol. Leave the RNA to precipitate at −20 °C for at least 2 h.
6. Pellet the RNA by centrifugation at 10 000*g* for 10 min at 5 °C.
7. Evaporate off the remaining ethanol using a stream of dry N_2.
8. Dissolve the RNA in 20 mM Tris–HCl (pH 8.0), 1 mM EDTA.

3. Differential pelleting of macromolecules

Macromolecules are relatively small with sedimentation coefficients ranging from less than 2 S to 100 S or more, although in some cases much larger complexes than this can be observed (e.g. chromatin and polysomes). Differential pelleting is used for the isolation of 'soluble' fractions of cells. Centrifugation of a whole cell homogenate in isotonic sucrose solution at 300 000*g* for 2 h at 5 °C should pellet all the cell components down to the size of ribosomes. The supernatant contains a wide variety of macromolecules, especially enzymes, that are involved in metabolic and synthetic processes of the cell. High-performance ultracentrifuge rotors should be used for pelleting macromolecules, and simple calculations based on sedimentation coefficients of the macromolecules and the *k*-factors of rotors will show the conditions required for pelleting (see Section 6 of Chapter 1).

Pelleting, followed by analysis of the pelleted macromolecules, has been used to determine the amount of water associated with different types of macromolecule (5), but such procedures do tend to be exceptional. In fact, only very large macromolecules (e.g. glycogen) or macromolecular complexes (e.g. chromatin) are routinely isolated by differential pelleting without some type of precipitation procedure. Usually proteins are precipitated by the addition of powdered ammonium sulphate (0.71 g/ml of protein solution); acetone and ethanol can also be used as protein precipitants at low temperature. Once precipitated, proteins can be readily pelleted by centrifugation at 10 000*g* for 10 min. When dissolved in an appropriate buffer precipitated proteins generally recover their native activity. In the case of nucleic acids, precipitation is achieved by addition of an equal volume of isopropranol or two volumes of ethanol in the presence of cations (e.g. 0.3 M ammonium or sodium acetate); in the absence of cations the aggregation required for precipitation does not occur. As in the case of precipitated proteins, only quite modest centrifugal forces are usually required to pellet precipitated nucleic aids. However, when pelleting microgram amounts of nucleic acids from ethanolic solutions it is necessary to centrifuge at 60 000*g* for 60 min in order to obtain high yields. When working with very small amounts of nucleic acids adsorption on to the tubes can be a problem; it can be reduced by siliconizing the tubes, rinsing them with protein solution, or by adding SDS to the solutions. Centrifugation can also be used for recovering samples from agarose gels. When pieces of gel containing proteins or nucleic acids are centrifuged in thick-walled tubes at 200 000*g* for 10 min the gel structure collapses leaving the macromolecule in the supernatant above the small pellet of gel matrix.

4. Separation of macromolecules on the basis of density

4.1 Introduction

Separating samples on the basis of density, by isopycnic centrifugation, remains the exclusive preserve of centrifugation. Even in the case of macromolecules

Table 2. Buoyant densities of macromolecules in different types of gradient media

Type of gradient medium	Observed buoyant densities (g/ml)					
	Native DNA	**Denatured DNA**	**RNA**	**Polysaccharides**	**Proteins**	**Proteoglycans**
CsCl	1.71	1.73	>1.9	1.62	1.3	—
Cs_2SO_4	1.43	1.45	1.64	—	1.3	1.46
CsTCA	1.58	—	1.75	—	1.5	—
CsTFA	1.62	1.70	1.8	—	1.4	1.65
RbCl	—	—	—	—	1.3	—
RbTCA	1.50	1.68	—	—	—	—
KBr	—	—	—	—	1.3	—
KI	1.49	1.52	1.62	—	—	—
NaBr	—	—	—	—	1.3	—
NaI	1.52	1.55	1.65	—	—	—
Metrizamide	1.12	1.14	1.17	1.28	1.27	—
Nycodenz	1.13	1.17	1.18	1.28	1.27	—

TCA = trichloroacetate.
CsTFA is caesium trifluoroacetate; CsTFA is a registered trademark of Pharmacia Biosystems AB.

where generally the sizes are more discrete than is the case for organelles, density separations have proved to be extremely valuable for the isolation and analysis of different classes of macromolecules because the separation is based on the nature of the particle alone, irrespective of its size; size only affects the speed with which the particle reaches its isopycnic position. The buoyant density of macromolecules can vary greatly depending on the type of gradient medium (*Table 2*); this is primarily a result of differences in the degree of hydration, that is the higher the degree of hydration the lower is the observed buoyant density. Molecules which are very highly hydrated in solution such as nucleic acids have more variable densities from one medium to another. As an example, DNA has a density of about 1.7 g/ml in CsCl gradients where there is little available water and 1.1 g/ml in Nycodenz gradients where the water activity is high. In contrast, those macromolecules which are less highly hydrated (e.g. proteins) have similar densities irrespective of the type of density gradient medium (*Table 2*).

4.2 General aspects of isopycnic centrifugation

4.2.1 Choice of rotor

As described in Section 4.3 of Chapter 2, the rotors of choice for the isopycnic separations of macromolecules are either vertical rotors or shallow fixed-angle rotors. The reasons for this are that the short pathlength of these rotors leads to the rapid formation of a pseudo-equilibrium gradient (see Section 4.2.3), and that the reorientation of the gradient that occurs greatly increases the

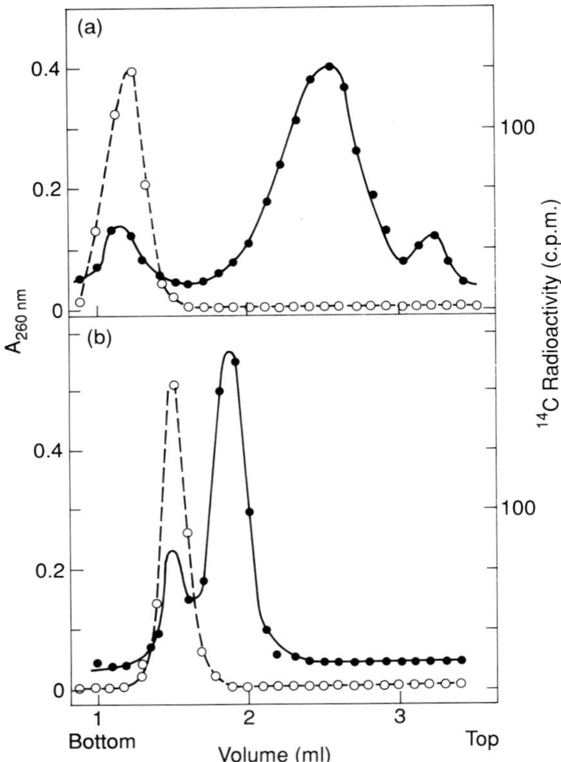

Figure 1. Comparison of the efficiency of fixed-angle and swing-out rotors for the isopycnic separation of DNA. A mixture of 50 µg of unlabelled mouse DNA and 3 µg of ^{3}H-DNA from *Escherichia coli* was centrifuged to equilibrium in CsCl gradients (initial density 1.710 g/ml) in either a fixed-angle rotor (a) or a swing-out rotor (b). At the end of the run the gradients were fractionated and the distributions of mouse (●—●) and *E. coli* (○---○) DNA determined by absorbance at 260 nm and by liquid scintillation counting, respectively.

sample band area and hence the capacity of the gradient (6). This reorientation effect also enhances the resolution of the gradient. *Figure 1* shows the difference in the gradient capacity and quality of the resolution between a fixed-angle rotor and a swing-out rotor. Swing-out rotors are not recommended for isopycnic separations of macromolecules; apart from the reasons listed above, the long pathlength of these rotors makes it all too easy to spin the rotor at such a speed that the gradient medium crystallizes at the bottom of the tube, overstressing the bucket which, in extreme cases, leads to catastrophic failure of the rotor.

4.2.2 Choice of gradient medium

The choice of gradient media depends on the type of macromolecule being separated and the nature of the separation needed. Gradient media vary not only

Table 3. Properties of gradient media used for isopycnic separations of macromolecules and nucleoproteins

Type of gradient	Maximum density	Slope of gradient[a]	Typical applications	Comments
CsCl	1.91	1.00	Plasmid DNA isolation; preparation of RNA; DNA base composition; separation of fixed nucleoproteins	RNA pellets
Cs_2SO_4	2.01	1.75	Separation of DNA, RNA, and proteins; isolation of proteoglycans banding nucleoproteins	Precipitates scintillation fluids
CsHCOO	2.10	0.81	Fractionation of RNA	More viscous than CsCl
CsTFA[b]	2.6	0.80	Separation of DNA, RNA, and proteins; RNA isolation	Chaotropic, UV-opaque
CsTCA[c]	—	0.97	Fractionation of DNA	Chaotropic, UV-opaque
KBr	1.37	0.34	Analysis of lipoproteins; analysing density-labelled proteins	
KI	1.72	0.66	Fractionation of DNA, and DNA:RNA hybrids	UV-opaque
NaBr	1.53	0.25	Analysis of lipoproteins	
NaI	1.90	0.39	Fractionation of DNA, RNA and DNA:RNA hybrids	UV opaque
RbCl	1.49	0.51	Analysing density-labelled proteins	
RbTCA[c]	—	0.86	Fractionation of DNA	
Metrizamide	1.5	—	Fractionation of nucleoproteins Analysis of ionic effects on DNA and RNA	Non-ionic, UV-opaque
Nycodenz	1.5	—	Fractionation of nucleoproteins Analysis of ionic effects on DNA and RNA	Non-ionic, UV-opaque

[a]Slope of gradient expressed relative to that of an equilibrium CsCl gradient with an initial density of 1.7 g/ml
[b]CsTFA is caesium trifluoroacetate; CsTFA is a registered trademark of Pharmacia Biosystems AB
[c]TCA = trichloroacetate

in terms of the characteristics of their gradient solutions but also in the nature of the gradient (for a more detailed discussion of the properties of different gradient media see Section 3 of Chapter 3). A summary of the properties and applications for the different gradient media for separating macromolecules and complexes of macromolecules are shown in *Table 3*. With the exceptions of metrizamide and Nycodenz, all of the other gradient media are highly ionic. In some cases the highly ionic nature of the gradient medium is essential, for example, in fractionating nucleic acids and for disrupting interactions between macromolecules such as acidic polysaccharides. On the other hand, macromolecules such as proteins and nucleoprotein complexes are likely to be affected by the high ionic strength of these media and certainly metrizamide and Nycodenz have proved to be ideal gradient media for the fractionation of intact nucleoprotein complexes.

4.2.3 Formation of gradients and determination of the length of centrifugation run

Gradients for separating macromolecules are usually self-forming, that is the centrifugal force sediments the gradient solute and at equilibrium this sedimentation is balanced by diffusion (see Section 5.3 of Chapter 1). The greater the centrifugal force the steeper is the gradient. However, as shown in *Table 3*, for any given centrifugal force some gradient media form steeper gradients than others. As a general rule, the steeper the gradient the poorer is the resolution of similar species that is achievable. A more detailed description of the theory of self-forming gradients is given in Section 5.3.1 of Chapter 1, and some of the practical aspects are covered in Section 5.7.1 of Chapter 3. The reader is referred to these chapters for detailed descriptions of these topics; here only some of the important points will be reiterated.

It is possible to calculate the time required for a self-forming gradient to reach true equilibrium, as described in Section 5.3.2 of Chapter 1 and Section 5.7.1 of Chapter 3, using the equation:

$$t(\text{hours}) = k(r_b - r_t)^2$$

where r_b and r_t are the distances (cm) from the centre of rotation of the bottom and top of the gradient, respectively. The constant, k, varies with the diffusion coefficient and viscosity of the gradient solute as well as the temperature of centrifugation. For CsCl solutions at 20 °C, k has a value of 5.6.

A calculation using this equation will indicate that a typical 12-ml gradient in a fixed-angle rotor will require about 100 h to reach equilibrium. Experimentally, it is found that perfectly adequate separations of DNA can be achieved in less than 30 h, depending on the type of macromolecule. Hence, it is possible to obtain good separations of macromolecules on what are effectively pseudo-equilibrium gradients. In this situation, the band of macromolecules is able to adjust its position in the gradient to match small changes in the gradient because the rate at which the particles band is much faster than the time required for the generation of equilibrium gradients. The length of time for particles to band in the gradient can be calculated using the following equation:

$$t(\text{hours}) = \frac{9.83 \times 10^{13} \times \beta^0 \times (\rho_p - 1)}{Q^4 \times r_p^2 \times s_{20,w}}$$

where β^0 is dependent on the gradient solute and its concentration (see *Table 5* of Chapter 3), ρ_p is the density of the particle, Q is the speed of centrifugation in revolutions/min, r_p is the radius at which the particles band, and $s_{20,w}$ is the sedimentation coefficient of the particles. The time required is inversely proportional to the size of the molecule and speed of centrifugation. Calculation shows that the time required to achieve true equilibrium is much longer than that usually considered necessary to band DNA; that is most separations are not carried out at true equilibrium. If the particles are allowed to reach true equilibrium then the width of the band reflects the molecular weight of the particle, but while it is a guide to the actual size of the particles, analysis of the width of bands is seldom used for calculating molecular weights.

Note that these equations assume that the initial sample is mixed throughout the tube and that the centrifugation conditions are constant. Layering the sample in the gradient can have a very beneficial effect in shortening the required time for isopycnic banding (7). In addition, the required centrifugation times can be shortened by increasing the centrifugal force. The limiting factors here are the maximum speed of the rotor (note that this often needs to be derated with dense solutions) and the degree of resolution required. The higher the centrifugal force, the steeper is the gradient, and this would appear to restrict the maximum speed which can be used. However, using the relaxation technique (9), the sample is banded rapidly by high-speed centrifugation, and then the speed is reduced to make the gradient shallower and so improve the resolution obtainable (see *Figure 15* of Chapter 3); usually the relaxation time needed at the end of the run is only about 1–2 h. This idea of using different speed profiles has been developed by manufacturers; for example, the ESP program of Beckman Instruments minimizes centrifugation time using a multi-step speed control profile which minimizes the centrifugation time required for gradients while avoiding the dangers of crystallization of CsCl at the bottom of the gradient. Software for the prediction and simulation of isopycnic separations is also available from other manufacturers of ultracentrifuges.

4.3 Fractionation of DNA

4.3.1 Isolation of plasmid DNA

The isolation of plasmid DNA has been one of the cornerstones of genetic engineering procedures and the method most frequently used for the isolation of plasmid DNA is based on centrifugation in CsCl gradients containing the intercalating agent ethidium bromide. The basis for this separation is that the ethidium bromide intercalates with DNA, reducing its density in CsCl gradients; less ethidium bromide binds to supercoiled DNA and so intact plasmid DNA which is supercoiled bands at a higher density than the linear chromosomal DNA. For example, the observed densities of linear and supercoiled SV40 DNA in CsCl-ethidium bromide gradients are 1.532 g/ml and 1.575 g/ml, respectively; in the absence of ethidium bromide both forms of DNA band at 1.694 g/ml (8). This definitive basis for the separation of plasmid DNA is at least part of the reason for the continued popularity of this method. In the author's experience the most reliable method for preparing plasmid DNA is that using alkaline lysis of the bacteria. A typical procedure for plasmid separation is given in *Protocol 2*.

Protocol 2. Purification of plasmids using CsCl-ethidium bromide gradients.

Solutions

Solution I: 50 mM glucose, 10 mM EDTA, 25 mM Tris–HCl (pH 8.0)
Solution II: 0.2 M NaOH, 1% SDS
Solution III: 3 M sodium acetate (pH 4.8)
TE buffer: 10 mM Tris–HCl (pH 7.4), 2 mM EDTA

Experimental protocol

1. Grow the bacteria under the appropriate selective conditions to maintain the plasmid; this will vary depending on the plasmid. This protocol assumes that you are starting with a litre of the bacterial culture grown to the end of the log phase; for lesser volumes or cell densities use correspondingly smaller amounts of reagents.
2. Centrifuge the culture at 4000 r.p.m. (2000*g*) for 20 min at 5 °C to pellet the bacteria.
3. Wash the bacteria in 100 ml of cold Solution I and re-pellet the bacteria as in Step 2.
4. Resuspend the bacteria in 20 ml of Solution I, ensuring that all of the pellet is evenly suspended throughout the medium.
5. When the cells are completely resuspended add lysozyme to a final concentration of 2 mg/ml and incubate the cell suspension at 0 °C in ice for 10 min.
6. Add 40 ml of Solution II to the lysozyme-treated bacterial suspension using a gentle swirling motion to complete the lysis and incubate in ice for 5 min.
7. Add 30 ml of Solution III and incubate in ice for 20 min; this will cause the precipitation of proteins and most of the potentially contaminating nucleic acids.

Protocol 1. *Continued*

8. Decant off the solution from most of the precipitate and remove any remaining precipitate by centrifuging the solution at 10 000 r.p.m. (10 000*g*) for 15 min at 5 °C in a high-speed centrifuge using a fixed-angle rotor.
9. Carefully pour off the supernatant without disturbing the pellet and precipitate the plasmid by the addition of 55 ml of isopropanol.
10. Allow the crude plasmid DNA to precipitate at −20 °C for at least 15 min and then pellet the plasmid by centrifugation at 12 000 r.p.m. (15 000*g*) for 5 min at 4 °C and discard the supernatant.
11. Briefly dry the pellet under vacuum to remove most of the isopropanol and dissolve the pellet in 18 ml of sterile TE buffer. Add 0.65 ml of a 1% ethidium bromide solution once the pellet has dissolved.
12. Add 1.0 g of CsCl to each millilitre of solution and dissolve completely; the refractive index should be 1.3890.
13. Fill 12 ml ultracentrifuge polyallomer tubes with the plasmid-containing CsCl solution, seal the tubes and centrifuge them either at 100 000*g* for at least 36 h at 20 °C in a fixed-angle rotor or at 350 000*g* for 6 h in a vertical rotor; for further details see the text following this protocol.
14. After centrifugation visualize the band(s) under UV light (it is best to use a wavelength of about 300 nm since this damages the DNA less). There is usually a fluorescent red pellet of RNA while the protein bands at the top.
15. Cut off the top of the tube and remove the plasmid DNA banding close to the centre of the gradient. If there is more than one band then take lower band which is the supercoiled plasmid DNA. Bands can be removed using a sterile Pasteur pipette or a hypodermic syringe.
16. Extract the ethidium bromide from the plasmid-containing CsCl solution by extracting a few times with isopropanol saturated with CsCl solution. Remove the CsCl by dialysis or by passage over a desalting column.

The choice of centrifugation conditions for Step 13 depends on a number of factors. If an ultracentrifuge is not available then it is possible to use one of the high-performance (28–30 000 r.p.m.) high-speed centrifuges which generate a maximum centrifugal force of about 100 000*g* and in this case centrifugation takes about 25–40 h depending on the method of loading the sample (10). However, if a short preparation time is required then it is necessary to use an ultracentrifuge (capable of greater than 75 000 r.p.m.) in combination with either a small volume vertical rotor or a shallow-angle fixed-angle rotor such as the NVT rotors of Beckman Instruments. The claimed shortest time required for plasmid DNA banding is 2 h using a vertical rotor if the sample is layered into the bottom of the gradient (11), but usually, with the sample mixed throughout the gradient, it is best to centrifuge the gradients for a minimum of 6–8 h, depending on the type of ultracentrifuge and rotor used. When using a NVT rotor it has been found to be important to include a final concentration of 0.01% Triton X-100 in the CsCl solution because this gives a more compact pellet of RNA separate from the plasmid DNA band (12), and in this publication it is recommended that the starting density of the CsCl solution should be 1.55 g/ml (refractive index of 1.3860). A typical plasmid separation is shown in *Figure 2*.

An alternative gradient medium which has been proposed for plasmid separation is caesium trifluoroacetate, which can separate linear and plasmid DNA in the absence of ethidium bromide (13); indeed, the presence of ethidium bromide minimizes the difference in density between the plasmid and linear DNA. However, in the absence of ethidium bromide the banded DNA is invisible and, furthermore, the trifluoroacetate ions absorb in the UV, making analysis of these gradients difficult. Hence CsCl–ethidium bromide gradients remain the method of choice for the isolation of plasmid DNA.

4.3.2 Separation of DNA on the basis of base composition

In gradients of concentrated salt solutions the extent of DNA hydration depends on the base composition of the DNA with A:T base pairs being more highly

Figure 2. Separation of plasmid DNA on CsCl–ethidium bromide gradients. Self-forming CsCl–ethidium bromide gradients prepared as described in *Protocol 1* were centrifuged in a shallow fixed-angle rotor at 250 000*g* for 24 h at 20 °C. After centrifugation the tubes were placed on a UV illuminator and photographed using a red filter.

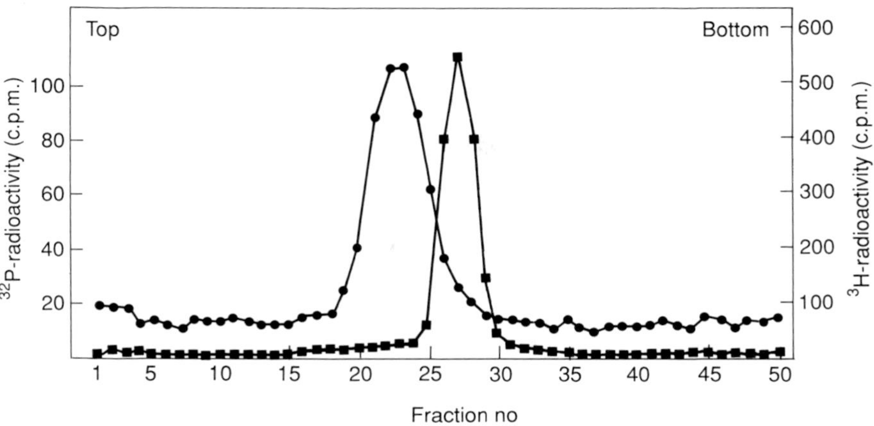

Figure 3. Separation of DNA on the basis of base composition on isopycnic CsCl gradients. A mixture of yeast nuclear ^{32}P-DNA (40% G + C) and *E. coli* ^{3}H-DNA (50% G + C) was centrifuged in a solution of CsCl (initial density 1.707 g/ml) at 100 000*g* for 66 h at 20 °C. The positions of the yeast DNA (●—●) and *E. coli* DNA (■—■) were determined by liquid scintillation counting of the diluted gradient fractions. The banding densities of the yeast and *E. coli* DNAs were found to be 1.699 g/ml and 1.710 g/ml, respectively.

hydrated, and this is reflected by the lower buoyant density of AT-rich DNA in these solutions. The usual gradient medium used is CsCl, primarily because it forms shallow enough equilibrium gradients to resolve DNAs which differ in their percentage of A + T by as little as 10%, as shown in *Figure 3*. This technique has been particularly useful in isolating DNA whose composition varies significantly from the bulk of the DNA, and examples include satellite DNA and the genes for rDNA. The method used is described in the following protocol.

Protocol 3. Separation of DNA on the basis of base composition using CsCl gradients

1. The density (D) of CsCl solutions can be calculated from the percentage (w/w) (P) of CsCl using the following formula:

$$D = 138.11/(137.48 - P)$$

For an initial density of 1.706 g/ml prepare the following mixture:

4.85 g CsCl (analytical grade)
50 µl 1.0 M Tris–HCl (pH 7.4)
25 µl 0.2 M EDTA (pH 7.4)
0.5 ml DNA sample[a]
3.3 ml distilled water

2. Check the density (D) of the initial gradient from its refractive index (RI) using the following equation:

$$D = 10.8601 \times \mathrm{RI} - 13.4974$$

3. The refractive index should be 1.4000, but if it is too high add more water; the amount required (V ml) for a gradient (G ml) can be calculated from the relationship:

$$V = 1.52 \times G(D_{obs} - D_{req})$$

If the gradient solution is not dense enough then add solid CsCl (W g)

$$W = 1.32 \times G(D_{req} - D_{obs})$$

4. Re-check the refractive index of the solution, fill the centrifuge tubes to the top and seal them; narrow-neck tubes are recommended for this.
5. Centrifuge the gradient solution at 150 000*g* for at least 48 h in a fixed-angle rotor. Alternatively, use a vertical rotor; in this case centrifuge the gradient at 300 000*g* overnight. The actual length of centrifugation time depends on the rotor pathlength[b]. In either case centrifugation should be at 20 °C.
6. After centrifugation fractionate the gradient into fractions. For a 5 ml gradient aim to divide it into about 50 fractions.
7. Read the refractive index of each fraction to an accuracy of 4 decimal places; remember to correct the observed refractive index for temperature by taking the refractive index of distilled water.
8. Add 1 ml of water to each fraction and read the optical density at 260 nm. Alternatively, if the DNA is radioactive then measure the radioactivity in a scintillation counter after the addition of water-soluble scintillation liquid.
9. Use the marker DNA of known buoyant density to check for the absence of significant diffusion of the gradient during the deceleration of the rotor and gradient fractionation. Calculate the base composition of the DNA using the formula:

$$\%(\mathrm{G}+\mathrm{C}) = \frac{D - 1.66}{0.098} \times 100$$

[a]As well as the sample DNA add a marker DNA, or preferably one lighter and one denser of known base compositions and buoyant densities.
[b]Do not use a swing-out rotor since they have a long pathlength, a lower maximum load and poorer resolution as compared with fixed-angle and vertical rotors (see *Figure 1*).

Note that this separation on CsCl gradients in the absence of ethidium bromide is not affected by the conformation of the DNA. Many gradient media are unsuitable because the gradients they form are too steep (e.g. Cs_2SO_4) or because they are non-ionic (e.g. Nycodenz); in the latter case differential hydration of the DNA does not occur. It is possible to enhance the separation of AT-rich and GC-rich DNA by including a DNA binding agent which is low density and binds preferentially to DNA which is AT- or GC-rich. An example of this is the addition of 6 µg/ml bisbenzamide (Hoechst 33258) to CsCl gradients. This compound binds specifically to AT-rich DNA (14); in this case the initial starting density is 1.463 g/ml. Using this method the G + C content of the DNA can be calculated from the equation:

$$\%GC = 351762.28 \times RI - 123778.66 \times (RI)^2 - 249789.47$$

where RI is refractive index of the gradient fraction corresponding to the peak of the DNA. The accuracy for most DNAs is within about 1.5% G + C. This selective separation of DNA on the basis of the binding of antibiotics can only be done in highly ionic media where the level of hydration is low (16).

Besides the variety of antibiotics which bind to DNA in a selective manner, silver and mercuric ions also interact specifically with DNA; the former bind to G:C base pairs and the latter to A:T base pairs. However, the chlorides of both these metals are insoluble and so it is necessary to use Cs_2SO_4 gradients. Using Cs_2SO_4 gradients containing either mercury or silver ions it is possible to fractionate sheared native DNA into a series of subfractions of DNA of slightly different base composition (39). It is also found that both silver and mercury ions bind stronger to denatured (single-stranded) DNA than native DNA and so the presence of these ions can enhance the degree of separation of native and denatured DNA in Cs_2SO_4 gradients.

In some cases alkaline CsCl gradients have been used to separate denatured DNA on the basis of base composition. This can be done because at pH 12.5 the increase in density of the DNA strands is dependent on the amount of T + G in each strand; the strand rich in T + G becomes denser than the complementary AC-rich strand (15). Alkaline CsCl gradients are adjusted to pH 12.5 by the addition of 50 mM K_3PO_4 or NaOH. Two of the best examples of this are the ability to separate mouse satellite DNA and the mitochondrial DNA of some organisms into H- and L-strands. As a general rule, single-stranded DNA has a higher density in most gradient media (*Table 2*).

Almost all work on the isopycnic fractionation of DNA has been done using the highly ionic media. However, with the availability of non-ionic media such as metrizamide and Nycodenz, it is possible to look at the behaviour of DNA in the presence of very low concentrations of ions or other compounds that can bind to DNA (16).

4.4 Isolation and fractionation of RNA

4.4.1 Isolation of RNA

In isolating RNA it is essential to minimize any degradation by inactivating the nucleases which are present in almost all types of biological samples. RNA is less stable than DNA not only in terms of sensitivity to enzymes but it is also chemically more labile. One effective method is the use of a combination of SDS and deproteinization with phenol (see *Protocol 1*). However, the RNA isolated in this way is often contaminated with DNA and polysaccharides. In addition, the toxic nature of phenol has led to concern over its use. One method which avoids these problems is to dissolve the sample in a highly denaturing, chaotropic solution such as 6 M guanidinium chloride, and then pellet the RNA, which is very dense, through a solution of CsCl. The details of this method are given in *Protocol 4*.

Protocol 4. Isolation of RNA by centrifugation through CsCl solution

Solutions

Solution I: 7.5 M guanidinium chloride, 50 mM sodium citrate (pH 7.0), 0.1 M 2-mercaptoethanol, 0.5% sarkosyl

TE buffer: 10 mM Tris–HCl (pH 7.4), 2 mM EDTA.

Experimental Protocol

1. Homogenize the tissue or cells in Solution I; use 10 ml of Solution I for each gram of cells.
2. Centrifuge the homogenate at 8000 r.p.m. (8000*g*) for 10 min at 5 °C in polyallomer or polypropylene tubes in a fixed-angle rotor and discard the pellet.
3. *Either* layer the supernatant over a 2 ml cushion of 5.7 M CsCl in TE buffer in a polyallomer tube if you are using a 14 ml swing-out rotor and centrifuge at 31 500 r.p.m. (185 000*g*) for 18 h at 10 °C, *or*, if you are using a fixed-angle rotor, then layer the supernatant over a 3 ml cushion of 5.7 M CsCl in TE buffer and centrifuge at 65 000 r.p.m. (410 000*g*) for 2 h at 10 °C.
4. After centrifugation, carefully remove the protein in the upper layer and the DNA at the interface leaving the pellet of purified RNA at the bottom of the tube.
5. Take up the pellet of RNA in 0.5 ml of Solution I and transfer to a microcentrifuge tube; rinse out the tube with another 0.5 ml of Solution I and add to the solution in the microcentrifuge tube. Add 50 µl of 1 M acetic acid followed by 500 µl of ethanol.
6. Allow the RNA to precipitate at −20 °C for at least 60 min and then pellet the RNA by centrifugation at 12 000 r.p.m. (11 000*g*) for 10 min. Discard the supernatant.
7. Dissolve the pelleted RNA in 400 µl of TE buffer and add 40 µl of 3 M sodium acetate (pH 5.0) followed by 880 µl of ethanol.
8. Allow the RNA to precipitate at −20 °C overnight and pellet the RNA by centrifugation as in Step 6, then wash the pellet twice with 70% ethanol in sterile distilled water.

9. Dissolve the RNA in TE buffer, determine the size of RNA by gel electrophoresis, and the concentration by spectrophotometry (1 mg/ml of RNA has an optical density of 25 at 260 nm). The RNA can be stored at −20 °C.

Notes
It is of paramount importance to avoid nuclease contamination of the RNA. Where possible all solutions, tubes, and glassware should be autoclaved. It is important to wear disposable plastic surgical gloves at all times to prevent contamination of glassware and solutions with skin ribonuclease.

The method described in *Protocol 4* is that commonly used for the purification of RNA. However, there are other dissociating media that have been recommended, such as using 4 M guanidinium isothiocyanate instead of guanidinium chloride (17), or changing the gradient medium to caesium trifluoroacetate (18). In this latter case the chaotropic effect of the trifluoroacetate ions inhibits nucleases much more effectively than other caesium salts.

4.4.2 Fractionation of RNA

RNA is too dense to band in CsCl gradients except at elevated temperatures (>40 °C) and, while RNA does band in Cs_2SO_4 gradients, it tends to aggregate and even precipitate unless steps are taken to prevent it. Mixed gradients of CsCl and Cs_2SO_4 (saturated CsCl solution, saturated Cs_2SO_4, and distilled water mixed in the ratio of 2:2:1 by vol) can be used to separate RNA without aggregation as long as the loading is kept to 2 µg/ml of RNA or less per gradient (7). Aggregation of RNA can be prevented by treating the RNA with formaldehyde before it is loaded on to the gradient, or by running Cs_2SO_4 gradients containing 1% formaldehyde (19). However, formaldehyde covalently modifies the amino groups of RNA. Alternatively, it is possible to minimize aggregation of the RNA by low pH or the inclusion of 4 M urea or 5% DMSO in the Cs_2SO_4 gradients. Using the last approach it is possible to separate single- and double-stranded RNA (20); addition of 5% DMSO to the mixed CsCl/Cs_2SO_4 gradients described previously also increases the sample capacity of the gradient (7). Caesium formate and acetate have been used for fractionating RNA but neither has achieved much popularity because these gradients are more viscous and RNA takes longer to band.

Apart from caesium salts, the other gradient media which have been widely used are gradients of KI (21) and NaI (7); the shallower gradients formed by these media avoid the precipitation problems seen with caesium salts, and so the amount of RNA which can be loaded is much greater. In the case of NaI gradients, it has been found to be very important to remove all heavy metal ions from the solutions to prevent precipitation of the RNA (7). A typical gradient for fractionating RNA is described in *Protocol 5*.

Protocol 5. Fractionation of RNA on KI gradients

1. Dissolve the RNA sample in 15 mM sodium citrate, 10 mM sodium bisulphite, 5 mM sodium phosphate (pH 7.0), to a final concentration not exceeding 50 μg/ml.
2. Add 1.0 g of KI to each millilitre of the RNA solution and dissolve to give a refractive index of 1.4290.
3. Fill 12 ml ultracentrifuge tubes with the KI solution, seal the tubes, and centrifuge at 150 000*g* for 60 h at 20 °C.
4. Unload the gradients by upward displacement. Locate the position of the RNA by chemical assays (Appendix 4) or, if the RNA is radioactive, by scintillation counting; in the latter case it is necessary to add a drop of 2-mercaptoethanol to each fraction to avoid the generation of free iodine in the presence of the scintillation fluid (7).

In KI gradients the buoyant density is independent of the G + C content of the RNA, but single- and double-stranded RNA can be separated (*Figure 4*). Using gradients of mixed KI and NaI (1:4, w/w) it is possible to separate RNAs on the basis of their base composition and secondary structure (23).

4.5 Separation of nucleic acid hybrids

Nucleic acid hybridization has proved to be one of the most powerful tools in the hands of the molecular biologists. Although many hybridization methods are now centred around filter techniques, solution hybridization still has a number of applications (24). From the previous paragraphs, it is obvious that single- and double-stranded DNA and RNA have markedly different buoyant densities in a number of different gradient media and so isopycnic centrifugation is well suited for the isolation of DNA:RNA hybrids. Both CsCl and NaI gradients have been used for separating hybrids and examples of the type of separation that can be achieved are shown in *Figure 5*; Cs_2SO_4 gradients are less ideal since the steep gradient slope gives a poorer separation for the other macromolecular species (7).

4.6 Fractionation of proteins

The partial specific volumes of amino acids vary (see *Table 6* of Chapter 5) and so one might expect variations in the buoyant density of proteins depending on amino composition of up to 0.1 g/ml (25); such variations have not been widely exploited. In contrast to nucleic acids, the level of protein hydration in most gradient media is fairly similar and so most proteins band in the range 1.27–1.40 g/ml. CsCl gradients have been used for separating proteins although experimental evidence indicates that better

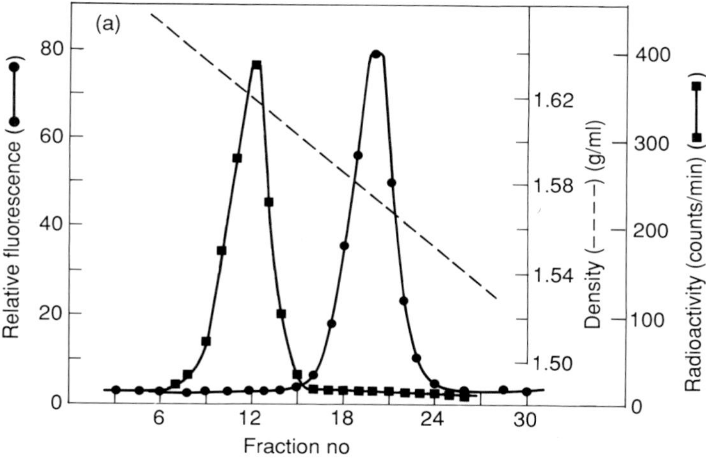

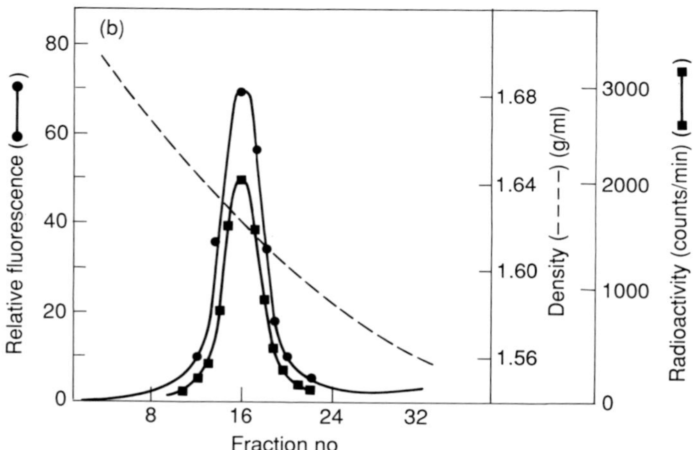

Figure 4. Banding of RNA in KI gradients: (a) banding of double-stranded reovirus RNA (●—●) and yeast ribosomal RNA (■—■); (b) banding of a mixture of yeast mRNA (■—■) and Ehrlich ascites tumour ribosomal RNA (●—●); the G + C content of each type of RNA was 40% and 62%, respectively. The amount of RNA loaded was up to 200 μg per 12-ml gradient. Gradients were centrifuged at 170 000*g* for 48–64 h at 20 °C.

results can be achieved with RbCl or KBr gradients because they form shallower gradients (*Table 3*). As described in Section 4.11, separations of density-labelled proteins have been well characterized. However, most work has been done on the separation of proteins modified in the cell by a variety of ways, such as by the addition of lipids or glycosylation.

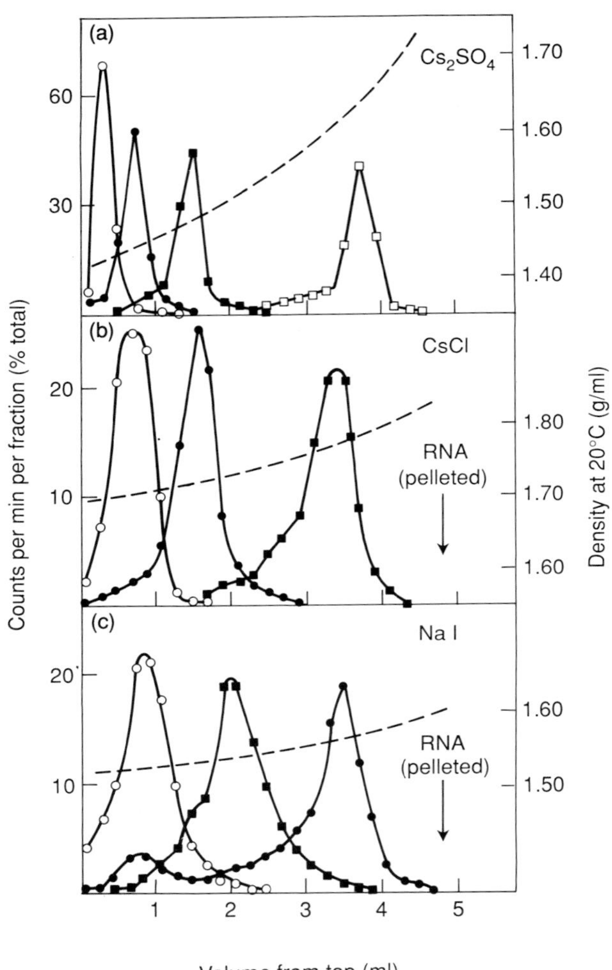

Figure 5. Comparison of different gradients for the separation of nucleic acid hybrids by isopycnic centrifugation. The figure shows the banding of native DNA (○—○), denatured DNA (●—●), DNA:RNA hybrids (■—■), and RNA (□—□) in (a) Cs_2SO_4 (1.54 g/ml), (b) CsCl (1.75 g/ml), and (c) NaI (1.55 g/ml) gradients. The gradients were centrifuged at 136 000*g* for 68 h at 25 °C in a fixed-angle rotor (from reference 7).

4.6.1 Fractionation of lipoproteins

The lipoproteins found in the blood are divided into low density lipoproteins (LDL) and very low density lipoproteins (VLDL). A typical protocol for the banding of blood lipoproteins is given in *Protocol 6*.

Protocol 6. Banding of blood lipoproteins in NaBr gradients (26)

Solutions

Dry NaBr crystals overnight at 210 °C and store in a dessicator. Prepare NaBr solutions with the following densities:

1.006 g/ml (9 g/litre)
1.019 g/ml (27 g/litre)
1.063 g/ml (95 g/litre)
1.210 g/ml (283.4 g/litre)
1.386 g/ml (524.8 g/ml)

All solutions also contain 0.05% EDTA (pH 7.0). Filter the solutions and store in brown bottles at 4 °C. Check the density of the solutions before use.

Experimental protocol

1. Remove the cells from the sample of blood by centrifugation (1000*g* for 10 min).
2. Mix 0.2 ml of the serum with 0.8 ml of the 1.386 g/ml NaBr solution.
3. Prepare a gradient by overlayering the diluted serum sample from Step 2 with the following solutions:

   3.2 ml of 1.21 g/ml NaBr solution
   3.8 ml of 1.063 g/ml NaBr solution
   3.3 ml of 1.016 g/ml NaBr solution
   1.2 ml of 1.006 g/ml NaBr solution

4. Centrifuge the gradients in a 14 ml swing-out rotor at 260 000*g* for 24 h at 14 °C in an ultracentrifuge. The lipoproteins float up through the gradient leaving the other serum proteins at the bottom of the tube.
5. Unload the gradients by upward displacement with Maxidens (Nycomed), fluorochemical FC40 or FC43 (Dupont or Sigma Chemical Co) through a continuous-flow UV monitor and collect 0.5 ml fractions.
6. The high optical density towards the bottom of the gradient reflects the presence of the serum proteins banding denser than the lipoproteins.
7. The serum lipoproteins are found banding in the gradient as follows:

Lipoprotein	Fraction no.	Volume from top (ml)
VLDL	1–2	0.5–1.0
IDL	3–4	1.5–2.0
LDL	5–9	2.5–4.5
HDL_1	10–12	5.0–6.0
HDL_2	13–17	6.5–8.5
HDL_3	18–20	9.0–10

Protocol 6. *Continued*

Notes on the method
High levels of triglyceride in serum samples can lead to the formation of a skin of triglyceride at the top of the gradient which must be aspirated off before unloading the gradient by upward displacement. Continuous monitoring of the gradient profile allows one to analyse directly the lipoprotein content of the blood.

An alternative procedure has been developed that uses a continuous gradient of NaBr from 1.003 g/ml to 1.200 g/ml (27); in this case the recommended centrifugation time is 95 000*g* for 2 h at 4 °C for flotation of the sample.

4.6.2 Fractionation of glycoproteins

Carbohydrates have different partial specific volumes than amino acids (for example, the disaccharide sucrose has a partial specific volume of 0.06 ml/g) and, in addition, carbohydrates are hydrated to different extents than proteins. Hence glycoproteins band at different densities to unmodified proteins, the actual density depending on the amount and the nature of the carbohydrate which is bound to the proteins. The gradient media of choice are the same as those recommended for other proteins; usually the medium of choice is 35% (w/w) RbCl. The centrifugation conditions used are also similar to those used for proteins, 250 000*g* at 5 °C with centrifugation times of 65 h for 12-ml gradients in a typical fixed-angle rotor. CsCl gradients can also be used although generally the degree of resolution is not as good as RbCl gradients.

4.7 Separation of polysaccharides and proteoglycans

Polysaccharides can be divided into neutral and charged molecules. Examples of the former are glycogen and starch that can be banded in a wide range of gradient media including CsCl gradients where they band close to 1.6 g/ml, while in Nycodenz gradients glycogen bands at 1.22 g/ml. Very little work has been done on the separation of neutral polysaccharides although isopycnic banding in metrizamide gradients has been used to show the association of specific enzymes with glycogen granules (28). In the case of proteoglycans and charged polysaccharides, a number of techniques have been devised for the isolation and characterization of polysaccharides from cartilage and other biomechanical structures. In this case the gradient needs to be designed to minimize macromolecular interactions. Typically, proteoglycans are extracted from the sample material by using 4 M guanidinium chloride followed by separation on CsCl gradients, either alone or in combination with 6 M guanidinium chloride (29), or Cs_2SO_4 gradients can be used (30). Proteoglycans extracted from detergent-lysed cells have also been fractionated on caesium trifluoroacetate gradients after removal of the bulk of the proteins by hot phenol extraction (31). Centrifugation conditions are typically 200 000*g* for 48–60 h at 20 °C. The proteoglycans are separated on the basis of their different compositions into their different classes.

4.8 Fractionation of nucleoproteins

Nucleoproteins can be divided into complexes of DNA and protein (e.g. chromatin and bacterial nucleoids), and ribonucleoproteins (e.g. ribosomes, informosomes, and spliceosomes). Although these complexes vary in terms of their relative amounts of nucleic acid and protein present as well as their stability to ionic environments, they all exhibit similar characteristics and so present similar problems for fractionation. The size diversity of many of these complexes (e.g. chromatin) makes isopycnic centrifugation an ideal method for purifying complexes from other subcellular components.

The major problem of separating nucleoproteins is that many types of complexes are readily dissociated by the ionic nature of some media, and in some cases by the hydrostatic pressure which is generated in the gradient by the centrifugal force. Of these two effects, the dissociative effects of highly ionic gradient media are usually the major problem. Some very stable complexes [e.g. transcription factor complexes (32)] can be banded without any prior fixation. However, usually such separations require the nucleoproteins to be fixed before fractionation. It is possible to 'glue' the complexes together by fixation of the complexes using formaldehyde or glutaraldehyde to cross-link the protein around the nucleic acids (33). Typically, samples are treated in dilute buffer (Hepes–NaOH or phosphate buffer are recommended) at neutral pH with a final concentration of 1% (w/v) formaldehyde (pH 7) for 24 h in ice. After fixation, formaldehyde should be removed by dialysis against dilute buffer. Alternatively, nucleoproteins can be cross-linked by exposure to UV light (34, 35) or by using one of the photo-activated psoralen derivatives (36). Two major problems of fixing any nucleoprotein complexes are that proteins may become adventitiously bound to the complex during fixation (22) and that analysis of the individual components of the complex after fractionation is difficult. Procedures for reversing the formaldehyde cross-linking of fixed complexes have been published (33, 37); however, they involve prolonged incubation of the sample at elevated temperatures and generally they have not been widely used. Cleavable cross-linkers can also be used (38) but again they have not been used extensively for nucleoproteins.

Problems associated with fixing samples, necessary for separating nucleoprotein complexes in ionic gradient media, can be avoided by using a non-ionic gradient medium such as Nycodenz or metrizamide which are completely non-ionic. These gradients have been widely used for separating almost all types of nucleoprotein complexes such as chromatin, bacterial nucleoids, and ribosomes, as well as organelles such as nuclei which contain nucleoprotein complexes (16, 43). In this case the complexes isolated reflect the state of the stable nucleoprotein complexes since any transiently bound proteins dissociate and separate from the complex during centrifugation. In the case of metrizamide gradients dissociation of transiently-bound proteins tends to be enhanced by the interaction of metrizamide with proteins (40); Nycodenz appears to interact less with protein and so this enhanced dissociation of proteins is much less pronounced. Alkaline, partially-denaturing metrizamide gradients containing triethylamine have also been used to study the distribution of histones associated with the DNA (41). *Table 4* summarizes the conditions used for the separation of different types of nucleoproteins on metrizamide and Nycodenz gradients.

The buoyant density of complexes generally reflects their overall composition in terms of nucleic acids and proteins. In the case of ionic gradient media the equation which can be applied is:

$$\text{Density} = F_n \rho_n + F_p \rho_p$$

Where F_n and F_p are the fractional amounts of nucleic acids and protein in the complex and ρ_n and ρ_p are the observed buoyant densities of the uncomplexed nucleic acids and proteins, respectively. This equation assumes that the interaction of protein and nucleic acid does not modify the extent of hydration of either type of macromolecule.

The situation for metrizamide gradients is more complex because, although for a fixed amount of sample on the gradient there is a linear relationship between the composition of nucleoproteins and their density (*Figure 6a*), it has been found that the density of chromatin is dependent on the amount of sample loaded on to the gradient (*Figure 6b*). This effect appears to be due to increased aggregation and hence decreased hydration of the complex as the band of chromatin becomes more concentrated; a similar effect is seen with some mitochondrial nucleoids. The extent of this concentration-dependent effect on buoyant density depends on the ionic content of the gradient medium (42). In Nycodenz gradients the buoyant density of chromatin appears to be less

Table 4. Banding of nucleoproteins in Nycodenz and metrizamide gradients

Nucleoprotein	**Density (g/ml)**	**Type of gradient[a] and rotor[b]**			**Centrifugation conditions (*g*/time)**
Chromosomes	1.29	25–60%	N	SO	16 000*g*/10 min
Chromatin (sheared)	1.19	36%	M/N	FA	100 000*g*/40 h
Chromatin (denatured)	1.22	27%	M	FA	80 000*g*/40 h
Histone:DNA complexes	1.15–1.25	35%	M	FA	63 000*g*/40 h
Bacterial nucleoids	1.22	33%	M	VT	140 000*g*/16 h
Mitochondrial nucleoids	1.26	20–55%	M	SO	100 000*g*/18 h
Ribosomal subunits	1.22				
Ribosomes	1.26	48%			
Ribosomes (Mg^{2+})	1.31		M	FA	100 000*g*/40 h
Polysomes (Mg^{2+})	1.35	30–50%	N	SO	80 000*g*/60 h
Messenger RNP (Mg^{2+})	1.21				

[a]N: Nycodenz; M: metrizamide.
[b]SO: swing-out; FA: fixed-angle; VT: vertical.

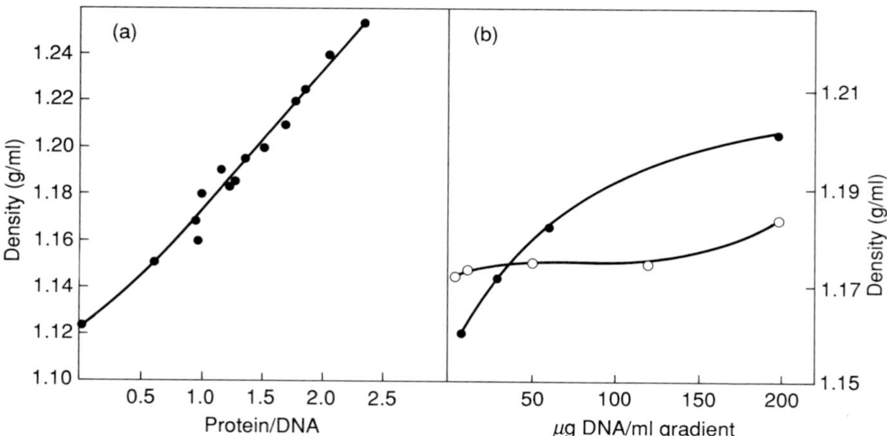

Figure 6. Banding of nucleoproteins in metrizamide and Nycodenz gradients: (a) relationship between the composition of complexes of DNA and protein (200 µg) and their buoyant density in metrizamide gradients; (b) effect of loading on the observed buoyant density of chromatin in metrizamide (●—●) and Nycodenz (○—○ gradients.

concentration-dependent (*Figure 6b*). Studies with ribonucleoproteins have shown that their buoyant densities are independent of the amount of sample on the gradient but their buoyant density can be affected by the ionic content of the gradient. Magnesium ions greatly increase the buoyant densities of ribosomes and polysomes to 1.33 g/ml while the density of mRNP is not greatly changed (43).

4.9 Use of isopycnic separations to study the interaction of macromolecules

Because high ionic strengths tend to disrupt most nucleoprotein complexes or stop them from forming, the frequently-used gradient media such as CsCl cannot be used for studying the interaction of macromolecules. It is for this reason that both metrizamide and Nycodenz have proved ideal for studying the binding of proteins to nucleic acids. The interaction of proteins with DNA can be of two types; the first is the essentially irreversible binding of basic proteins (e.g. histones), and second there are the weaker interactions of non-histone proteins. The two types of binding demand different gradient separation procedures.

Complexes of histones or other basic proteins will bind to DNA to form stable complexes and these can be banded in metrizamide or Nycodenz gradients very much in the same way as chromatin, with the banding patterns obtained depending on the ionic strength of the gradients (44). In the case of proteins which bind more transiently, then it is difficult to obtain a good separation using self-forming gradients with the sample distributed throughout the gradient. However, loading the sample into the bottom of the gradient and allowing the

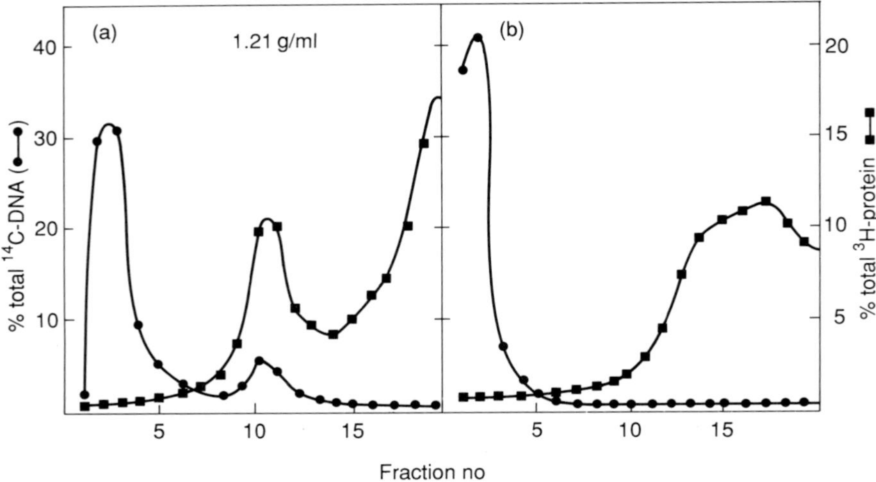

Figure 7. Use of non-ionic gradients to study the binding of non-histone proteins to DNA. A non-histone protein fraction from mouse cell nuclei was mixed with sheared mouse DNA in 2 M NaCl, 5 M urea, and dialysed into 0.2 M NaCl. The mixture was then made up to 38% metrizamide and 1 ml was layered under 4 ml of 35% metrizamide solution also containing 0.2 M NaCl. Gradients were centrifuged at 60 000*g* for 40 h at 2 °C. After centrifugation the distribution of protein (■) and DNA (●) was determined (a). Control gradients containing either DNA or protein alone in 0.2 M NaCl were loaded and centrifuged in the same way. The banding patterns of these control gradients are shown in (b). Separate experiments revealed that under these conditions the proteins did not bind to bacterial DNA. (Data derived from reference 45.)

DNA and DNA–protein complexes to float up differentially is capable of separating different DNA fractions (45). An example of this type of fractionation is shown in *Figure 7*; the actual binding pattern obtained depends on the ionic composition of the gradient.

4.10 Separation of mixtures of macromolecules

Since most types of macromolecules are present as complexes bound to other macromolecules in the cell, separation of these different types of molecules usually requires a dissociative medium which is usually ionic or chaotropic, or both. In addition, because the different types of macromolecules have widely different densities usually it is better to use a gradient medium which forms a steep gradient at equilibrium (see *Table 3*). In the author's experience, it is possible to obtain good separations of proteins, DNA, and RNA using Cs_2SO_4 gradients containing 6 M urea and 0.4 M guanidinium chloride (46). An example of the procedure for the separation of macromolecules is given in *Protocol 7*.

Protocol 7. Separation of DNA, RNA, and protein on Cs_2SO_4 gradients containing urea and guanidinium chloride.

1. Dissolve the sample in a solution of 6 M urea (analytical grade, deionized before use), 0.4 M guanidinium chloride, 20 mM dithiothreitol (DTT), 1 mM EDTA, and 20 mM Tris–HCl (pH 8.0). If the solution appears very viscous at this stage then it is better to reduce the sample concentration rather than shear the DNA which is the cause of the viscosity.
2. Add 0.64 g of Cs_2SO_4 (analytical grade) to each millilitre of the sample solution and make sure that it is dissolved.
3. Load the samples into thick-walled tubes in a 6 × 4 ml swing-out rotor and centrifuge for 65 h at 100 000*g* at 15 °C.
4. After centrifugation, first check that there is no pellicle of protein floating on the surface of the gradient then unload the solution by upward displacement. If a pellicle has formed on the top of the gradient it can be removed using a stainless steel spatula, or the gradient can be unloaded by dripping it out of the bottom of the tube.

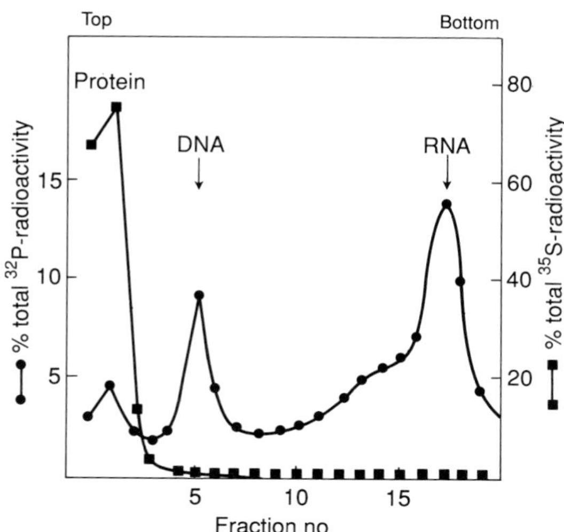

Figure 8. Isopycnic separation of protein, DNA, and RNA on Cs_2SO_4 gradients. Yeast nuclear chromatin labelled *in vivo* with ^{35}S-sulphate and ^{32}P-phosphate was fractionated on Cs_2SO_4 gradients containing 6 M urea and 0.4 M guanidinium chloride as described in *Protocol 7*. The distribution of ^{35}S (■—■) and ^{32}P (●—●) radioactivity, incorporated primarily into proteins and nucleic acids, respectively, was measured by liquid scintillation counting.

In this case the use of a swing-out rotor is used to avoid the problems associated with unloading of gradients after reorientation of the solution as does occur in fixed-angle and vertical rotors. However, the capacity of the gradients is rather limited. An example of such a separation is shown in *Figure 8*.

An alternative method which has been suggested is to use a procedure similar to that given in *Protocol 4* but using guanidinium isothiocyanate. Since this compound is both highly ionic and chaotropic it is a powerful denaturing agent, and so it is not necessary to add any other denaturant (36). The sample is mixed with the solution to give 4 M guanidinium isothiocyanate, 0.5% lauryl sarcosine, 20 mM DTT, and 20 mM citrate buffer pH 7.0. This solution is layered over a 5 ml cushion of 5.7 M CsCl in a 12-ml centrifuge tube and centrifuged at 150 000*g* for 18 h at 20 °C. The advantage of this method is that the DNA is found at the interface between the two solutions and the RNA is pelleted. The proteins which are found floating in the upper part of the tube can be used for blotting, and other studies suggest that they may retain some biological activity after separation. It has also been reported that gradients of caesium trifluoroacetate can be used to separate proteins, DNA, and RNA on a single gradient with banding densities of 1.4 g/ml, 1.63 g/ml, and 1.83 g/ml, respectively (47).

4.11 Separation of density-labelled macromolecules

The buoyant density of macromolecules can be modified by the incorporation of stable isotopes such as ^{15}N-nitrogen or ^{13}C-carbon or deuterium; the maximum density increases for nucleic acids if these isotopes replace the normal isotopes are 0.016 g/ml, 0.036 g/ml, and 0.035 g/ml, respectively (7, 25, 48). Alternatively, it is possible to incorporate dense analogues (e.g. bromodeoxyuridine) for density labelling DNA and RNA (7, 49), and in this case an increase in density up to 0.08 g/ml can be achieved. It is also possible to incorporate analogues which can then be modified subsequently. An example of this latter method is the incorporation of mercury-containing nucleotides into nucleic acids (50), which produce nucleic acids which are denser as a result of the presence of mercury atoms. The extent to which the density is modified depends on the method of labelling and the percentage incorporation of the density label. This methodology has proved useful for looking at the synthesis of macromolecules both *in vivo* and *in vitro*. For separating density-labelled macromolecules it is best to use a gradient medium which forms fairly shallow gradients. The original gradient media used were CsCl for DNA, caesium formate for RNA, and RbCl for proteins (7). However, it has been shown that mixed CsCl/Cs_2SO_4 gradients give good separations of RNA (see Section 4.4.2), and that for proteins KBr gradients which are very shallow give better resolution (48) than CsCl gradients (*Figure 9*). It has also been shown that mixed gradients of potassium and caesium thiocyanates are suitable for separating density-labelled proteins. Iodinated media such as metrizamide and Nycodenz can also be used for the fractionation of density-labelled nucleic acids (16, 50) and proteins (51).

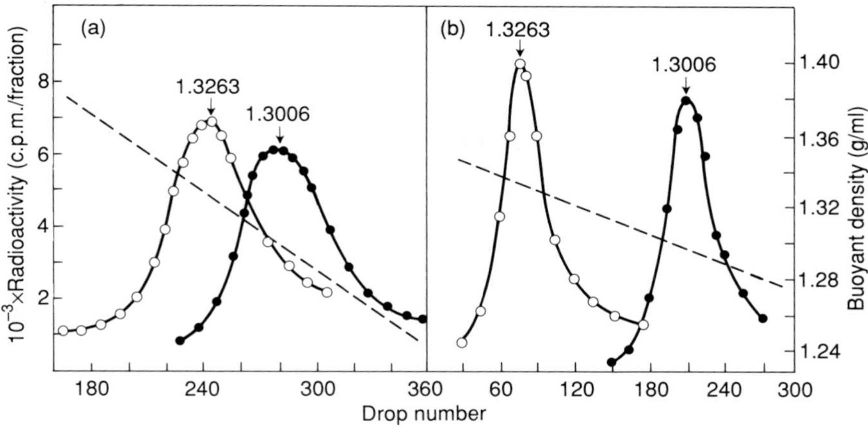

Figure 9. Separation of density-labelled proteins on CsCl and KBr gradients. Proteins from plants grown for one week on 50% D_2O and labelled for 24 h with ^{14}C-leucine (○—○) were mixed with ^{3}H-labelled protein extracts from control plants (●—●) and separated by isopycnic centrifugation on either (a) CsCl or (b) KBr gradients. The distributions of the proteins were determined by liquid scintillation counting. (Data derived from reference 48.)

Density labelling has been very important for investigating the synthesis of macromolecules and in making it possible to isolate newly-synthesized macromolecules. However, care must be taken to ensure that density labelling itself does not disrupt the normal synthetic processes. As an example, growth of organisms can be markedly inhibited by 50% deuterium oxide (D_2O). Analogues can be even more disruptive in terms of both synthesis and the catabolism of macromolecules since they appear as different types of molecule within the cell. Hence care must be exercised when interpreting these experiments. The other problem which arises is that the highly ionic nature of the gradient can inhibit some enzymes (25) or cause salt-induced aggregation of the proteins. In an attempt to avoid the problems associated with these highly ionic media, metrizamide/D_2O gradients have been used and appear to offer an alternative methodology (16, 51), especially for proteins which are very susceptible to highly ionic solutions.

5. Rate-zonal separations of macromolecules

5.1 Introduction

The development of versatile and sophisticated electrophoretic techniques which separate molecules on the basis of their migration through agarose or polyacrylamide gels (1, 2) has tended to reduce the use of rate-zonal ultracentrifugation for the analytical determination of macromolecule sizes. In addition, electrophoretic techniques have also become the primary method of looking at the interaction of macromolecules (2). However, some parameters such as partial specific volume can not be calculated by electrophoresis. In addition, the weakness of electrophoresis is that the method is limited in terms of the ionic conditions during the separation, while centrifugation can be carried out in the absence of any ions or in concentrated salt solutions as required. It is for this reason that a significant part of this chapter is devoted to separations based on the size of macromolecules and macromolecular complexes as determined by their sedimentation rate. The theoretical aspects of rate-zonal centrifugation have been described in Section 5.2 of Chapter 1 and the quantitative aspects of rate-zonal centrifugation are described in detail in Chapter 5.

5.2 General aspects of rate-zonal centrifugation

The rate of sedimentation of particles in rate-zonal gradients is dependent on the size, density, and conformation of the particles. Of these three parameters, the conformation of particles is most variable since the shape of particles can change as a result of ionic effects of the solution in which the samples are suspended or, in the case of DNA, as a result of the centrifugal force used for the separation. It is important to appreciate the extent to which biological particles and their ability to interact with each other can be modified by centrifugation and this, in turn, can affect the sedimentation properties of the particles.

5.2.1 Choice of rotor

In order to obtain optimal rate-zonal separations particles should move through the gradient without interacting with each other or the sides of the tube. As might be expected, zonal rotors have the optimum geometry for rate-zonal separations because there are no wall effects, that is, as the particles move along the radial lines of centrifugal force they do not strike the internal vanes of the rotor. However, zonal rotors are designed primarily for large-scale separations and tend to be inconvenient for the average person to set up and use. Of the tube rotors, swing-out (swinging bucket) rotors are the rotors of choice for almost all rate-zonal separations. The most popular rotors are the so called long-bucket swing-out rotors (e.g. Beckman SW41) although the high-performance rotors generating 350 000*g* can be used to obtain very rapid separations. For rapid rate-zonal separations the rotors of choice are vertical rotors because these have very short pathlengths compared to swing-out rotors of equivalent volume. In vertical rotors, however, wall effects do become significant once the particles cross the half-way point of the tube and the walls begin to converge (see *Figure 9* of Chapter 2). The band capacity of gradients is related to the size of band and so, as a result of liquid reorientation which occurs in vertical rotors, the capacity is much greater than a gradient of similar size in a swing-out rotor. However, the much greater band area means that the diffusion of the band is much greater and this can result in the formation of much broader bands, particularly if the sample has a molecular weight of 10^5 or less. Because of this bands can be much broader than those in swing-out rotors, even if the time of centrifugation is much shorter (see *Figure 2* of Chapter 3). As described in

Section 4.3 of Chapter 2, fixed-angle rotors are not generally suitable for rate-zonal separations (52), although they do seem to be suitable for the rate-zonal separation of small proteins $< 10S$ and for sedimentation equilibrium studies (for a review see ref. 53).

5.2.2 Choice of gradient medium

In contrast to the wide range of different gradient media used for isopycnic separations, sucrose is the medium of choice for almost all rate-zonal separations of macromolecules. However, the quality of sucrose used must be sufficiently high as sucrose from some sources absorbs in the UV much more than others. Once a good source has been located keep to that source; note that the highest price does not always indicate the highest quality. For separations of nucleic acids it is advantageous to use beet sugar rather than cane sugar since the levels of nucleases are generally less. Although some people have advocated various purification procedures for sucrose, it is best to buy the best quality sucrose. A detailed description of the properties and characteristics of sucrose are given in Section 3.1.2 and Appendix A of Chapter 3. The extensive use of sucrose gradients over the years means that there are descriptions of the use of sucrose gradients for almost every type of separation. In addition, the physicochemical properties of sucrose have been fully characterized and this has allowed the use of sucrose gradients for accurate determinations of sedimentation coefficients of macromolecules as described in Chapter 5.

In the few cases where sucrose is not appropriate, because, for example, the sample is a protein that interacts with sucrose, the alternative gradient medium used is glycerol. Glycerol has the added advantage for protein separations that it can help to preserve protein structure against denaturation. The physicochemical characteristics of glycerol are also well documented, making it possible to use these gradients for calculating sedimentation coefficients also. Gradients of Cs_2SO_4 have been used for the rate-zonal fractionation of proteoglycans (74) and gradients of guanidinium hydrochloride have been used for the fractionation of proteins in their denatured state (68). However, these applications are very much the exception to the general rule that sucrose is the medium of choice for rate-zonal separations.

5.2.3 Choice of gradient solvent and buffer

Most non-denaturing gradients are aqueous and buffered close to neutral pH with one of the common buffers, and they contain low concentrations of salt and EDTA. These substances are added to ensure the optimal conditions of the separation although the actual choice of the concentrations of these components is sometimes rather arbitrary. Some thought should be given to what is actually present; for example, EDTA not only chelates free metal ions which can modify nucleic acid sedimentation but EDTA can also remove the metal ions from metalloproteins which often results in their inactivation. The conformation and hence sedimentation of RNA is particularly sensitive to the presence of both monovalent and divalent cations. The major problem of separating macromolecules is that their poly-electrolyte nature leads to interactions between them which can lead to aggregation and, in extreme cases, precipitation. It is possible to denature samples by, for example, heating or treating the sample with denaturants before loading them onto the gradients. However, such treatments tend to have only a short-term effect and the molecules begin to interact again as they sediment down the gradient.

It is also possible to minimize aggregation by blocking active groups. As an example, in the case of RNA, treatment with formaldehyde blocks the amino groups which are involved in hydrogen bonds which are responsible for the secondary structure (20).

Alternatively, separations can be carried out in denaturing gradients. The choice of medium for denaturing gradients depends on the nature of the macromolecule. As an example, in the case of DNA, sucrose gradients containing 0.1 M NaOH can be used, but alkali rapidly degrades RNA and so these gradients cannot be used for this purpose. Alternatively, gradients based on non-aqueous solvents can be used; for nucleic acids one can use sucrose gradients in 99% DMSO or 85% formamide, both of which ensure complete denaturation of hydrogen-bonded molecules such as RNA and DNA. When using these non-aqueous solvents one encounters the problem that sucrose has a fairly low solubility in these solvents and so gradients must necessarily be rather shallow. It is also important to choose the correct type of tube when using gradients containing DMSO or formamide since these solvents attack some plastics used for tubes, especially polycarbonate.

5.2.4 Formation of gradient shape and centrifugation conditions

In terms of gradient shape, it is usual to use linear gradients since they are easiest to prepare by diffusion or using a simple two-chamber gradient maker (see Section 5.3 of Chapter 3), and usually little is gained by the use of more complex profiles such as exponential gradients which are in any case more inconvenient to prepare. Discontinuous gradients, often used for isopycnic separations, should not be used. Indeed, discontinuities in gradient can spoil rate-zonal separations and the presence of a density plateau within the gradients makes their capacity very low.

The choice of gradient concentration is also important; in some cases it is limited by the solvent as in the case of formamide and DMSO gradients but this is not a restriction with aqueous gradients. However, the choice of the sucrose concentration at the top and the bottom is not quite as arbitrary as it may seem. For many years people have tended to use 5–20% sucrose gradients because they are essentially isokinetic, that is particles move at the same speed as they sediment down the gradient because the greater centrifugal force on the particle as it moves away from the centre of rotation is balanced by the increasing viscosity of the more concentrated sucrose solution further down the gradient. Using isokinetic gradients enables you to use markers of known sedimentation coefficients to calculate the sedimentation coefficients of sample particles on the basis of the relative distances moved through the gradient (see Section 3 of Chapter 5). For higher-performance rotors it is often beneficial to use steeper, more viscous gradients which will induce zone-sharpening (see Section 6.4 of Chapter 3) and so improve the resolution of the gradient. For the very-high-performance rotors, such as the Beckman SW60, 10–60% sucrose gradients have been used. However, in using these high concentrations of sucrose it is important to ensure that at no time does the density of the gradient exceed that of the sample, otherwise particles will simply stop at their isopycnic positions.

As a general rule, the centrifugation conditions should be adjusted so that the time required is as short as possible, so that diffusion and possible changes to the sample (e.g. enzymic degradation) are minimal. Centrifuging at a low temperature also helps to preserve the native properties of samples, although it is often the composition of the gradient that determines the temperature of centrifugation. As an example, it is usual to include the detergent SDS in sucrose gradients for fractionating RNA as a nuclease inhibitor, but SDS precipitates at low temperature and so these gradients must be centrifuged at room temperature (~20 °C). The advantage of centrifugation at higher temperatures is that the viscosity of gradient solutions is less and so the time of centrifugation can be shorter. As a general rule, one should use the rotor which generates the most centrifugal force because this will allow the fastest separation and best resolution of the particles. However, some samples can be affected either by the centrifugal force itself (e.g. high-molecular-weight DNA) or by the hydrostatic pressure generated by the centrifugal force (e.g. some macromolecular complexes). These effects are discussed in detail in Section 9.1 of Chapter 5.

Provided that you have mastered the art of preparing and loading gradients as described in Chapter 3, then there is no reason why you should not achieve good results. Probably the most likely error is to load too much sample on to gradients; the symptoms indicative of overloading have been discussed in Section 2.1.3 of Chapter 3 and this topic is discussed further in Section 9.3 of Chapter 5. There is no simple guide to the amount of sample that should be loaded, but as a guide try loading 20–40 μg of sample per millilitre of gradient. The length of time of centrifugation often can be deduced from reading through the work of others as given in scientific papers; however, care is needed because such descriptions are sometimes not very accurate. If you are using 5–20% sucrose gradients at 5 °C and the k^*-factor (i.e. k'-factor) of the rotor is known, then this can be used to calculate the approximate centrifugation time required. It is also possible to obtain accurate estimates of the time required using simulation programs such as those described in Chapter 5. Similar types of simulation program are available commercially and the source codes of similar programs have been published (54, 55).

It is important to appreciate that good results depend not only on the centrifugation conditions but also on the care taken when unloading the gradient. It is all too easy to spoil separations as a result of a poor gradient fractionation technique. Section 7.6 of Chapter 3 describes various techniques which have been devised; as a general rule upward displacement should be the method of choice unless a pellicle is present on the top of the gradient.

5.3 Rate-zonal separations of nucleic acids

5.3.1 Fractionation of RNA

The defined sizes of almost all types of RNA molecule from 80–8000 nucleotides in length are well suited to rate-zonal separations on sucrose gradients. Good separations should be obtained using 10–40% (w/w) sucrose gradients in long-bucket swing-out rotors (e.g. Beckman SW41) or 10–60% (w/w) sucrose gradients in a high-performance swing-out rotor (e.g. Beckman SW60). Gradient solutions are usually buffered to neutral pH and contain 0.1 M NaCl. EDTA is also added (~1 mM) in order to chelate metal ions, and this is particularly important because metal ions not only activate many nucleases but also they can affect the conformation (and hence observed sedimentation coefficient) of RNA molecules. In some cases 0.5% SDS is added as a nuclease inhibitor and if so then gradients must be centrifuged at 20 °C, otherwise gradients should be centrifuged at 4 °C.

The major problem with RNA is its tendency to aggregate. Heating samples at 65 °C for 10 min disrupts aggregates but they tend to re-form on cooling. Hence the only way to ensure the absence of aggregation is to prevent it by covalent modification of the groups involved in aggregation prior to loading the sample on to the gradient, or to separate the RNA on gradients which denature the RNA and maintain it in a denatured state as it sediments down the gradient.

Hydrogen bonds involving the amino groups of RNA bases are responsible for the aggregation of RNA and so modification of these groups will ensure that aggregation does not occur. To do this RNA can be denatured and treated with 1% formaldehyde; this has the advantage that the reaction can be reversed by heating (19, 33). Alternatively, the RNA can be glyoxylated by incubation with 1 M glyoxal, 50% DMSO, at pH 7.0, for 60 min at 50 °C (56). Hydrogen bonding is completely disrupted by denaturants such as formamide and DMSO. Hence the other approach has been to separate RNA on sucrose gradients made up in either 99% DMSO or 85% formamide (57, 58); only minimal amounts of water, added as buffer or EDTA solution, are present. The gradients should be centrifuged at 25 °C and contain minimal amounts of salt in order to maximize the denaturation of the RNA. *Figure 10* illustrates how hidden nicks in the RNA can be revealed by centrifugation under denaturing conditions. DMSO and formamide both absorb in the UV at 260 nm, but gradients can be monitored at 280 nm.

5.3.2 Fractionation of DNA

DNA isolated from most sources other than viruses and some cell organelles is of very high molecular weight and, because DNA has a stiff, rod-like nature, it is very susceptible to degradation as a result of physical shearing or enzymic

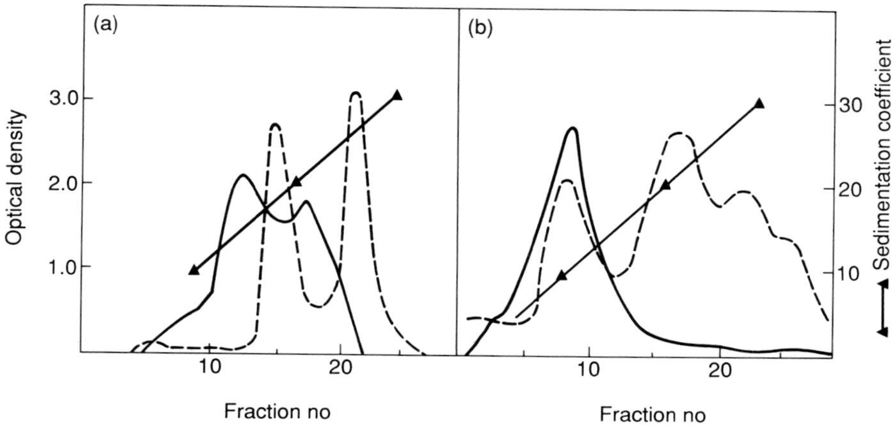

Figure 10. Rate-zonal separation of RNA in non-denaturing and denaturing sucrose gradients. Samples of (a) ribosomal RNA and (b) globin 9S mRNA were centrifuged in either sucrose (-----) non-denaturing gradients or formamide–sucrose (——) denaturing gradients at 25 °C in a swing-out rotor and the sedimentation pattern of the RNA was monitored. (Data derived from reference 58.)

degradation. In order to minimize degradation samples can be lysed after they have been loaded on to the gradient (59). However, one of the features of DNA sedimentation in rate-zonal gradients is that the apparent sedimentation coefficient of linear DNA and bacterial nucleoid structures depends on the centrifugation conditions (60, 61). This effect has been explained in terms of the centrifugal force causing conformational changes in the DNA (62, 63) and this is reflected in the increase in sedimentation coefficient with increasing rotor speed. In spite of these potential artefacts, work has been done on the rate-zonal separation of DNA especially in terms of DNA fragments and small bacteriophage DNAs (64, 65).

Denaturing gradients have been used to study DNA damage caused by various treatments. Unlike RNA, DNA is stable to alkali and so it is possible to separate DNA in alkaline sucrose gradients containing 0.1–0.5 M alkali instead of buffer (63, 66). However, in the case of mitochondrial DNA, it has been found that the DNA is sensitive to alkali, probably as a result of ribonucleotides remaining in the DNA after replication (59). It is possible that this problem would be seen in other situations where DNA is undergoing active replication. In the same study it was shown that supercoiled, relaxed, circular, and nicked mitochondrial DNA could be separated on a 5–15% sucrose gradient in formamide; centrifugation was for 140 000*g* for 2 h at 4 °C. Damage to DNA has also been examined using a combination of rate-zonal and isopycnic centrifugation on neutral and alkaline pre-formed NaI gradients (67).

5.4 Rate-zonal separations of proteins

Rate-zonal centrifugation has been widely used to study proteins and changes in their characteristics as a result of modification of their shape and/or molecular mass. Both sucrose and glycerol gradients have been used to fractionate and characterize proteins of many different types; changes in proteins arising from proteolysis or changes in conformation are readily detectable. Glycerol gradients have been widely used to study steroid hormone receptor proteins and the reputation of glycerol as an aid to the prevention of protein denaturation has resulted in these gradients being used for a wide range of enzymes. A 5–20% glycerol gradient has similar density and viscosity profiles as a 5–20% sucrose gradient (see Appendices A and B of Chapter 3) and so the run conditions for both types of gradient are also very similar.

Very little work has been done on the development of denaturing gradients for proteins. In principle, sucrose gradients could be prepared containing any of the accepted protein denaturants that do not bind to proteins; urea and guanidinium chloride would both be suitable for this purpose. As an alternative approach, it has been reported that proteins can be separated in their denatured state by centrifuging them in a gradient of 5–8 M guanidinium chloride containing 20 mM dithiothreitol (68). Marker proteins of known size can be used to determine the molecular mass of the sample proteins. Estimates of the size of proteins using this method gave values to within 6% of the true values.

The rate-zonal fractionation of lipoproteins is also an established technique. The interest in this case is that for optimal resolution it is best to use differential flotation; that is, the sample of lipoproteins is loaded into the bottom of the gradient and separated on the basis as to how quickly they float up the gradient (69, 70). In order for this separation to work the proteins must be less dense than the medium, and thus sucrose is not suitable for these gradients. Instead gradients are usually preformed from NaBr or KBr and the sample, mixed with the appropriate dense solution, is carefully layered into the bottom of the tube. It seems that the wall effects which can disrupt rate-zonal sedimentation fractionations are much less important in differential flotation and so this type of separation can be carried out in fixed-angle rotors; best results are obtained using shallow angle (20 ° or less) rotors.

5.5 Rate-zonal separations of nucleoproteins

Polysomes, ribosomes, and ribosomal subunits are all routinely fractionated on sucrose gradients, primarily for the isolation and characterization of their components, especially the proteins which are tightly bound to the RNA but whose function remains unclear. Similarly, rate-zonal separations are used to isolate and characterize nuclear ribonucleoproteins associated with both the nucleolus and the transcriptionally-active genes; analyses of these nuclear complexes is extremely important for investigations into the post-transcriptional processing of RNA in the nucleus. Rate-zonal fractionation of these complexes does not pose any real problems and the separations obtained are as good as those obtained by electrophoretic methods. However, it appears that ribosomes and, by implication, possibly other ribonucleoproteins can be disrupted by the hydrostatic pressure created by very large centrifugal forces (71). Hence it is better to use a moderate centrifugal force and a slightly longer centrifugation time rather than use the highest performance rotor to obtain a rapid separation.

In terms of the rate-zonal separation of DNA–protein complexes, most of these (e.g. chromatin) are very polydisperse and so are not really suitable for this type of separation. However, nucleosomes, derived from nuclease digestion of chromatin, have been separated on sucrose gradients without any problems. In addition, a method for the rate-zonal separation of metaphase chromosomes on hexylene glycol gradients has been devised (72); however, the interest in this type of technique is not very significant.

5.6 Rate-zonal separation of proteoglycans

Proteoglycans are generally very polydispersed and in addition they have a tendency to aggregate in solution. For this reason it is necessary to separate the proteoglycans on sucrose gradients containing 0.5 M NaCl to minimize intermolecular interactions; typically the 4 M guanidinium chloride extract of cartilage is centrifuged on a 10–50% sucrose gradient containing 0.5 M NaCl at 240 000*g* at 20 °C in a long-bucket swing-out rotor (73). Isokinetic Cs_2SO_4 gradients have also been used for fractionating proteoglycans (74) and there has been a great deal of theory developed for the rate-zonal separation of proteoglycans in this type of gradient (75). This same group has developed a sector-shaped cell that will fit into a normal swing-out rotor; however, it remains unclear as to whether such esoteric approaches actually offer real benefits.

5.7 Analyses of macromolecular interactions

A range of methods have been devised for studying the interaction between macromolecules; some of these are based on electrophoretic methods (2) but centrifugal separations have also been used. As described in Section 4.9, isopycnic centrifugation can be used to study the formation of stable complexes. However, quite often the interactions between macromolecules are only transient and it is in these situations that rate-zonal centrifugation can be used to study macromolecular interactions.

One of the main areas of study are the interactions of nucleic acids with proteins. In such experiments one of the keys to successful experiments is to ensure that the size of the nucleic acid component is correct; the preferred size will depend on the nature of the experiment. Nuclear DNA as isolated is normally larger than 15S while the size of RNA can vary from 4S to greater than 50S. If small molecules are required then the size of RNA can be reduced by partial alkaline digestion at pH 9–10 or by limited enzyme digestion. DNA can be randomly sheared by sonication or needle shearing or it can be degraded more

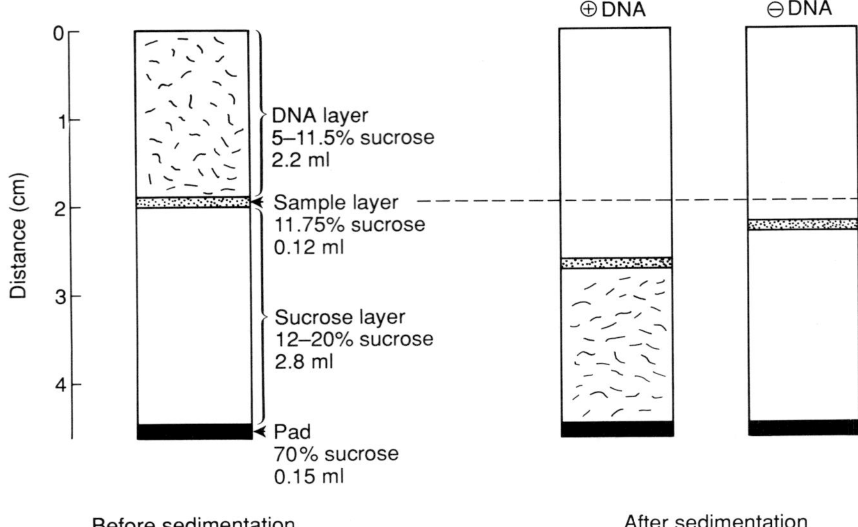

Figure 11. An illustration of the principle of the use of sedimentation partition chromatography to study the interaction of proteins with nucleic acids. (Derived from reference 77.)

specifically by restriction nucleases. The DNA sequence of interest can usually be cloned into a suitable vector and this will provide milligram amounts of DNA which can then be digested to the correct size as required.

A boundary method similar to that used for analytical centrifuges has been devised (76) for studying the interaction of proteins with nucleic acids, but it is limited in that it cannot be used if the binding constant is greater than 10^7 M^{-1} and it requires a fairly large amount of sample. However, two other methods have been devised which involve zone sedimentation. The first of these was devised to investigate the binding of steroid hormone receptor proteins to DNA. The method measures the displacement of a band of the smaller hormone receptor protein (5S) in the gradient as much larger DNA (18–20S) is centrifuged through it; this has been termed as 'sedimentation partition chromatography' (77). *Figure 11* shows the principle of the method and the procedure is given in *Protocol 8*.

Protocol 8. Sedimentation partition chromatography for studying the interaction of DNA with steroid receptor protein

1. Prepare purified DNA, and for the purposes of this procedure its size should be much larger than that of the protein ligand; DNA 20S in size was used in the original method.

2. Label the protein ligand, and in this case it can be done by labelling the protein with ^{3}H-estradiol. Add sucrose to the solution to give a final concentration of 11.75% (determine this from the refractive index).
3. Using a simple two-pot gradient maker, prepare a 2.2 ml 5–11.5% linear sucrose gradient light-end first and containing 80 μg/ml of the 20S DNA in a 5-ml tube for a swing-out rotor. At the same time prepare a control gradient which does not contain any DNA.
4. Layer 0.12 ml of the labelled protein ligand from Step 2 at the bottom of each gradient (care is required since the density difference is not very great).
5. Using the gradient maker layer a 2.8 ml 12–20% sucrose gradient under the radioactive layer of the protein ligand, and again care is required as the density difference is not very great. Finally, place a pad of 0.15 ml of 70% sucrose at the bottom of each gradient.
6. Centrifuge the gradients at 100 000*g* for 15 h at 5 °C; note that the exact time required will depend on the sizes of the DNA and ligand.
7. Fractionate the gradients by upward displacement into at least 30 fractions.
8. Measure how much the band of ligand has moved in the gradient containing the DNA and in the control gradient. At the same time determine the average DNA concentration in the lower part of the gradient from the A_{260} of each fraction. A typical result is shown in *Figure 12*.
9. Calculate the binding affinity from the relative distances moved by the DNA (X) and the distance moved by the protein ligand as a result of its interaction with the DNA (Y) from the equation:

$$K_{RD} = (X/Y) \times (\text{DNA sites})$$

An example of the calculation is shown in the legend to *Figure 12*.

The second approach has been to study the sedimentation of the complex of protein and DNA as it sediments through a sucrose gradient (78). In this case the DNA and proteins are mixed in the required amounts, in a suitable ionic environment, and loaded on to a 5–20% sucrose gradient. For this method to be valid the concentration of protein must be much lower than the number of binding sites on the DNA and so this must be taken into account when deciding on the ratio of protein-to-DNA. When the gradients are centrifuged the complex begins to sediment out of the sample zone leaving the majority of the protein behind. The higher the affinity of the protein for the DNA then the further it will migrate down the gradient bound to the DNA. As long as the complexes do not sediment faster than the time for the binding equilibrium to be established, then it can be shown that the concentration of protein in the sample (P_0) and in the nth fraction (P_n) are related by the equation:

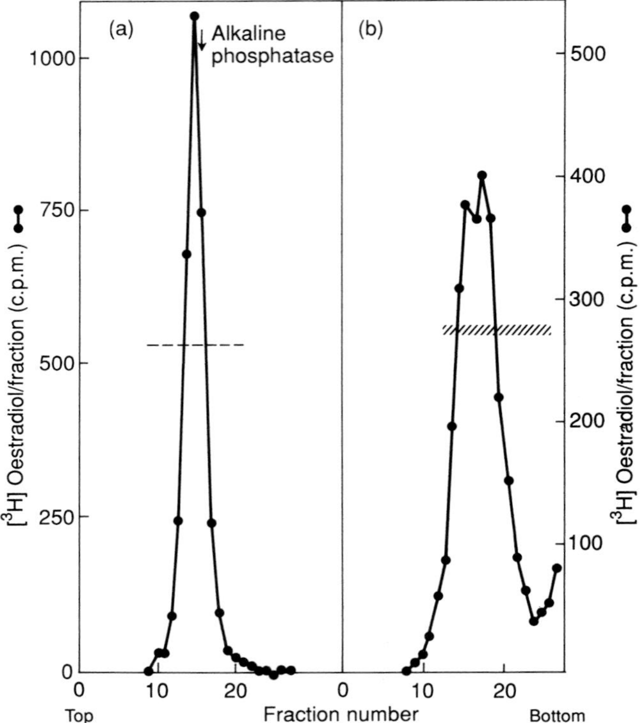

Figure 12. Use of experimental data derived from sedimentation partition chromatography. The figure shows the experimental data obtained from using the procedure given in *Protocol 8*. Comparison of the control gradient (a) containing no DNA with the experimental gradient (b) shows that the presence of DNA has caused the band of receptor protein (●—●) to sediment faster. In the data shown the receptor protein has sedimented an average of 3.3 fractions while the DNA has sedimented 14 fractions. Hence on average the receptor was bound to the DNA for 3.3/14 (i.e. 23.5%) of the time. Since the association constant, K_{RD}, is defined by: $K_{RD} = (\text{free/bound}) \times (\text{DNA sites})$, then if one site is one base pair then 1 µg/ml of DNA is equivalent to 1.5 µM of sites. Substituting in the data gives $K_{RD} = (76.5/23.5) \times 80 \times 1.5 = 4 \times 10^{-4}$M. (Data derived from reference 77.)

$$\frac{P_n}{P_0} = \left(\frac{KL}{1+KL}\right)^n$$

where K is the association constant and L is the concentration of free nucleic acid binding sites. K can be calculated if the equation is rearranged to:

$$KL = \left[\left(\frac{P_n}{P_0}\right)^{-1/n} - 1\right]^{-1}$$

5.8 Determination of partial specific volumes of macromolecules

Edelstein and Schachman (79) using an analytical centrifuge first showed that it was possible to determine the partial specific volume of macromolecules by comparing their sedimentation in H_2O and D_2O. Subsequent work showed that it was also possible to obtain accurate calculations of partial specific volumes using sucrose gradients (80, 81).

The method used is described in *Protocol 9*.

Protocol 9. Determination of partial specific volumes by sedimentation in sucrose gradients

1. Prepare 5–20% sucrose gradients in either H_2O or D_2O; take great care to ensure that the linearity of both types of gradient is as similar as possible.
2. Carefully load the sample on to each of the gradients and centrifuge for an appropriate time and speed that will ensure that the particles sediment between one-third and two-thirds of the way down the gradient.
3. Fractionate the gradients into at least 25 fractions, and read the refractive index of each to determine the sucrose concentration. Locate the peak of the sample material by an appropriate assay.
4. Calculate the distances sedimented by the sample in sucrose/H_2O and sucrose/D_2O gradients, r_H and r_D, respectively.
5. Determine the average density of the medium through which the particles have sedimented, ρ_H and ϱ_D, respectively. This can be done by taking an average of the densities at the top of the gradient and at the position of the sample at the end of centrifugation (equivalent to the density at position $r/2$ in the gradient).
6. The partial specific volume of the particles ($\bar{v}$) is related to the measured parameters by the equation:

$$\frac{r_H}{r_D} = \frac{k_H(1 - \bar{v}\rho_H)}{k_D(1 - \bar{v}\rho_D)}$$

   where k_H and k_D are constants for H_2O and D_2O, respectively, and so it is possible to calculate the partial specific volume of the sample.

The major advantage of this method is that it is not necessary to have a pure sample in order to determine the partial specific volume. As can be seen from the protocol, all that is required is to be able to detect the position of the macromolecule in the gradient, and this can often be done on the basis of enzymic activity or labelled antibodies.

References

1. Hames, B. D. and Rickwood, D. (ed.) (1990). *Gel electrophoresis of proteins: a practical approach* (2nd edn) IRL Press, Oxford.
2. Rickwood, D. and Hames, B. D. (ed.) (1990). *Gel electrophoresis of nucleic acids: a practical approach* (2nd edn) IRL Press, Oxford.
3. Howe, C. J. and Ward, E. S. (ed.) (1989). *Nucleic acid sequencing: a practical approach*. IRL Press, Oxford.
4. Findlay, J. B. C. and Geisow, M. J. (ed.) (1989). *Protein sequencing: a practical approach*. IRL Press, Oxford.
5. Lundh, S. (1985). *Arch. Biochem. Biophys.* **241**, 265.
6. Flamm, W. G., Birnstiel, M. C., and Walker, P. M. B. (1972). In *Subcellular components: preparation and fractionation* (ed. G. D. Birnie) p. 279. Butterworths, London.
7. Birnie, G. D. (1978). In *Centrifugal separations in molecular and cell biology* (G. D. Birnie and D. Rickwood ed.) p. 169. Butterworths, London.
8. Bauer, W. and Vinograd, J. (1968) *J. Mol. Biol.* **33**, 141.
9. Anet, R. and Stryer, D. R. (1969). *Biochem. Biophys. Res. Commun.* **34**, 328.
10. Abeyasekera, G. and Rickwood, D. (1991). *Biotechniques* **10**, 460.
11. Stougaard, P. and Molin, S. (1981). *Anal. Biochem.* **118**, 191.
12. Little, S. E. (1990). *Beckman Instrument Application Sheet DS-770*. Beckman Instrument Inc, Palo Alto, CA, USA.
13. Andersson, K. and Hjorth, R. (1985). *Plasmid* **13**, 78.
14. Karlovsky, P. and de Cock, A. W. A. M. (1991). *Anal. Biochem.* **194**, 192.
15. Kasamatsu, H., Robberson, D. L., and Vinograd, J. (1971). *Proc. Nat. Acad. Sci.* **68**, 2252.
16. Ford, T. C. and Rickwood, D. (1983). In *Iodinated density gradient media: a practical approach* (ed. D. Rickwood) p. 23. IRL Press Ltd, Oxford.
17. Chirgwin, J. M., Przybyla, A. E., MacDonald, R. J., and Rutter, W. J. (1979). *Biochemistry* **18**, 5294.
18. Okayama, H., Kawaich, M., Brownstein, M., Lee, F., Yokota, T., and Arai, K. (1987). *Methods in Enzymol.* **154**, 3.
19. Lozeron, H. A. and Szbalski, W. (1966). *Biochem. Biophys. Res. Commun.* **23**, 612.
20. Williams, A. E. and Vinograd, J. (1971). *Biochim. Biophys. Acta* **228**, 423.
21. De Kloet, S. R. and Andrean, B. A. G. (1971). *Biochim. Biophys. Acta* **247**, 519.
22. Hinton, R. H. and Dobrota, M. (1976). *Density gradient centrifugation*. North Holland, Amsterdam.
23. Andrean, B. A. G. and De Kloet, S. R. (1973). *Arch. Biochem. Biophys.* **156**, 373.
24. Young, B. D. and Anderson, M. L. M. (1985). In *Nucleic acid hyridisation: a practical approach* (ed. B. D. Hames and S. J. Higgins). IRL Press Ltd, Oxford.
25. Hu, A. S., Bock, R. M., and Halvorson, H. O. (1962). *Anal. Biochem.* **4**, 489.
26. Kelley, J. L. and Kruski, A. W. (1986). *Methods in Enzymol.* **128**, 170.
27. Hinton, R. H., Al-Tamer, Y., Mallinson, A., and Marks, V. (1974). *Clin. Chim. Acta* **53**, 355.
28. Guenard, D., Morange, M., and Buc, H. (1977). *FEBS Lett.* **76**, 262.
29. Crawford, T. (1988). *Biochim. Biophys. Acta* **964**, 183.
30. Webber, C., Glant, T. T., Roughley, P. J., and Poole, A. R. (1987). *Biochem J.* **248**, 735.

31. Lyon, M., Greenwood, J., Sheehan, J. K., and Nieduszynski, I. A. (1983). *Biochem J.* **213**, 355.
32. Fukuda, R. and Ishihama, A. (1972). *Biochem. Biophys. Res. Commun.* **45**, 1255.
33. Jackson, V. (1978). *Cell* **15**, 945.
34. Schouten, J. P. (1985). *J. Biol. Chem.* **260**, 9916.
35. Blanco, J., Kimura, H., and Mueller, G. C. (1987). *Anal. Biochem.* **163**, 537.
36. Coombs, L. M., Pigott, D., Proctor, A., Eydman, M., Denner, J., and Knowles, M. A. (1990). *Anal. Biochem.* **188**, 203.
37. Ip, Y. T., Jackson, V., Meier, J., and Chalkley, R. (1988). *J. Biol. Chem.* **263**, 14044.
38. Nielsen, P. E., Hansen, J. B., and Buchardt, O. (1984). *Biochem. J.* **223**, 519.
39. Wilson, V. L., Rinehart, F. P., and Schmid, C. W. (1976). *Anal. Biochem.* **73**, 350.
40. Rickwood, D. and Jones, C. (1981). *Biochim. Biophys. Acta* **654**, 26.
41. Russev, G. and Hancock, R. (1981). *Nucleic Acids Res.* **4**, 129.
42. Kondo, T., Nakajima, Y., and Kawakami, M. (1979). *Biochim. Biophys. Acta* **561**, 526.
43. Houssais, J. F. (1983). In *Iodinated density gradient media: a practical approach* (ed. D. Rickwood), p. 43. IRL Press Ltd, Oxford.
44. Rickwood, D., Birnie, G. D., and MacGillivray, A. J. (1975). *Nucleic Acids Res.* **2**, 723.
45. Rickwood, D. and MacGillivray, A. J. (1977). *Expl. Cell Res.* **104**, 287.
46. Adameitz, P. and Hilz, H. (1976). *Hoppe Seyler's Z. Physiol. Chem.* **357**, 527.
47. Mirkes, P. E. (1985). *Anal. Biochem.* **148**, 376.
48. Boudet, A., Humphrey, T. J., and Davies, D. D. (1975). *Biochem. J.* **152**, 409.
49. Phillips, W. H., Peterson, C. A., and Grainger, R. M. (1989). *Anal. Biochem.* **177**, 333.
50. Mory, Y. and Gefter, M. (1978). *Nucleic Acids Res.* **5**, 3899.
51. Huttermann, A. and Guntermann, U. (1975). *Anal. Biochem.* **64**, 360.
52. Castaneda, M., Sanchez, R., and Santiago, R. (1971). *Anal. Biochem.* **44**, 381.
53. Minton, A. P. (1989). *Anal. Biochem.* **176**, 209.
54. Steensgaard, J. and Rickwood, D. (1985). *Microcomputers in biology: a practical approach* (ed. C. Ireland and S. P. Long). IRL Press, Oxford.
55. Young, B. D. and Krumlauf, R. (1981). *Anal. Biochem.* **115**, 97.
56. Nash, M. A., Johnson, M., Knesek, J. E., Chan, J. C., and East, J. L. (1981). *Anal. Biochem.* **111**, 376.
57. Imaizumi, T, Diggelman, H., and Scherrer, K. (1973). *Proc. Nat. Acad. Sci. USA* **70**, 1122.
58. MacNaughton, M., Freeman, K. B., and Bishop, J. O. (1974). *Cell* **1**, 117.
59. Ellis, C. N., Conley-Hixon, S., and Blakemore, W. S. (1987). *Anal. Biochem.* **163**, 200.
60. Chia, D. and Schumaker, V. N. (1974). *Biochem. Biophys. Res. Commun.* **56**, 241.
61. Kovacic, R. T. and Van Holde, K. E. (1977). *Biochemistry* **16**, 1490.
62. Pettijohn, D., Hecht, R. M., Stimpson, D., and Van Scoyk, S. (1978). *J. Mol. Biol.* **119**, 353.
63. Korba, B. E., Hayes, J. B., and Boehmer, S. (1981). *Nucleic Acids Res.* **9**, 4403.
64. Serwer, P., Graef, P. R., and Garrison, P. N. (1978). *Biochemistry* **17**, 1166.
65. El-Gewely, M. R. and Helling, R. B. (1980). *Anal. Biochem.* **102**, 423.
66. Hell, A., Birnie, G. D., Slimming, T. K., and Paul, J. (1972). *Anal. Biochem.* **48**, 369.
67. James, M., Mansbridge, J., and Kidson, C. (1983) *Anal. Biochem.* **128**, 60.

68. Patterson, B. W., Schumaker, V. N., and Fisher, W. R. (1984). *Anal. Biochem.* **136**, 347.
69. Patsch, W., Patsch, J. R., Kostner, G. M., Sailer, S., and Braunsteiner, H. (1978). *J. Biol. Chem.* **253**, 4911.
70. Patsch, W., Schonfeld, G., Gotto, A. M., and Patsch, J. R. (1980). *J. Biol. Chem.* **255**, 3178.
71. Hauge, J. G. (1971). *FEBS Lett.* **17**, 168.
72. Stubblefield, E. and Wray, W. (1978). *Biochem. Biophys. Res. Commun.* **83**, 1404.
73. Franzen, A., Bjornsson, S., and Heinegaard, D. (1982). *Anal. Biochem.* **120**, 38.
74. Pita, J. C., Mueller, F. J., and Pezon, C. F. (1985). *Biochemistry* **24**, 4250.
75. Mueller, F. J., Pezon, C. F., and Pita, J. C. (1989). *Biochemistry* **28**, 5276.
76. Jensen, D. E. and Von Hippel, P. H. (1977). *Anal. Biochem.* **80**, 267.
77. Yamamoto, K. R. and Alberts, B. (1974). *J. Biol. Chem.* **249**, 7076.
78. Draper, D. E. and Von Hippel, P. H. (1979). *Biochemistry* **18**, 753.
79. Edelstein, S. J. and Schachman, H. K. (1967). *J. Biol. Chem.* **242**, 306.
80. Meunier, J. C., Olsen, R. W., and Changeux, J. P. (1972). *FEBS Lett.* **24**, 63.
81. O'Brien, R. D., Timpone, C. A., and Gibson, R. E. (1978). *Anal. Biochem.* **86**, 602.

5

Measurements of sedimentation coefficients

JENS STEENSGAARD, STEVEN HUMPHRIES, and S. PETER SPRAGG

1. Introduction

The sedimentation coefficient (s-value) of a particle is, by definition, its sedimentation velocity in a unit gravitational field (1). It is a commonly-used measure of particle size, especially when a molecular mass cannot be determined or when it is not physically meaningful as in the case of ribosomal subunits or other subcellular components. The sedimentation coefficient (given in Svedberg units, S) is a useful measure of particle size and it is an important parameter for the optimization of centrifugal procedures and the characterization of biological particles. This chapter describes how sedimentation coefficients can be calculated in practice.

Data for the calculation of sedimentation coefficients can be obtained in two different ways. The first of these requires a dedicated analytical ultracentrifuge with direct recording of the sedimentation processes in the spinning rotor fitted with optical cells; this type of instrument represents a major capital cost for most laboratories. The second approach is based on rate-zonal centrifugation in sucrose gradients using a preparative ultracentrifuge and it is this latter technique which will be the main topic of this chapter.

It should be stressed that the usefulness of ultracentrifugation today lies mainly in the preparative field. Approximate molecular weights can be estimated quickly and usually with greater accuracy by using gel electrophoresis or chromatographic methods or in some cases by protein or DNA sequence analysis. However, some proteins having unusual sizes, charge, or carbohydrate content are obvious candidates for the calculation of molecular size and macromolecular conformation by ultracentrifugation; examples include glycoproteins, histones, and lipoproteins. An important feature is that individual particles can be followed by tracing a specific property such as an enzymic activity or specific label, so making it possible to estimate sedimentation coefficients without prior purification of the preparation. The same theoretical basis for the calculation of sedimentation coefficients is also at the core of the simulation programs for optimizing centrifugal experiments (2, 3). This is an important reason to appreciate the theory of centrifugation. A number of programs suitable for IBM-PC compatible computers have been published previously and are now commercially available (4).

2. Choice of gradient medium and gradient solvent

As described in Chapter 3, the medium of choice for rate-zonal separations is sucrose. The reason for this is that sucrose has all the desirable characteristics of an ideal gradient medium for rate-zonal separations. In addition, because of its frequent use, the relationships between concentration, density, viscosity, and temperature have been fully investigated and polynomial expressions linking these different parameters have been devised (5). Similarly, the usual solvent for sucrose gradients is water or a dilute aqueous buffer. Because aqueous sucrose gradients have become the standard for rate-zonal separations there has been a great deal of work done on devising computational methods for the simulation of experiments and the calculation of sedimentation coefficients from experimental data, and these are described in the later sections of this chapter. The major part of this chapter will concentrate on calculation of sedimentation coefficients using this type of gradient.

While aqueous sucrose gradients are standard for most rate-zonal separations other gradient media are also used, the most frequent being glycerol. Glycerol gradients have been used for separations where either sucrose is not suitable or where their use has been traditional. Glycerol gradients are widely used for separating proteins, especially enzymes. In the absence of any programs for calculating sedimentation coefficients for glycerol gradients it is necessary to use isokinetic gradients with markers of known size. For the rate-zonal separation of proteoglycans, isokinetic Cs_2SO_4 have been devised (see Section 5.6 of Chapter 4).

In most cases, aqueous sucrose gradients separate macromolecules without changing their native conformation (for exceptions see Section 9.1). However, it may be desirable to separate molecules in their denatured form and for this it is necessary to change the nature of the gradient. DNA can be separated in its denatured form by adding alkali to sucrose gradients to increase the pH to greater than pH 12, at which point the DNA becomes single stranded. In the case of RNA, it is best to change the solvent from water to formamide or DMSO (see Section 5.2.3 of Chapter 4); in both cases only fairly shallow sucrose gradients can be formed. Very little work has been done on the separation of denatured proteins but there has been a report of the sedimentation of denatured proteins in gradients of guanidinium chloride containing 1 mM dithiothreitol (6). In all cases where non-standard conditions are used for separations, the sedimentation coefficients of the macromolecules can only be determined by using markers of known size in gradients which are essentially isokinetic. Note that in denaturing conditions the conformation and hence sedimentation coefficient (see Section 10) of markers will change. Hence, in denaturing

Table 1. Sedimentation coefficients of marker molecules

Particle	Molecular weight	Sedimentation coefficient
Ribonuclease	13 683	1.6
Ovalbumin	45 000	3.6
Haemoglobin	68 000	4.3
Ceruloplasmin	134 000	7.1
Catalase	250 000	11.4
Urease	480 000	18.6

gradients, markers and sample molecules appear to sediment according to their molecular weight as long as both are completely denatured.

3. Use of isokinetic gradients for calculating sedimentation coefficients

This is a simple way for estimating sedimentation coefficients of particles, making it useful for novices or for those without any access to a microcomputer, or in cases where it is necessary to use non-standard conditions (e.g. sucrose–formamide gradients). Shallow sucrose gradients (above 5 °C) can be designed so that they are isokinetic, that is, particles in such gradients sediment with a constant velocity under a variety of different conditions. Isokinetic gradients have been designed for swing-out rotors (7, 8) and zonal rotors (9). In an analytical ultracentrifuge with water as the sedimentation medium the particles accelerate as they sediment in the optical cell. In contrast, in a strictly isokinetic gradient, the increasing viscosity and density of the sucrose medium will brake the particles just enough to give them a constant sedimentation velocity. A 5–20% (w/w) linear sucrose gradient is essentially isokinetic and is widely used for the separation of macromolecules. This type of gradient can be used experimentally to calculate sedimentation coefficients by comparing the sedimentation distance of a particle with a known sedimentation coefficient with the sedimentation distance of the test particles. The relationship is very simple, namely:

$$\frac{s_1}{s_2} = \frac{r_1}{r_2}$$

where s_1 and s_2 are the sedimentation coefficients of the two particles and r_1 and r_2 are the respective distances that each particle sediments. If possible, it is desirable to use two markers of known size, one larger and one smaller than the particle of interest, since this will give a more accurate estimate of the particle size and check that the gradient is behaving as a truly isokinetic gradient. *Table 1* gives a list of markers which are likely to prove useful for people wishing to use markers of known size. Note that only small amounts of markers should be loaded to ensure that the gradient is not overloaded (see Section 9.3); for this reason the use of radioactively-labelled markers is recommended. The method used is described in the following protocol.

Protocol 1. Use of isokinetic sucrose gradients to determine sedimentation coefficients

1. Prepare identical linear gradients of 5–20% (w/w) sucrose using one of the methods described in Sections 5.2 or 5.3 of Chapter 3. Make at least two more gradients than you are likely to need.
2. To prepare the sample dissolve it in the gradient buffer without sucrose. If the chosen markers are radioactive and can be added to the sample containing the particles under study mix them with the sample before it is loaded onto the gradient. The total concentration of solutes in the sample should be less than 3 mg/ml.
3. When the gradients are ready *and* equilibrated to the correct temperature, carefully layer the samples on the top of the gradient using one of the techniques described in Section 7.3 of Chapter 3. The volume of the sample should be less than 4% of the gradient volume; take care not to overload the gradient (see Section 9.3).
4. Load the gradients into a temperature-equilibrated swing-out rotor and place the rotor in the ultracentrifuge, also at the correct temperature.
5. Carry out centrifugation for the appropriate time; an estimate of this can be obtained from the k^*-factor of the rotor (see Section 6 of Chapter 1).
6. After centrifugation fractionate the gradient into at least 25 fractions using the upward displacement method as described in Section 7.6 of Chapter 3.
7. Measure the refractive index of each fraction to check the linearity of the gradient and locate the positions of the sample and markers.
8. Calculate the sedimentation coefficient as described in the text.

Isokinetic sucrose gradients tend to be very stable even during long periods of centrifugation. This is because sucrose has a low molecular weight and its solutions have a high viscosity and so very little sedimentation of sucrose occurs and the shape of the gradient is essentially constant, undisturbed by diffusion processes.

4. Principles learned from analytical ultracentrifugation

It can be seen from the definition of the sedimentation coefficient (s) that the primary requirement is to determine the sedimentation velocity (v)

and the corresponding gravitational field (g); for a centrifugal field g depends on the speed of rotation (ω) given as radians/sec. They are defined as follows:

$$v = \mathrm{d}r/\mathrm{d}t \qquad [1]$$

and

$$g = \omega^2 r \qquad [2]$$

Thus

$$s = v/g \qquad [3]$$

From Equation 3 it can be seen that a sedimentation coefficient has time (sec) as its dimension. Sedimentation coefficients in absolute units are typically numbers of an inconvenient order of magnitude, and they are better referred to in terms of Svedberg units, which are equal to 10^{-13} sec. The exponent, -13, gives practical sedimentation coefficients a convenient numerical size, invariably larger than 1, and particles with sedimentation coefficients near 1 are also the smallest particles that can be sedimented in an ultracentrifuge.

By using the analytical ultracentrifuge the radial position of the sample boundary can be recorded directly at different times. The average sedimentation velocity during the largest possible time is easily obtained by re-writing Equations 1 and 2 to:

$$s\,\omega^2 \mathrm{d}t = \mathrm{d}r/r \qquad [4]$$

Integration of both sides gives:

$$\ln(r) = s\,\omega^2 t \qquad [5]$$

Thus plotting $\ln(r)$ against t gives a straight line having a slope of s/ω^2. Regression analysis is a simple and convenient way to determine the slope, but the plot in itself may be very informative if it shows deviations from linearity, for instance if relatively high concentrations are used. This is because during centrifugation a concentration gradient of the particles will be formed in the centrifuge cell. Sedimentation coefficients are concentration sensitive, and increasing concentrations may lead to association or dissociation processes. *Figure 1* gives an example of the type of data that can be obtained using an analytical ultracentrifuge.

The standard or normalized sedimentation coefficient ($s_{20,\mathrm{w}}$) is defined as that equivalent to centrifugation in water at 20 °C, when both the density and viscosity of the medium surrounding the particles are uniquely defined. A higher viscosity and/or higher density of the medium will act as a brake on the

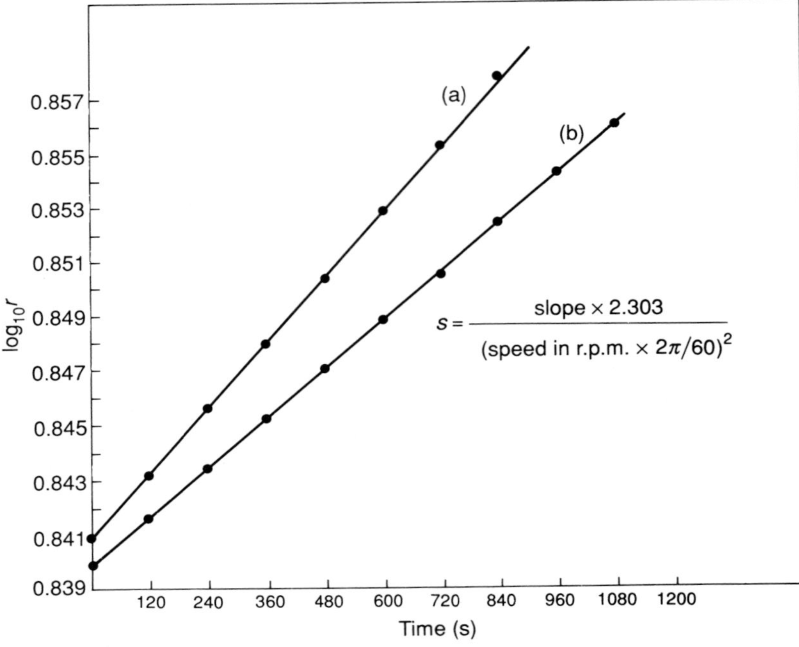

Figure 1. Determination of sedimentation coefficients of (a) T7 DNA and (b) SV40 DNA at 39 000 r.p.m. at 20 °C. (Reproduced from reference 10 with permission.)

sedimenting particles; the magnitude of this effect depends on the density, viscosity and temperature of the medium. Sedimentation coefficients ($s_{T,m}$), obtained experimentally in a medium of density $\rho_{T,m}$ and a viscosity of $\eta_{T,m}$ at a temperature T, can be corrected to the standard conditions using the equation:

$$s_{20,w} = s_{T,m} \frac{\rho_p - \rho_{20,w}}{\eta_{20,w}} \times \frac{\eta_{T,m}}{\rho_p - \rho_{T,m}} \qquad [6]$$

5. Sedimentation theory applied to sucrose gradient centrifugation

The techniques used for rate-zonal gradient centrifugation are clearly different from conventional analytical centrifugation because the sample particles sediment as discrete bands or zones and because the centrifugation medium is a supporting gradient for the sample band. A typical experiment is one where a sucrose gradient is pre-formed in a centrifuge tube or a zonal rotor. The sample is carefully layered onto the top of the gradient to form a thin distinct layer. During centrifugation the zones, each approximating to the shape of a Gaussian distribution curve, moves according to the individual properties of the particles. The theory behind analytical ultracentrifugation and sucrose gradient centrifugation are nevertheless closely related as described in Chapter 1 because the mass centre of a zone can be treated theoretically as the pivotal point of a sedimenting boundary.

The special feature of gradient centrifugation is that only the situations at the beginning and at the end of an experiment are known. This holds for variables which are affected by geometrical considerations and for the gravitational forces. All these aspects are coupled together in the following expression:

$$s_{20,\mathrm{w}} \int_0^t \omega^2 \,\mathrm{d}t = \frac{\rho_\mathrm{p} - \rho_{20,\mathrm{w}}}{\eta_{20,\mathrm{w}}} \int_i^r \frac{\eta_{\mathrm{T,m}}}{\rho_\mathrm{p} - \rho_{\mathrm{T,m}}} \times \frac{r}{\mathrm{d}r} \qquad [7]$$

If this equation is integrated between the limits from time zero to a fixed time t, and that during the interval the particles in question move from an initial position i (cm) to r (cm), then both integrals can be evaluated. Unfortunately these integrals cannot be solved analytically, but they can be determined numerically by an iterative procedure. In using this procedure it is assumed the gradient measured at the end of the run has not changed during centrifugation, and experimental data show that sucrose fulfils this requirement.

Thus because the calculations are stepwise in nature (naturally, each step corresponds to a fraction collected on completion of the run) a table is constructed in which the sedimentation coefficients are equated to fractions. An example is given in *Figure 2*. It will be known which fractions contain the particles of interest from the measurements of their concentration in each fraction. Hence the required sedimentation coefficients can be related to those fractions containing maximum concentrations of the sample.

The overall problem can be divided into a series of less complicated procedures which are the basis for the computer program described in this chapter. The procedures involve:

(a) determining the force–time integral

(b) determining radial movement of the particles

(c) analysing the effect of the gradient on sedimentation

(d) standardizing the estimated coefficients to water at 20 °C

5.1 The force–time integral

The force–time integral is displayed constantly during centrifugation on most modern ultracentrifuges as an option and some centrifuges can be programmed to stop a run at the correct time equivalent to a pre-set integral. If this facility is not available then the integral must be calculated. The complete force–time integral comprises of contributions from the period of acceleration (A), the actual run period (P), and the period of deceleration (D). The run period contribution is given by:

$$\omega_\mathrm{p}^2(60P) = (2\pi Q/60)^2\,(60P) = \frac{(\pi Q)^2 P}{15} \qquad [8]$$

Rotor speed:	47 000 r.p.m.	Run duration:	300 min
Acceleration period:	15 min	Deceleration period:	10 min
Sample volume:	10 ml	Overlay volume:	100 ml
Particle density:	1.4 g/ml	Rotor temperature:	8.0 deg

No	Vol	Suc	Avol	Rad	Dens	Viscos	*S*-val	Act
1	45	0						0.
2	45	0						0.
3	45	4.6	135	3.56	1.0183	1.5786	3.2	0.
4	45	7.1	180	3.95	1.0283	1.7072	7.4	0.
5	45	9.2	225	4.30	1.0369	1.8313	11.4	0.
6	45	11.1	270	4.62	1.0449	1.9585	14.8	0.
7	45	12.4	315	4.92	1.0504	2.0548	18.0	0.
8	45	13.6	360	5.20	1.0555	2.1513	21.1	0.025
9	45	14.6	405	5.46	1.0598	2.2377	23.9	0.180
10	45	15.5	450	5.71	1.0637	2.3207	26.7	0.270
11	45	16.3	495	5.94	1.0672	2.3989	29.3	0.070
12	45	17.0	540	6.17	1.0703	2.4709	31.8	0.008
13	45	18.1	585	6.39	1.0752	2.5915	34.3	0.
14	45	19.1	630	6.60	1.0797	2.7097	36.7	0.
15	45	20.2						
16	45	21.3						

Figure 2. Printout from calculation of sedimentation coefficients after zonal centrifugation with data corresponding to *Figure 3*.

where P is run time in minutes and Q is the average rotor speed (r.p.m.) during the run period.

Assuming that the accleration is constant, that is, there is a linear relationship between speed and time during this phase, the following formula for describing the angular velocity during acceleration, ω_a, can be deduced:

$$\omega_a = k_a t \tag{9}$$

where k_a is the accleration in radians/sec. As ω^2 is linear during the time $60A$ (A is the acceleration period in min), the acceleration must be:

$$k_a = \frac{\omega_p}{60A} = \frac{2\pi Q/60}{60A} = \frac{2\pi Q}{60^2 A} \tag{10}$$

Combining these equations gives:

$$\int_0^{60A} \omega_a^2 dt = \int_0^{60A} (k_a^2 t^2) dt = \left(\frac{2\pi Q}{60^2 A}\right)^2 \int_0^{60A} t^2\, dt = \tag{11}$$

$$\frac{(2\pi Q)^2}{60^4 A^4} \times \frac{(60A)^3}{3} = \frac{\pi Q^2 A}{45}$$

By analogy

$$\int_0^{60D} \omega^2 \mathrm{d}t = \frac{(\pi Q)^2 D}{45} \qquad [12]$$

Combination of all of these equations finally gives:

$$\int_0^{t} \omega^2 \mathrm{d}t = \frac{(\pi Q)^2 \times [P + (A + D)/3]}{15} \qquad [13]$$

It is relatively safe to assume that acceleration and deceleration are constant with respect to time. Most ultracentrifuges are electronically designed to achieve this and any deviations from linearity will be minimized because the run period typically is far longer than the periods of acceleration and deceleration. The indicated rotor speeds of direct drive ultracentrifuges are very accurate, usually to within 10 r.p.m. and readings can be taken directly. If using one of the older types of ultracentrifuge then it will be necessary to measure the rotor speed accurately from tachometer readings and a stop watch. The equations listed here are useful for checking the integrator against the displayed rotor speed by dividing the integral value by the run time (or *vice versa*) thus making these equations a way for checking some basic features of expensive ultracentrifuges.

5.2 Determining the radial movement of the particles

It is clear from Equation 7 that it is necessary to know the radial position of the sample zone mass centre at the beginning of the centrifugation and at the end. Also, one must be able to describe the gradient in terms of a concentration–radius relationship. The radius–volume relationship depends primarily on the type of rotor in use, whether it is a swing-out (swinging bucket) rotor, a vertical rotor, or a zonal rotor. The principal mathematical formulae for converting volumes to radii are given in Chapter 1 and can be used directly in combination with dimensions of the rotors and tubes given in the instruction manual. It is important to remember that the dimensions in the rotor manual should be corrected for the thickness of the wall of the tubes. The plastic of a typical centrifuge tube has a volume of 1–2 ml which means that serious inaccuracies can arise unless the correct dimensions are included in the calculations.

To illustrate the general approach an example involving zonal rotors will be given. The practical use of zonal rotors is discussed in Chapter 8 and only a brief description will be given here. A zonal rotor is a hollow cylinder with four or six internal vanes. It is loaded with a pre-formed gradient from the edge, while the sample and overlay are loaded from the centre. The situation in the rotor is illustrated in *Figure 3*, where a 640-ml zonal rotor is schematically divided into concentric rings each containing 45 ml (apart from the outermost). A thin sample zone (2 ml) is placed under 100 ml of overlay. Upon centrifugation the

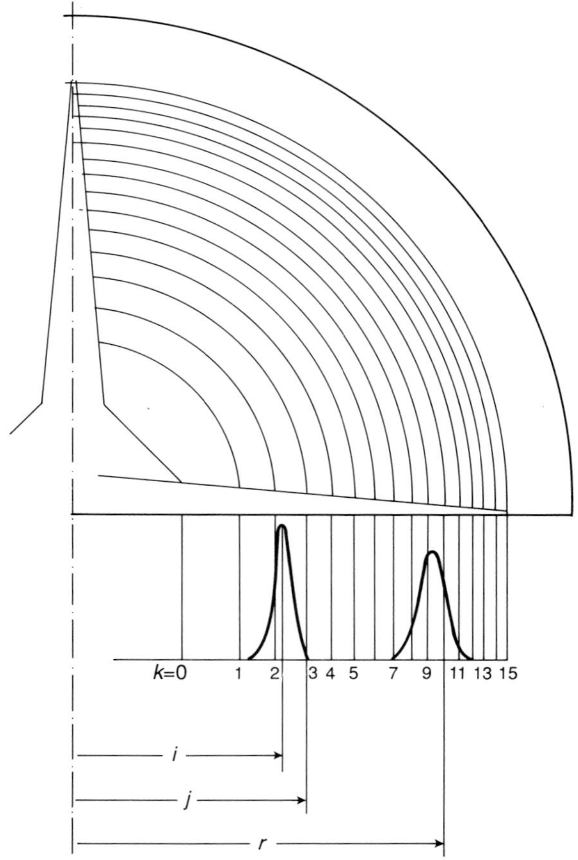

Figure 3. A sector of a BXIV zonal rotor with a diagrammatic representation of the initial and final distribution of the sample. The upper part is a horizontal cross section showing the radii equivalent to 45-ml increments in volume. The lower part shows the sample distribution against radius, *r*. The fraction numbers (*k*) are shown at the end of each fraction and *j* depicts the fraction containing the initial mass centre. (From reference 11 with permission.)

mass centre of the sample zone is located at a radius corresponding to 420 ml from the centre.

Thus, the starting position is given by the volume of the overlay plus half of the sample volume, which defines the position of the mass centre of the sample. The relationship between volume (V) and radius (R) for the B14 type zonal rotor (640 ml) is given by:

$$R = 0.5947 + (1.394 + 0.05495V)^{1/2} \quad [14]$$

or

$$V = 18.20R^2 - 21.34R - 18.93 \quad [15]$$

In case of the larger B15 zonal rotor with the original design (12) the corresponding relationships are:

$$R = 0.5126 + (1.480 + 0404V)^{1/2} \quad [16]$$

or

$$V = 24.75R^2 - 25.37R - 30.12 \quad [17]$$

It should be noted that the sedimentation geometry of zonal rotors is ideal since particles sediment without any wall effects (see Section 3.2.1 of Chapter 1). The angle in each sector is actually larger than 90°. Moreover, as zonal rotors contain fairly large volumes, several fractions can be obtained with perfect volumetric control, providing data for accurate calculation of sedimentation coefficients. General rules for establishing relationships between radius and volumes for tubes are given in Section 5.1.1 of Chapter 1. These include swing-out rotors and vertical rotors with flat or hemispherical top (see also refs 11 and 13). Fixed-angle rotors are not suitable for rate-zonal separations because of the wall effects experienced by the particles during centrifugation.

5.3 Analysis of the gradient

Upon completion of a gradient centrifugation experiment the contents of the tube or zonal rotor must be divided into a number of fractions. The volume of each fraction must be measured and a relationship between accumulated volume and the radius of the last drop of each fraction must be established. In addition, the concentrations of gradient material and of the particles being studied in each fraction must be measured. There are a number of ways in which this can be done. For zonal rotors where the volume of fractions is large it is simplest to collect the fractions in weighed tubes and weigh the tubes again after the fraction has been collected. In the case of smaller fractions, the density of each can be measured either by refractive index or, for samples larger than 0.5 ml, by using a densitometer as described in Section 7.7 of Chapter 3. Since the density is calculated as a part of the following computations inputs such as weight or refractive index can be included in the corresponding computer program.

Having calculated these parameters it is now possible to evaluate the integral of Equation 8 by stepwise approximation. Each fraction represents the natural radial increment in this procedure making it necessary to calculate the density and viscosity of each fraction from the concentration (or measured refractive index).

Barber (5) has worked out a very accurate set of polynomials relating density and viscosity of sucrose solutions to concentration and temperatures. The following detailed description of these polynomials follows the original description given by Barber quite closely. The original data cover a very wide range of concentrations from 0–75% (w/w) sucrose for temperatures ranging from 0–30 °C. Two polynomials were derived from the experimental data; one matching the original data with a maximum deviation of 0.07%, and extrapolated values with a maximum deviation from tabulated values of 0.4% (it also seems to give reliable values in the range 30–60 °C). This first polynomial, being entirely empirical, has the following form:

$$\rho = (B_1 + B_2T + B_3T^2) + (B_4 + B_5T + B_6T^2)Y + (B_7 + B_8T + B_9T^2)Y^2 \qquad [18]$$

where ρ is the density (g/ml), the values for B are given in *Table 2*, T is the temperature (°C), and Y is the weight fraction of sucrose. This is the recommended polynomial within the stated temperature range.

The second formula is based on more classical chemical concepts arguing that the density of a solution can be evaluated as:

$$\rho_{\mathrm{T,m}} = \frac{yM_1 + (1-y)M_2}{yC(T) + (1-y)A(T)} \qquad [19]$$

where $C(T)$ and $A(T)$ are temperature dependent constants (expressing the molar volumes of liquid sucrose and liquid water at the temperature in question, respectively) and y is the mole fraction of solute. M_1 and M_2 are the molecular weights of sucrose (342.30) and water (18.032), respectively. $C(T)$ can be evaluated as:

$$C(T) = C_1 + C_2T + C_3T^2 \qquad [20]$$

and

$$A(T) = A_1 + A_2T + A_3T^2 \qquad [21]$$

using the constants given in *Table 3*.

The mole fraction of sucrose (y = weight fraction of sucrose, g/g of solution) is given by:

$$y = \frac{y/M_1}{y/M_1 + (1-y)/M_2} \qquad [22]$$

Table 2. Constants for empirical density calculation

Constant	**Value**[a]
B_1	1.0003698
B_2	3.9680504×10^{-5}
B_3	$-5.8513271 \times 10^{-6}$
B_4	0.38982371
B_5	$-1.0578919 \times 10^{-3}$
B_6	1.2392833×10^{-5}
B_7	0.17097594
B_8	4.7530081×10^{-4}
B_9	$-8.9239737 \times 10^{-6}$

[a] Values are given to 8 figures for machine calculations; use of the first 5 figures would be sufficient for hand calculations.

Table 3. Constants for extrapolatable density function

Constant	**Value**[a]
A_1	18.027525
A_2	4.8318329×10^{-4}
A_3	7.7830857×10^{-5}
M_1	342.30
M_2	18.032
C_1	212.57059
C_2	0.13371672
C_3	$-2.9276449 \times 10^{-4}$

[a] Values are given to 8 figures for machine calculations; use of 5 figures would be sufficient for hand calculations.

In spite of the chemical appeal of the second formula the first is strongly recommended below 30 °C because of its numerical simplicity and accuracy.

To calculate the viscosity of sucrose solutions (expressed in centipoise) Barber devised the following approach, again using a numerical method based on thermodynamic considerations. The equations cover the enormous range from 0–75% (w/w) sucrose and up to 80 °C. Calculated individual values only rarely deviate more than 0.3% from standard tabulated values. The basic equation has the form:

$$\log \eta_{\mathrm{T,m}} = \frac{(A+B)}{(T+C)} \qquad [23]$$

where A and B are polynomials of the form:

$$A = D_0 + D_1 y + D_2 y^2 + \ldots D_n y^n \qquad [24]$$

These coefficients are given in *Tables 4* and *5*. The temperature dependent coefficient C (°C) is given by:

$$C = G_1 - G_2 \left[1 + \left(\frac{y}{G_3} \right)^2 \right]^{1/2} \qquad [25]$$

where G_1, G_2, and G_3 have the following values:

$$G_1 = 146.06635;\ G_2 = 25.251728;\ G_3 = 0.070674842$$

The equations given in detail here appear at first glance to be slightly cumbersome for repeated use, but they can be easily converted into a computer program as shown in Appendix D in this chapter.

Since it is possible to calculate the density and viscosity of any sucrose solution at all sensible temperatures, the term:

$$\frac{\eta_{T,m}}{\rho_p - \rho_{T,m}}$$

can be evaluated providing that the density of the particle itself in sucrose solutions is known. A typical value is 1.3 g/ml for most proteins and DNA; RNA is slightly denser at about 1.7 g/ml. Lipoproteins and membrane fragments have lower densities depending on their lipid content. More precise values of density are not required since it does not have a large effect on the calculated values of sedimentation coefficient. This can be tested if you have access to a computer program for the calculation of sedimentation coefficients.

The final step in evaluation of the expression inside the integral sign of Equation 7 is dr/r. The term dr is by definition the radial increment determined by the volume of the fraction in question, and may safely be used as such. The radial position of the particle, r, determines the centrifugal force acting on the particle, and it may be taken as the average radius of the fraction as long as the fractions have reasonably small volumes.

5.4 Standardization

Equivalent or standard sedimentation coefficients are sedimentation coefficients recalculated to a theoretical sedimentation process in water at 20 °C. The viscosity of water at 20 °C is 0.9982 centipoise and the density is 1.005 g/ml (or equivalent). Note that the standardization term is inverted compared with the previous expression for individual fractions and so it is less important

Table 4. Coefficients[a] for calculation of the limiting viscosity or *A*-constant as a function of mole fraction sucrose

Coefficients[a]	Range of equation, weight percent	
	0–48	48–75
D_0	-1.5018327	-1.0803314
D_1	9.4112153	-2.0003484×10^{1}
D_2	-1.1435741×10^{3}	4.6066898×10^{2}
D_3	1.0504137×10^{5}	-5.9517023×10^{3}
D_4	-4.6927102×10^{6}	3.5627216×10^{4}
D_5	1.0323349×10^{8}	-7.8542145×10^{4}
D_6	-1.1028981×10^{9}	
D_7	4.5921911×10^{9}	

[a]Numbers as given for machine calculation.
[b]Coefficient subscript indicates the exponent of the composition by which the coefficient is to be multiplied.

Table 5. Coefficients[a] for calculation for the activation energy or *B*-constant as a function of mole fraction sucrose

Coefficients[b]	Range of equation, weight percent	
	0–48	48–75
D_0	2.1169907×10^{2}	1.3975568×10^{2}
D_1	1.6077073×10^{3}	6.6747329×10^{3}
D_2	1.6911611×10^{5}	-7.8716105×10^{4}
D_3	-1.4184371×10^{7}	9.0967578×10^{5}
D_4	6.0654775×10^{8}	-5.5380830×10^{6}
D_5	-1.2985834×10^{10}	1.2451219×10^{7}
D_6	1.3532907×10^{11}	
D_7	-5.4970416×10^{11}	

[a]Numbers as given for machine calculations.
[b]Coefficient subscript indicates the exponent of the composition by which the coefficient is to be multiplied.

to know the particle densities precisely. The temperature during centrifugation is a particularly important parameter. The density of sucrose-containing solutions is not crucially dependent on temperature, but the viscosity of sucrose solutions is to a significant extent dependent on the temperature of the solution during centrifugation. An accuracy of the temperature control to less than 0.5 °C is recommended.

6. Steps in writing a computer program to calculate sedimentation coefficients

The first computer programs for calculating sedimentation coefficients from zonal centrifugation data were written by Barbara S. Bishop (13) along with the development of the zonal rotors at the Oak Ridge National Laboratory (14). Although written in a most labyrinthine computer language, FORTRAN II, it is an intricately stringent program that continues to be used unchanged in corresponding programs today. Appendices A, B, and C of this chapter give examples written in Basic. The general design of these programs is given in *Protocol 2*.

Protocol 2. Steps for computation of sedimentation coefficients

1. Input force–time integral or calculate it from rotor speed, run duration, periods of acceleration and deceleration (Equation 17).
2. Input the known or expected particle density and calculate the standardization term $(\rho_p - \rho_{20,w})/\eta_{20,w}$.
3. Input temperatures and volumes required to calculate the radius of the mass centre at the beginning of the centrifugation. For zonal rotors this is given by the radial position (Equation 14 or 16) corresponding to the overlay volume plus half of the sample volume. Set accumulated volume and integration term to zero.
4. Input a description of the experiment. The minimum is clearly an experiment identification number and the date, but a more detailed description is usually functional and may conveniently serve as a laboratory diary. From a programming point of view the information can be printed out immediately as it is not used in the calculations. At this point the required information on the physical aspects of the experiment are obtained. The following step (Step 5) is repeated for all of the gradient fractions until the total volume of the tube or rotor is reached.
5. Input the volume (or weight) and sucrose concentration for the current fraction and calculate the corresponding density and viscosity. If fraction weight is used as input, then the volume is obtained using the calculated density.
6. Calculate the radius and radial increment for the fraction in question.
7. Calculate the complete incremental value and add it to the previous accumulated value to give the actual value of the integral.
8. Convert the intermediate values to Svedberg units.
9. Print for each fraction number, the volume (especially if calculated from fraction weight and density), radius (last drop), sucrose concentration, density, viscosity, and the sedimentation coefficient.

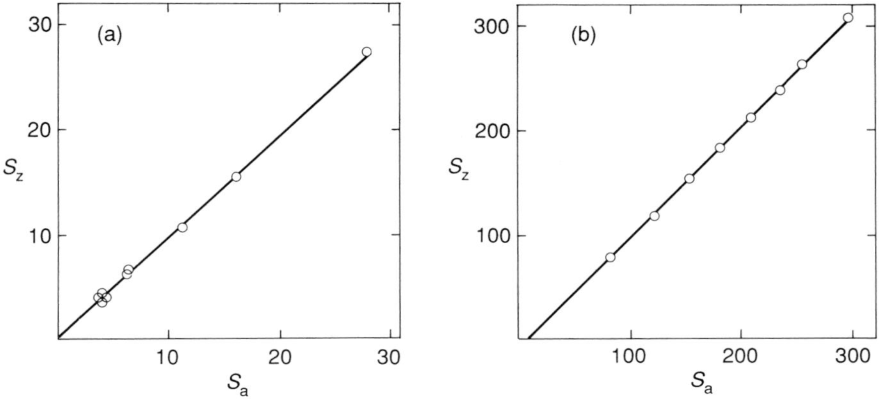

Figure 4. Relationship between sedimentation coefficients obtained by analytical ultracentrifugation (S_a) and by rate-zonal centrifugation in a zonal rotor (S_z). The values on the left (a) show the relationship for proteins and those on the right (b) are for polysomes. (Values from references 19 and 20, reproduced from reference 12 with permission.)

It is important in this approach to work in the direction of the sedimentation process. Using this method data does not necessarily have to be stored as they are used only once per fraction. If data are collected in the opposite way they must all be read in and stored in order to be used as described previously. However, storage of the original data is strongly recommended for several reasons. It is useful to be able to compare experiments from different runs without having to input all of the data again. The original data can be extended to cover measured activities as well as providing data for computer-based graphics.

7. Manual methods for calculating sedimentation coefficients

Before the advent of personal computers these calculations were very cumbersome, but some alternative approaches based on semi-graphical principles or the use of tabulated values (12, 15–18) have been published. Although these mainly reflect the difficulties at that time of doing substantial numerical calculations, these papers are still useful because they contain a host of hints for aiding such calculations.

8. Comparison of data obtained from analytical ultracentrifugation and sucrose gradients

The concepts of sedimentation coefficients and other hydrodynamic properties of biological particles were historically developed using the analytical ultracentrifuge as the one and only available technique. Norman (19) and Steensgaard (20) have compared results obtained by the analytical and gradient ultracentrifugation. Examples are shown in *Figure 4* and they demonstrate fairly good agreement between the two sets of results using proteins and polysomes as test particles covering a range of 4–300 Svedberg units. The accuracy of values obtained by analytical ultracentrifugation is considered to be better as they can be accurate to the second decimal place. Sedimentation coefficients calculated from sucrose gradient centrifugation should be given without decimals unless the value is the mean from a series of experiments, when it is justifiable to express values to one decimal place. In practice, sedimentation coefficients calculated from sucrose gradient centrifugation are accurate enough for most purposes.

In the estimation of sedimentation coefficients some of the sources of inaccuracy are common to both analytical ultracentrifugation and sucrose gradient centrifugation. The particle density and temperature control are examples of uncertainty and their importance in affecting calculated values should be checked by running the program several times with data reflecting limiting values of these two variables. Gradient centrifugation, volumetric control, estimation of sucrose concentrations, and overloading of the gradient all play important roles in determining the accuracy of the experimental data. The main factors which can cause inaccuracies in the measurements of sedimentation coefficients are described in the next section.

9. Sources of inaccuracy in the determination of sedimentation coefficients

9.1 Characteristics of the sample

Variations in the evaluated sedimentation coefficients may reflect changing characteristics of the sample which are caused by centrifugation. An example of this is the observation that the apparent sedimentation coefficient of large DNA molecules varies depending on the centrifugal force applied (21). It appears that in this case the centrifugal force changes the conformation of the DNA, possibly leading to some unwinding of the DNA helix (see Section 5.3.2 of Chapter 4). A further factor to consider is the hydrostatic pressure created in gradients by the centrifugal force (see Section 3.2.2 of Chapter 1) which can cause the dissociation of some macromolecular complexes such as ribosomes and multimeric proteins (22).

9.2 Instrument inaccuracies

As has been noted previously in this chapter, it is extremely important to control the centrifugal conditions as precisely as possible. As a general rule, the latest models of centrifuge are the best because the trend is to produce instruments which can control centrifugal conditions in a very precise way; for example, the latest direct drive machines, capable of controlling the speed to within

10 r.p.m., are to be preferred over the much less accurate speed control of the earlier types of centrifuge. Exact knowledge of the temperature of the gradient during the run is also extremely important. Use a centrifuge which can control temperature as accurately as possible; some can control temperature to within less than 0.5 °C while others only control it to within 2 °C. Temperature is very important because small changes in temperature have a large effect on the viscosity of sucrose solutions. It is good practice to run a blank gradient, stop the centrifuge half way through the run period, and measure the temperature of the solution in the tube manually using a sensitive thermocouple.

9.3 Operator error

Assuming that the operator has mastered the skills of preparing, loading, and fractionating gradients using the methodologies which are described in Chapter 3, the other main source of error comes from overloading the gradient. There is an almost irresistible urge to load too much sample on to a gradient; more sample, bigger peaks, easier detection, and recovery of samples for an experiment. However, succumbing to such temptations will almost always result in inaccurate quantitation of sedimentation coefficients.

Some theoretical aspects of the capacity of a gradient have been discussed in Section 5.2 of Chapter 1 and in Section 2.1.3 of Chapter 3. In very pragmatic terms a gradient (comprising the contributions from the gradient material and the sample material together) is only a gradient as long as the density increases in the direction of the gravitational field. If the density of the sample zone exceeds the density of the top of the gradient it is bound to create turbulence at the interface, causing the sample zone to tip over and to move deeper into the gradient (see *Figure 4* of Chapter 3). On completion of the run the results suggest the particles have moved faster than expected and their calculated sedimentation coefficients are accordingly larger. An example is given in *Figure 5* showing how thin, well-separated particle zones become very broad and outwardly dislocated as a result of simple overloading. A measured relationship between the calculated sedimentation coefficients of four different particles and an increasing sample load is shown in *Figure 6*. The effects of overloading are quite dramatic, and it can clearly be seen that these effects work in the opposite direction to the concentration dependence seen in analytical ultracentrifugation. Without calculations and control experiments, overloaded zones simply appear broad and distorted in shape.

There is no way to estimate the maximum possible sample load of a gradient. A useful guide is that the concentration (w/w) of material in the sample should not exceed 10% of the concentration (w/w) at the top of the gradient (e.g. <5 mg/ml of sample in water on a 5–20% gradient). A more accurate method is to include a radioactive marker with a known sedimentation coefficient (see *Table 1*) in the sample and then to calculate sedimentation coefficients as described previously. If the standard marker seems to give too high a

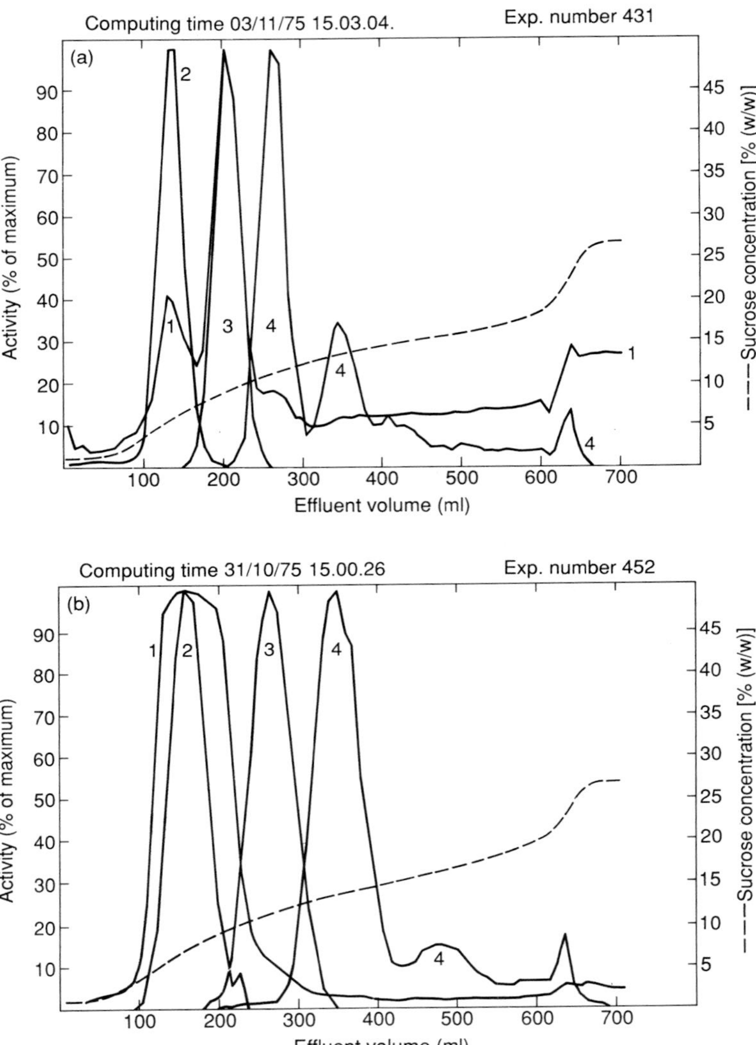

Figure 5. Computer-drawn effluent patterns of a multi-component sample centrifuged in a zonal rotor: (a) correct loading, and (b) overloading of the sample; the latter condition caused by the addition of human serum albumin. The experiments were carried out in a BXIV rotor containing an isokinetic 3–23% (w/w) sucrose gradient with a 2 ml sample and 100 ml overlay. Centrifugation was at 8 °C at 47 000 r.p.m. until a force–time integral of 4.5×10^{11} rad^2/sec was achieved (~5 h). The sample consisted of 0.3 mg of ^{131}I-labelled human serum albumin (alkylated and monomerized), 0.5 mg of beef liver catalase, 0.1 mg of β-galactosidase. In (b) the sample also contained 100 mg of unlabelled human serum albumin. The curves (1) shows optical density at 220 nm, (2) radioactivity, (3) catalase, and (4) β-galactosidase. The dotted line shows the sucrose gradient. (From reference 12 with permission.)

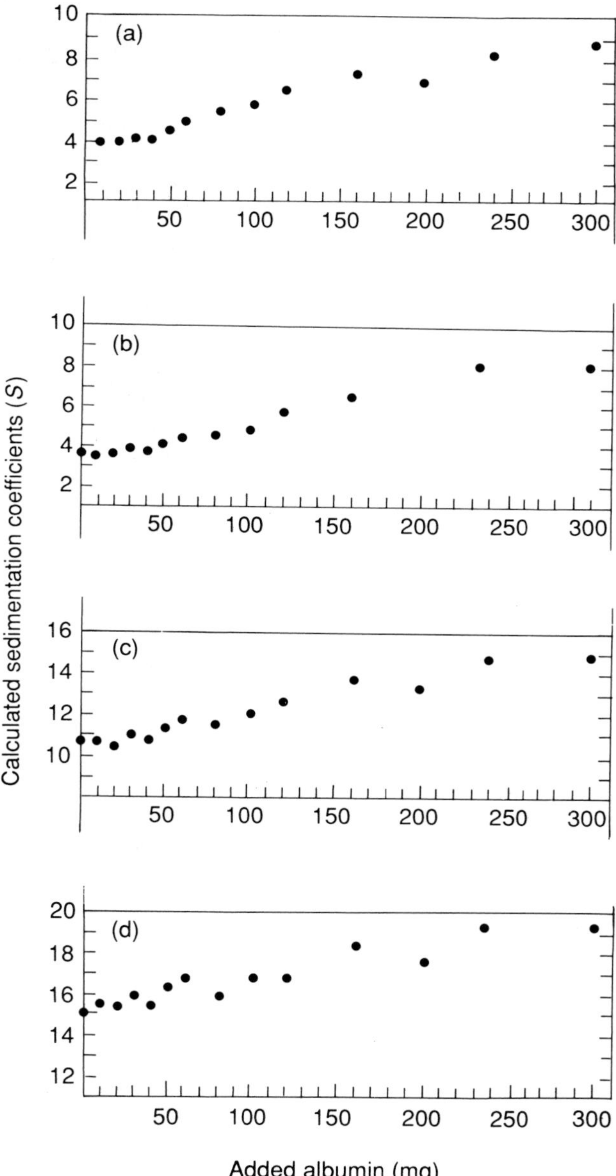

Figure 6. The relationship between calculated sedimentation coefficients and the amount of sample loaded onto gradients. (a) shows the value for human serum albumin based on UV absorption, (b) shows the sedimentation of labelled human serum albumin, and (c) and (d) show the sedimentation of catalase and β-galactosidase, respectively, with increasing amounts of human serum albumin. (Derived from reference 12.)

Table 6. Calculation of the partial specific volume of a protein (ribonuclease) from its amino acid composition (27)

Amino acid residue	Number of residues/ molecule[a]	W_i (% by weight[a] of residue)	V_i (specific volume of residue)	V_iW_i (% by volume of residue)
Aspartic acid	7	5.69	0.59	3.36
Asparagine[b]	9[b]	7.31[b]	0.60	4.39
Glutamic acid	4	3.63	0.66	2.40
Glutamine[b]	8[b]	7.27[b]	0.67	4.87
Glycine	3	1.25	0.64	0.80
Alanine	12	6.12	0.74	4.53
Valine	9	6.34	0.86	5.45
Leucine	2	1.74	0.90	1.57
Isoleucine	3	2.30	0.90	2.07
Serine	15	9.44	0.63	5.95
Threonine	10	7.56	0.70	5.29
Half cystine	8	5.95	0.63	3.75
Methionine	4	3.52	0.75	2.64
Proline	5	3.32	0.76	2.52
Phenylalanine	3	3.13	0.77	2.41
Tyrosine	6	6.84	0.71	4.86
Histidine	4	3.73	0.67	2.50
Lysine	10	9.22	0.82	7.56
Arginine	4	4.43	0.70	3.10
Tryptophan	0	0.00	0.74	0.00
Total =	126	$\Sigma W_i = 98.79$		$\Sigma W_iV_i = 70.02$

$$\text{Partial specific volume } (\bar{v}) = \frac{\Sigma W_iV_i}{\Sigma W_i} = \frac{70.02}{98.79} = 0.709 \text{ ml/g}$$

[a] The number of each amino acid residue per molecule of ribonuclease, which gives a molecular weight of 13 895 for this enzyme, was obtained from reference 28.
[b] To take into account the seventeen amide groups per molecule of RNase, it was assumed that these were approximately equally distributed between the glutamic and aspartic residues.

sedimentation coefficient, then it is likely that the gradient has been overloaded.

10. Relationships between sedimentation coefficients and other molecular parameters

There is a natural interest in the molecular masses of particles and over the years a number of empirical equations have been derived for converting sedimentation

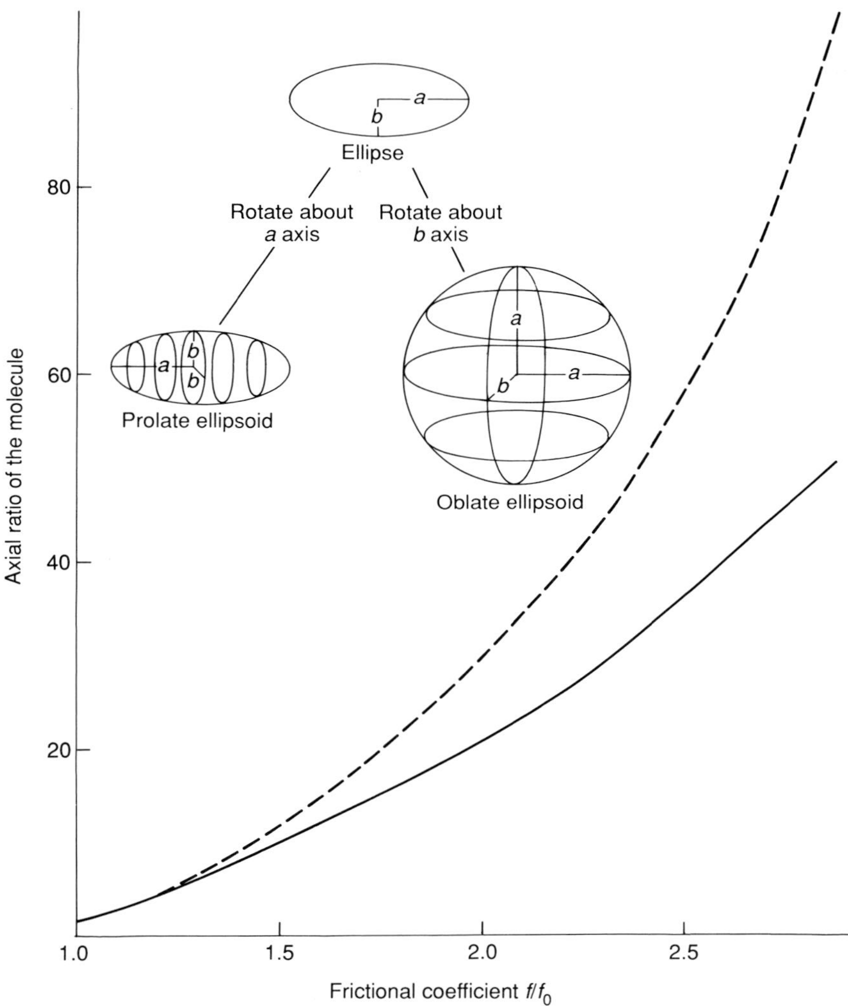

Figure 7. Relationship between the frictional coefficient of macromolecules and their axial ratio for oblate (---) and prolate (—) molecules. (Data derived from reference 26.)

coefficients to molecular masses. A list of the various equations that have been used has been published (23).

However, the sedimentation coefficient is a function of particle size, density, and shape. For macromolecules the size is given by its molecular mass (M), the density by the partial specific volume ($\bar{v}$), and the shape by the frictional coefficient (f/f_0); the last parameter reflects how much the molecule deviates

Table 7. Molecular parameters of macromolecules

Protein	**Molecular mass**	$\bar{v}$	f/f_0
Lysozyme	14 100	0.688	1.32
Ovalbumin	45 000	0.748	1.17
Haemoglobin	68 000	0.749	1.14
Catalase	250 000	0.730	1.25
Fibrinogen	330 000	0.710	2.34
Collagen	345 000	0.695	6.80
Urease	480 000	0.73	1.20
Myosin	493 000	0.728	3.53

Data from reference 26.

from the shape of a perfect sphere. These four parameters are physically correctly inter-related as shown below:

$$s = M^{2/3}\left(\frac{1-\bar{v}\rho}{\sqrt[3]{\bar{v}}}\right) / [(N\pi 6\eta\sqrt[3]{0.75/(N\pi)}\,(f/f_0)] \qquad [26]$$

Hence, if any three of these are known then the fourth can be calculated (24); a program for this is given in Appendix E of this chapter. There are now very accurate ways of determining the molecular mass of macromolecules by sequencing methods, gel electrophoresis, and HPLC. In the case of the partial specific volume of a macromolecule, this can be measured by pycnometry (25) or by sedimentation in solvents of different density (see Section 5.8 of Chapter 4); alternatively, it can be calculated from the composition as the sum of its individual components. In the case of proteins the partial specific volume can be calculated using the following steps:

(a) Determine the amino-acid composition of the protein in terms of weight per cent (if the composition is given as molar per cent then this must be converted using the molecular mass of the amino acids less the molecular mass of water, which is lost on peptide bond formation).

(b) Multiply the partial specific volume of each amino acid (*Table 6*) by its proportional (not percentage) contribution to the protein composition.

(c) Sum the individual contributions of each amino acid to obtain the calculated specific volume.

Calculation of partial specific volumes of proteins which contain other components (e.g. carbohydrates) must take into account the contribution of the other components and so it is necessary to know both the proportion of material relative to the total amino acids and its partial specific volume; sucrose is about 0.66 ml/g.

If the partial specific volume and molecular mass of any macromolecule are known and the sedimentation coefficient is measured, then it is possible to determine the hydrodynamic shape of the macromolecule. *Figure 7* shows the relationship between the value of the frictional coefficient (f/f_0) and the axial ratio of the molecule. Hence the determination of the sedimentation coefficient of particles should not be seen to be just another way of finding molecular weight but rather it allows the determination of other hydrodynamic properties of particles. *Table 7* gives an indication of the partial specific volumes and frictional coefficients for different types of protein.

References

1. Svedberg, T. and Pederson, K. O. (1940). *The ultracentrifuge*. Oxford University Press, Oxford.
2. Steensgaard, J. and Moller, N. P. H. (1978). *Subcellular Biochemistry* **6**, 117.
3. Sartory, W. K., Halsall, H. B., and Breillat, J. P. (1976). *Biophys. Chem.* **5**, 107.
4. Steensgaard, J. and Rickwood, D. (1985). in *Microcomputers in biology: a practical approach* (ed. C. Ireland and S. P. Long). IRL Press Ltd, Oxford.
5. Barber, E. J. (1966). *Nat. Cancer Inst. Monogr.* No **21**, 219.
6. Patterson, B. W. (1984). *Anal. Biochem.* **136**, 347.
7. Martin, R. G. and Ames, B. N. (1961). *J. Biol. Chem.* **236**, 1372.
8. Noll, H. (1967). *Nature* **215**, 360.
9. Steensgaard, J. (1970). *Eur. J. Biochem.* **16**, 66.
10. Eason, R. and Campbell, A. M. (1978). In *Centrifugal separations in molecular and cell biology* (ed. G. D. Birnie and D. Rickwood), p. 251. Butterworths, London.
11. Young, B. D. and Rickwood, D. (1981). *J. Biochem. Biophys. Methods* **5**, 95.
12. Steensgaard, J., Møller, N. P. H., and Funding, L. (1978). In *Centrifugal separations in molecular and cell biology* (ed. G. D. Birnie and D. Rickwood). Butterworths, London.
13. Bishop, B. S. (1966). *Nat. Cancer Inst. Monogr.* No **21**, 175.
14. Anderson, N. G. (1966). *Nat. Cancer Inst. Monogr.* No **21**, 175.
15. Halsall, H. B. and Schumaker, V. N. (1969). *Anal. Biochem.* **30**, 368.
16. McEwen, C. R. (1967). *Anal. Biochem.* **20**, 368.
17. Young, B. D. (1972). *MSE Application Information, A6/6/72*. MSE Instruments Ltd, Crawley, UK.
18. Fritsch, A. (1973). *Anal. Biochem.* **55**, 57.
19. Norman, M. R. (1971). In *Separations with zonal rotors* (ed. E. Reid). Longmans, London.
20. Steensgaard, J. (1974). *Separation og analyse af makromolekyler ved s-zoneultra-centrifugering*. Aarhus.
21. Pettijohn, D., Hecht, R. M., Stimpson, D., and Van Scoyk, S. (1978). *J. Mol. Biol.* **119**, 353.
22. Marcum, J. M. and Borisy, G. G. (1978). *J. Biol. Chem.* **253**, 2852.
23. Young, B. D. (1984). In *Centrifugation: a practical approach*, 2nd edn (ed. D. Rickwood). IRL Press Ltd, Oxford.

24. Jacobsen, O. and Steensgaard, J. (1979). *Immunochem.* **16**, 571.
25. Eason, R. (1984). In *Centrifugation: a practical approach*, 2nd edn (ed. D. Rickwood). IRL Press Ltd, Oxford.
26. Tanford, C. (1961). *Physical chemistry of macromolecules.* John Wiley, New York.
27. Cohn, E. J. and Edsall, J. T. (1943). In *Proteins, amino acids and peptides*, p. 372. Reinold, New York.
28. Hirs, C. H. W., Stein, W. H., and Moore, S. (1954). *J. Biol. Chem.* **211**, 941.

Appendix A. A program for calculating sedimentation coefficients in swing-out rotors

```
10      SCREEN 2: KEY OFF: CLS
20      REM     THIS PROGRAM CALCULATES SEDIMENTATION COEFFICIENTS
30      REM     FOR SUCROSE GRADIENTS IN SWING-OUT ROTORS.
40      REM
100     DIM P(80), M(80)
110     DIM L(80), A$(10)
130     CLS
135     GOSUB 140: GOTO 160
140     LOCATE 3, 23: PRINT "S-VALUES IN SWING-OUT ROTORS"
150     RETURN
160     Z1 = 0
170     DATA 0.0528,0.4
175     DATA 0.0882,0.346
176     MEN = 0
177     GOSUB 10000
180     LOCATE 23, 3: PRINT "Is sample DNA (Y/N)                              "
182     LOCATE 23, 3: INPUT "Is sample DNA (Y/N)"; C$
183     IF C$ <> "n" AND C$ <> "N" AND C$ <> "y" AND C$ <> "Y" THEN
            IF C$ <> "q" AND C$ <> "Q" THEN BEEP: GOTO 180
        END IF
185     IF C$ = "q" OR C$ = "Q" THEN CLS : SYSTEM
187     CLS : LOCATE 3, 23: PRINT "S-VALUES IN SWING-OUT ROTORS"
190     IF INSTR("Nn", C$) <> 0 THEN
          LOCATE 8, 3
          PRINT "Sample DNA", , "N"
          GOTO 250
        END IF
191     MEN = 1
192     GOSUB 10000
195     LOCATE 7, 3: PRINT "Sample DNA", , "Y"
200     LOCATE 23, 3: PRINT "Alkaline gradient (Y/N)                              "
201     LOCATE 23, 3: INPUT "Alkaline gradient (Y/N)"; C$
202     IF C$ = "q" OR C$ = "Q" OR C$ = "l" OR C$ = "L" THEN GOTO 210
203     IF C$ = "n" OR C$ = "N" OR C$ = "y" OR C$ = "Y" THEN GOTO 210
204     BEEP: GOTO 200
210     RESTORE 170
215     IF C$ = "l" OR C$ = "L" THEN RESTORE: GOTO 130
217     IF C$ = "Q" OR C$ = "q" THEN RUN
220     IF INSTR("Yy", C$) <> 0 THEN
           LOCATE 8, 3
           PRINT "Alkaline gradient", "Y"
           GOTO 240
        END IF
225     LOCATE 8, 3: PRINT "Alkaline gradient", "N"
230     RESTORE 175
240     READ Z1, Z2
250     NO = 10
```

```
260     REM     THE FOLLOWING DATA STATEMENTS CONTAIN THE
270     REM     BASIC INFORMATION FOR EACH ROTOR, i.e. DISTANCE
280     REM     FROM CENTRE OF ROTATION TO TUBE BOTTOM AND
290     REM     INTERNAL TUBE RADIUS (CM).
300     REM
310     DATA 12.0,0.5: A$(1) = "BECKMAN SW 60Ti"
320     DATA 11.3,0.6: A$(2) = "CENTRIKON TST 55.5"
330     DATA 16.0,0.65: A$(3) = "CENTRIKON TST 41.14"
340     DATA 15.9,0.65: A$(4) = "HITACHI RPS40T"
350     DATA 16.0,1.2: A$(5) = "BECKMAN SW 28"
360     DATA 8.8,0.625: A$(6) = "BECKMAN SW 65Ti"
370     DATA 12.1,0.5: A$(7) = "CENTRIKON TST 60.4"
380     DATA 15.2,0.65: A$(8) = "BECKMAN SW 41Ti"
390     DATA 15.8,0.65: A$(9) = "BECKMAN SW 40Ti"
400     DATA 15.3,1.6: A$(10) = "HITACHI RPS25-2"
410     FOR N = 1 TO NO: LOCATE 8 + N, 2: PRINT " "; N, A$(N): NEXT
405     REM
412     MEN = 1
415     GOSUB 10000
420     LOCATE 23, 3
430     INPUT "Which rotor do you require"; N1$
435     IF N1$ = "L" OR N1$ = "l" THEN 20032 ELSE N1 = VAL(N1$)
437     IF N1$ = "q" OR N1$ = "Q" THEN RUN
438     N1 = INT(N1)
440     IF N1 > NO OR N1 < 1 THEN 420
442     LOCATE 9, 1: FOR N = 1 TO NO: PRINT "                                        ": NEXT
443     GOSUB 10000
445     LOCATE 9, 3: PRINT "Rotor:", , A$(N1) + " "
450     ON N1 GOTO 2000, 2010, 2020, 2030, 2040, 2050, 2060, 2070, 2080, 2090
460     READ R1, R2
470     LOCATE 23, 3: PRINT "Total volume (ml) =                              "
472     LOCATE 23, 3: INPUT "Total volume (ml) = ", U1$
475     IF U1$ = "L" OR U1$ = "l" THEN 405 ELSE U1 = VAL(U1$)
477     IF U1$ = "q" OR U1$ = "Q" THEN RUN
478     LOCATE 10, 3
        PRINT "Total volume (ml)", U1$; "                  "
480     LOCATE 23, 3: PRINT "Sample volume (ml) =                             "
482     LOCATE 23, 3: INPUT "Sample volume (ml) = ", U2$
485     IF U2$ = "L" OR U2$ = "l" THEN 470 ELSE U2 = VAL(U2$)
487     IF U2$ = "q" OR U2$ = "Q" THEN RUN
488     LOCATE 11, 3: PRINT "Sample volume (ml)", U2$; "                      "
490     LOCATE 23, 3: PRINT "Fraction volume (ml) =                           "
492     LOCATE 23, 3: INPUT "Fraction volume (ml) = ", U3$
495     IF U3$ = "L" OR U3$ = "l" THEN 480 ELSE U3 = VAL(U3$)
497     IF U3$ = "q" OR U3$ = "Q" THEN RUN
498     LOCATE 12, 3: PRINT "Fraction volume (ml)", U3$; "                    "
500     N3 = U1 / U3
510     PRINT
520     LOCATE 23, 3: PRINT "Acceleration time (minutes) =              "
522     LOCATE 23, 3: INPUT "Acceleration time (minutes) = ", T1$
525     IF T1$ = "L" OR T1$ = "l" THEN 490 ELSE T1 = VAL(T1$)
527     IF T1$ = "q" OR T1$ = "Q" THEN RUN
528     LOCATE 13, 3: PRINT "Acceleration time (min.)", T1$; "              "
530     LOCATE 23, 3: PRINT "Run time (hours) =              "
532     LOCATE 23, 3: INPUT "Run time (hours) = ", T4$
535     IF T4$ = "L" OR T4$ = "l" THEN 520 ELSE T4 = VAL(T4$)
537     IF T4$ = "q" OR T4$ = "Q" THEN RUN
538     LOCATE 14, 3: PRINT "Run time (hours)", T4$; "                   "
540     T2 = T4 * 60
550     LOCATE 23, 3: PRINT "Deceleration time (minutes) =              "
552     LOCATE 23, 3: INPUT "Deceleration time (minutes) = ", T3$
555     IF T3$ = "L" OR T3$ = "l" THEN 530 ELSE T3 = VAL(T3$)
557     IF T3$ = "q" OR T3$ = "Q" THEN CLS : SYSTEM
```

```
558     LOCATE 15, 3: PRINT "Deceleration time (min.)", T3$; "                    "
560     LOCATE 23, 3: PRINT "Average speed (RPM) =                            "
562     LOCATE 23, 3: INPUT "Average speed (RPM) = ", Q$
565     IF Q$ = "L" OR Q$ = "l" THEN 550 ELSE Q = VAL(Q$)
567     IF Q$ = "q" OR Q$ = "Q" THEN RUN
568     LOCATE 16, 3: PRINT "Average speed (RPM)", Q$; "                         "
570     LOCATE 23, 3: PRINT "Temperature (deg.C) =                            "
572     LOCATE 23, 3: INPUT "Temperature (deg.C) = ", T5$
575     IF T5$ = "L" OR T5$ = "l" THEN 560 ELSE T5 = VAL(T5$)
577     IF T5$ = "q" OR T5$ = "Q" THEN RUN
578     LOCATE 17, 3: PRINT "Temperature (deg.C)", T5$; "                         "
580     LOCATE 23, 3: PRINT "Particle density (g/ml) =                         "
581     LOCATE 22, 3: PRINT "                                                  "
582     LOCATE 23, 3: INPUT "Particle density (g/ml) = ", P5$
585     IF P5$ = "L" OR P5$ = "l" THEN 570 ELSE P5 = VAL(P5$)
587     IF P5$ = "q" OR P5$ = "Q" THEN RUN
588     LOCATE 18, 3: PRINT "Particle density (g/ml)", P5$; "                     "
590     I = 15 / ((T2 + (T1 + T3) / 3) * (PI * Q) ^ 2)
600     P$ = "Input % sucrose of the ## fractions"
605     IF N3 = 1 THEN
           LOCATE 22, 3
           PRINT "Input % sucrose of the one fraction"
           GOTO 620
         END IF
607     IF N3 < 1 THEN 690
610     LOCATE 22, 3: PRINT USING P$; N3
620     J = 1
621     LOCATE 23, 3: PRINT "                                                  "
630     LOCATE 23, 3: PRINT J,
640     INPUT PE$
645     IF PE$ = "l" OR PE$ = "L" THEN
             J = J - 1
             IF J < 1 THEN 580 ELSE 621
         END IF
646     P(J) = VAL(PE$)
650     IF PE$ = "q" OR PE$ = "Q" THEN RUN
660     FOR MAR = J + (J > 1) TO J
665        LOCATE 19 + MAR - J - (J <> 1), 3
           PRINT "Fraction"; MAR; CHR$(29); "  ", CHR$(29);
           PRINT P(MAR); "           "
670     NEXT
680     J = J + 1: IF J <= N3 THEN 621: '                         NEXT J
685     LOCATE 23, 3: PRINT "DATA OK                                     "
686     LOCATE 23, 3: INPUT "DATA OK"; JA$
        IF JA$ = "y" OR JA$ = "Y" THEN 690
688     J = J - 1: GOTO 621
690     J = 1
700     L(J) = LOG(R1 - R2 + (J * U3 - U1 + 2 * PI * (R2 ^ 3) / 3) / (PI * R2 ^ 2))
710     IF EXP(L(J)) > R1 - R2 THEN 740
720     J = J + 1
730     GOTO 700
740     L(J) = 0
750     J = J - 1
760     M(1) = L(1) - LOG(R1-R2+(U2/2-U1+2*PI*(R2^3)/3)/(PI*R2^2))
770     FOR J1 = 2 TO J
780     M(J1) = L(J1) - L(J1 - 1)
790     NEXT J1
800     GOSUB 805: GOTO 890
805     CLS : MEN = 0: GOSUB 10000: GOSUB 140
810     LOCATE 24, 50: PRINT "                         ";
820     P$ = "& AT ##### RPM FOR ## HOURS AT ## DEG.C  P.D. = #.#"
830     LOCATE 7, 3: PRINT USING P$; A$(N1); Q; T4; T5; P5
840              '
```

```
850     IF Z1 = 0 THEN 880
860     LOCATE 9, 3: PRINT "FRACTION NO   % SUCROSE   S VALUE   MOL. WT."
870     GOTO 885
880     LOCATE 9, 3: PRINT "FRACTION NO   % SUCROSE   S VALUE"
885     RETURN
890     SO = 0
900     FOR J1 = 1 TO J
910     TEMP = T5: SUC = P(J1): DENS = P5: GOSUB 1470
920     SO = SO + M(J1) * E
930     IF Z1 = 0 THEN 970
940     P$ = "    ##          ##.##      ###.##     ##.###^^^^"
950     LOCATE 10 - INT((J1 - 1) / 10) * 10 + J1, 3
        PRINT USING P$; J1; P(J1); SO*I* 1E+13; EXP((LOG(SO*I*1E+13/ Z1))/ Z2)
960     GOTO 985
970     P$ = "    ##          ##.##      ###.##"
980     LOCATE 10 + J1 - INT((J1 - 1) / 10) * 10, 3
        PRINT USING P$; J1; P(J1); SO * I * 1E+13
985     IF J1 / 10 = INT(J1 / 10) AND J1 <> J THEN GOSUB 25000
990     NEXT J1
1005    LOCATE 24, 3: PRINT "Press Q then ENTER to QUIT                    ";
1007    EDDY$ = INKEY$: IF EDDY$ = "" THEN 1007
1015    IF EDDY$ = "q" OR EDDY$ = "Q" THEN
           LOCATE 24, 31
           PRINT "Q";
           GOSUB 40000
           GOTO 1007
         END IF
1016    GOTO 1007
1020    END
1030    REM SUBROUTINE TO CALCULATE DENSITY (D), VISCOSITY (V)
1040    REM FROM TEMPERATURE (T) AND SUCROSE PERCENTAGE (P)
1050    DATA 1.0003698,3.9680504E-5,-5.8513271E-6
1060    DATA 0.38982371,-1.0578919E-3,1.2392833E-5
1070    DATA 0.17097594,4.7530081E-4,-8.9239737E-6
1080    DATA 18.027525,4.8318329E-4,7.7830857E-5
1090    DATA 342.3,18.032
1100    DATA 212.57059,0.13371672,-2.9276449E-4
1110    DATA 146.06635,25.251728,0.070674842
1120    DATA -1.5018327,9.4112153,-1.1435741E3
1130    DATA 1.0504137E5,-4.6927102E6,1.0323349E8
1140    DATA -1.1028981E9,4.5921911E9
1145    DATA -1.0803314
1150    DATA -2.0003484E1,4.6066898E2,-5.9517023E3
1160    DATA 3.5627216E4,-7.8542145E4,0,0
1170    DATA 2.1169907E2,1.6077073E3,1.6911611E5
1180    DATA -1.4184371E7,6.0654775E8,-1.2985834E10
1190    DATA 1.3532907E11,-5.4970416E11
1195    DATA 1.3975568E2
1200    DATA 6.6747329E3,-7.8716105E4,9.0967578E5
1210    DATA -5.5380830E6,1.2451219E7,0,0
1220    RESTORE 1050
1230    READ B1, B2, B3, B4, B5, B6, B7, B8, B9
1240    READ A1, A2, A3, M1, M2, C1, C2, C3
1250    READ G1, G2, G3
1260    Y1 = P / 100
1270    Y = (Y1 / M1) / (Y1 / M1 + (1 - Y1) / M2)
1280    IF T > 30 THEN 1310
1290    D = B1 + B2*T + B3*T^2 + (B4+B5*T+B6*T^2)*Y1 + (B7+B8*T+B9*T^2) * Y1^2
1300    GOTO 1330
1310    SDO = Y * M1 + (1-Y) * M2
1320    D = SDO/(Y*(C1+C2*T+C3*T^2) + (1-Y)*(A1+A2*T+A3*T^2))
1330    RESTORE 1120
1340    IF P <= 48 THEN 1360
```

```
1350    RESTORE 1145
1360    READ D0, D1, D2, D3, D4, D5, D6, D7
1370    A = D0 + D1*Y + D2*Y^2 + D3*Y^3 + D4*Y^4 + D5*Y^5 + D6*Y^6 + D7*Y^7
1380    RESTORE 1170
1390    IF P <= 48 THEN 1410
1400    RESTORE 1195
1410    READ D0, D1, D2, D3, D4, D5, D6, D7
1420    A = D0 + D1*Y + D2*Y^2 + D3*Y^3 + D4*Y^4 + D5*Y^5 + D6*Y^6 + D7*Y^7
1430    C = G1 - G2 * SQR(1 + (Y/G3)^2)
1440    V = 10 ^ ((A + B) / (T + C))
1450    RETURN
1470    REM SUBROUTINE TO CALCULATE SEDIM (E) FROM PARTICLE
1480    REM DENSITY (DENS), TEMPERATURE (T) AND % SUCROSE (SUC)
1490    T = 20: P = 0: GOSUB 1030
1500    H2 = D
1510    V2 = V
1520    T = TEMP: P = SUC: GOSUB 1030
1530    H3 = D
1540    V3 = V
1550    E = ((DENS - H2) / (DENS - H3)) * (V3 / V2)
1560    RETURN
2000    RESTORE 310: GOTO 460
2010    RESTORE 320: GOTO 460
2020    RESTORE 330: GOTO 460
2030    RESTORE 340: GOTO 460
2040    RESTORE 350: GOTO 460
2050    RESTORE 360: GOTO 460
2060    RESTORE 370: GOTO 460
2070    RESTORE 380: GOTO 460
2080    RESTORE 390: GOTO 460
2090    RESTORE 400: GOTO 460
10000 ' window
10010   PI = 103.638 / 32.989
10195   IF MEN = 1 THEN
            LOCATE 23, 50
            PRINT "Press Q then ENTER to QUIT";
            GOTO 10205
         END IF
10200    LOCATE 24, 50: PRINT "Press Q then ENTER to QUIT";
10202    RETURN
10205    LOCATE 24, 46: PRINT "Press L then ENTER for last value";
10210    RETURN
20000    CLS
20010    LOCATE 8, 12
         PRINT "Y O U R   D A T A   A R E   I M P O S S I B L E ! ! !"
20030    LOCATE 10, 31: PRINT "Please try again..."
20032    CLEAR
20040    GOTO 135
25000    LOCATE 23, 2: PRINT "Press space bar to continue.";
25010    A$ = INKEY$: IF A$ <> " " THEN 25010
25020    LOCATE 23, 2: PRINT "                              ";
25030    GOSUB 805
25040    RETURN
40000    LET A$ = INKEY$: IF A$ = "" THEN GOTO 40000
40010    LET A = ASC(A$): IF A = 13 THEN RUN
40020    IF A = 8 THEN LOCATE 24, 31: PRINT " "; : RETURN
40030    BEEP: GOTO 40000
```

Appendix B: A program for calculating sedimentation coefficients in vertical rotors

```
10      SCREEN 2: KEY OFF: CLS
20      DIM P(80), M(80), L(80), A$(10)
30      MLIN = 9
35      GOSUB 40: GOTO 50
40      LOCATE 3, 26
41      PRINT "S-VALUES FOR VERTICAL ROTORS"
42      RETURN
50      Z1 = 0
60      DATA 0.0528,0.4
65      DATA 0.0882,0.346
67      MEN = 0
68      GOSUB 18000
70      LOCATE 23, 3: PRINT "IS SAMPLE DNA? (Y/N)                "
71      LOCATE 23, 3: INPUT "IS SAMPLE DNA? (Y/N) ", CED$
72      IF CED$ = "q" OR CED$ = "Q" THEN CLS : SYSTEM
73      IF CED$ = "" THEN 71
75      LOCATE 7, 3
        PRINT "Sample DNA", , CHR$(ASC(CED$) AND 223); "                    "
80      IF INSTR("Nn", CED$) <> 0 THEN 140
82      IF CED$ <> "y" AND CED$ <> "Y" THEN 70
84      MEN = 1
85      GOSUB 18000
90      LOCATE 23, 3: PRINT "ALKALINE GRADIENT? (Y/N)                "
92      LOCATE 23, 3: INPUT "ALKALINE GRADIENT? (Y/N) ", C$
93      IF C$ = "q" OR C$ = "Q" OR C$ = "l" OR C$ = "L" THEN RUN
94      IF C$ <> "y" AND C$ <> "Y" AND C$ <> "n" AND C$ <> "N" THEN 92
95      LOCATE 8, 3
        PRINT "Alkaline gradient", CHR$(ASC(C$) AND 223); "                    "
100     RESTORE 60
110     IF INSTR("Yy", C$) <> 0 THEN 130
120     RESTORE 65
130     READ Z1, Z2
140     NO = 10
145     REM LIST OF ROTORS AND THEIR DIMENSIONS
150     DATA 4.7,0.6,7.83: A$(1) = "SORVALL TV-865 (8x6ml)"
160     DATA 2.8,1.2,7.2: A$(2) = "SORVALL TV-865B (8x18.5ml)"
170     DATA 6.4,1.2,7.2: A$(3) = "SORVALL TV-850 (8x38.5ml)"
180     DATA 5.7,0.75,7.2: A$(4) = "CENTRIKON TV-850 (13.5ml adaptors)"
190     DATA 4.4,0.5,7.2: A$(5) = "CENTRIKON TV-850 (4.4ml adaptors)"
200     DATA 7.217,1.232,7.4: A$(6) = "CENTRIKON TVT-50.38"
210     DATA 5.7,0.75,7.4: A$(7) = "CENTRIKON TVT-50.38 (13.5ml adaptors)"
220     DATA 4.7,0.6,7.4: A$(8) = "CENTRIKON TVT-50.38 (6ml adaptors)"
230     DATA 4,0.57,6.45: A$(9) = "BECKMAN VTi80 (8x5ml)"
240     DATA 4,0.57,8.13: A$(10) = "BECKMAN VTi65.2 (16x5ml)"
250 '
260     FOR N = 1 TO NO: LOCATE N + 8, 2: PRINT N, A$(N): NEXT N
270     LOCATE 23, 3: PRINT "WHICH ROTOR DO YOU REQUIRE?                    "
271     LOCATE 23, 3: INPUT "WHICH ROTOR DO YOU REQUIRE? ", N1$
275     IF N1$ = "l" OR N1$ = "L" THEN 85 ELSE N1 = VAL(N1$)
277     IF N1$ = "q" OR N1$ = "Q" THEN RUN
280     IF N1 > NO OR N1 < 1 THEN 270
282     FOR N = 1 TO NO: LOCATE N + 8, 2: PRINT "   ", "                    ": NEXT N
283     LOCATE 9, 3: PRINT "Rotor", , A$(N1)
285 '
290     ON N1 GOTO 5000, 5001, 5002, 5003, 5004, 5005, 5006, 5007, 5008, 5009
299     A$ = "N"
300     READ L, R1, R2
301     BO = 1
```

```
302     LOCATE 23, 3: PRINT "HEMISPHERICAL TOP (Y/N)                    "
303     LOCATE 23, 3
        INPUT "HEMISPHERICAL TOP (Y/N) ", A$
        IF A$ = "q" OR A$ = "Q" THEN RUN
304     IF A$ = "" THEN 303
305     IF A$ = "l" OR A$ = "L" THEN
            GOTO 250
        ELSE
            LOCATE 10, 3
            PRINT "Hemispherical top", CHR$(ASC(A$) AND 223); "            "
        END IF
306     IF INSTR("Nn", A$) <> 0 THEN
            LOCATE 23, 3
            PRINT "TOTAL VOLUME =              "
            GOTO 308
         END IF
307     IF A$ <> "y" AND A$ <> "Y" THEN
            GOTO 302
        ELSE
            BO = BO + 1
            LOCATE 23, 3
            PRINT "TOTAL VOLUME =                   "
        END IF
308     P = 2: GOSUB 1000
        LOCATE 23, 3
        INPUT "TOTAL VOLUME = ", V$: V9 = VAL(V$)
310     IF V$ = "l" OR V$ = "L" THEN 302
312     IF V$ = "q" OR V$ = "Q" THEN RUN
314     LOCATE 11, 3
        PRINT "Total volume", CHR$(29); V9; "                 "
320     LOCATE 23, 3
        PRINT "SAMPLE VOLUME =              "
322     LOCATE 23, 3
        INPUT "SAMPLE VOLUME = ", U2$
325     IF U2$ = "l" OR U2$ = "L" THEN 306 ELSE U2 = VAL(U2$)
326     IF U2$ = "q" OR U2$ = "Q" THEN RUN
327     LOCATE 12, 3:
        PRINT "Sample volume", CHR$(29); U2; "                "
330     LOCATE 23, 3
        PRINT "FRACTION VOLUME =            "
332     LOCATE 23, 3
        INPUT "FRACTION VOLUME = ", U3$
335     IF U3$ = "l" OR U3$ = "L" THEN 320 ELSE U3 = VAL(U3$)
336     IF U3$ = "q" OR U3$ = "Q" THEN RUN
337     LOCATE 13, 3
        PRINT "Fraction volume", CHR$(29); U3; "              "
340     N3 = V9 / U3
350
360     LOCATE 23, 3
        PRINT "ACCELERATION TIME (MINUTES) =        "
362     LOCATE 23, 3: INPUT "ACCELERATION TIME (MINUTES) = ", T1$
365     IF T1$ = "l" OR T1$ = "L" THEN 330 ELSE T1 = VAL(T1$)
366     IF T1$ = "q" OR T1$ = "Q" THEN RUN
367     LOCATE 14, 3
        PRINT "Acceleration time", CHR$(29); T1; "                "
370     LOCATE 23, 3
        PRINT "RUN TIME (MINUTES) =                  "
372     LOCATE 23, 3: INPUT "RUN TIME (MINUTES) = ", T4$
375     IF T4$ = "L" OR T4$ = "l" THEN 360 ELSE T4 = VAL(T4$)
376     IF T4$ = "q" OR T4$ = "Q" THEN RUN
377     LOCATE 15, 3
        PRINT "Run time", , CHR$(29); T4; "                 "
380     T4 = T4 / 60: T2 = T4 * 30
```

```
385     '
390     LOCATE 23, 3
        PRINT "DECELERATION TIME (MINUTES) =          "
392     LOCATE 23, 3: INPUT "DECELERATION TIME (MINUTES) = ", T3$
394     IF T3$ = "l" OR T3$ = "L" THEN 370 ELSE T3 = VAL(T3$)
396     IF T3$ = "q" OR T3$ = "Q" THEN RUN
397     LOCATE 16, 3
        PRINT "Deceleration time", CHR$(29); T3; "                    "
400     LOCATE 23, 3: PRINT "AVERAGE SPEED (RPM) =                   "
402     LOCATE 23, 3: INPUT "AVERAGE SPEED (RPM) = ", Q$
404     IF Q$ = "l" OR Q$ = "L" THEN 390 ELSE Q = VAL(Q$)
406     IF Q$ = "q" OR Q$ = "Q" THEN RUN
407     LOCATE 17, 3
        PRINT "Average speed", CHR$(29); Q; "                  "
410     LOCATE 23, 3: PRINT "TEMPERATURE (DEG.C) =                   "
412     LOCATE 23, 3: INPUT "TEMPERATURE (DEG.C) = ", T5$
414     IF T5$ = "l" OR T5$ = "L" THEN 400 ELSE T5 = VAL(T5$)
416     IF T5$ = "q" OR T5$ = "Q" THEN RUN
417     LOCATE 18, 3: PRINT "Temperature", , CHR$(29); T5; "                   "
420     LOCATE 23, 3: PRINT "PARTICLE DENSITY =                      "
422     LOCATE 23, 3: INPUT "PARTICLE DENSITY = ", P5$
424     IF P5$ = "l" OR P5$ = "L" THEN 410 ELSE P5 = VAL(P5$)
425     PI = 4 * ATN(1)
426     IF P5$ = "q" OR P5$ = "Q" THEN RUN
427     LOCATE 19, 3
        PRINT "Particle density", CHR$(29); P5; "                  "
430     I = 15 / ((T2 + (T1 + T3) / 3) * (PI * Q) ^ 2)
435     CLS : GOSUB 18000: GOSUB 40
440     P$ = "INPUT % SUCROSE OF THE ## FRACTIONS"
450     LOCATE 22, 3: PRINT USING P$; N3
460     J = 1
461     LOCATE 23, 3: PRINT J,
470     INPUT P$: '                          P(J)
471     IF P$ = "q" OR P$ = "Q" THEN RUN
472     IF P$ = "l" OR P$ = "L" THEN J = J - 1: GOTO 18500
475     LOCATE 23, 2: PRINT "                          "
476     P(J) = VAL(P$)
480     FOR EDDY = J + (J > 1) + (J > 2) + (J > 3) + (J > 4) TO J
481        P$ = STR$(EDDY): IF LEN(P$) = 2 THEN P$ = " " + P$
482        LOCATE 12 + EDDY - J, 3
           PRINT "The % sucrose of fraction"; P$; " is"; P(EDDY); "  ";
485     NEXT EDDY
489     J = J + 1: IF J <= N3 THEN 461
490     J = 1
495     V1 = J * U3
496     IF V1 > V9 THEN 540
497     GOSUB 1200
500     L(J) = LOG(R2 - R1 + P * R1)
510     IF EXP(L(J)) > R1 + R2 THEN 540
520     J = J + 1
530     GOTO 495
540     L(J) = 0
550     J = J - 1
552     V1 = U2 / 2: GOSUB 1200
555     M(1) = L(1) - LOG(R2 - R1 + P * R1)
570     FOR J1 = 2 TO J
580     M(J1) = L(J1) - L(J1 - 1)
590     NEXT J1
600     REM
610     MEN = 0: CLS : GOSUB 18000: GOSUB 40: LOCATE 24, 65: PRINT "        ";
620     P$ = "& AT ##### RPM FOR ## HOURS AT ## DEG.C  P.D.= #.#"
630     LOCATE 7, 3: PRINT USING P$; A$(N1); Q; T4; T5; P5
640 '
```

```
650     IF Z1 = 0 THEN 680
660     LOCATE 9, 3: PRINT "FRACTION NO   % SUCROSE   S VALUE   MOL. WT.";
670     GOTO 690
680     LOCATE 9, 3: PRINT "FRACTION NO    % SUCROSE   S VALUE";
690 '
695     GOTO 2000
998     REM SUBROUTINE TO CALCULATE VOLUME (V) FROM
999     REM RADIAL DISTANCE (P) ACROSS THE TUBE
1000    IF P = 2 THEN
            V = 0
        ELSE
          V = 1.570796
          V = V - ATN((1 - P) / SQR(2*P - P*P)) - (1 - P) * SQR(2*P - P*P)
        END IF
1010    V = V + BO * (PI * P * P * (3 - P) * R1) / (6 * L)
1020    V = V * R1 * R1 * L
1030    RETURN
1198    REM SUBROUTINE TO CALCULATE RADIAL DISTANCE (P)
1199    REM ACROSS THE TUBE FROM VOLUME (V)
1200    P = 2 * V1 / V9
1210    GOSUB 1000: V2 = V: IF ABS(V1 - V2) / (V1) < .001 THEN 1250
1220    P = P * .99: GOSUB 1000: D = (V2 - V) / (.01 * P)
1230    P = P / .99: P = P - (V2 - V1) / D
1240    GOTO 1210
1250    RETURN
2000     SO = 0
2010     FOR J1 = 1 TO J
2020     TEMP = T5: SUC = P(J1): DENS = P5: GOSUB 2590: ' calculatings
2030     SO = SO + M(J1) * E
2040     IF Z1 = 0 THEN 2080
2050     P$ = "     ##          ##.##      ###.##    ##.###^^^^"
2060     LOCATE 11 + NLIN, 3
         PRINT USING P$; J1; P(J1); SO * I * 1E+13;
         PRINT EXP((LOG(SO * I * 1E+13 / Z1)) / Z2)
         NLIN = NLIN + 1
2065    IF NLIN > MLIN AND J1 <> J THEN GOSUB 25000
2070     GOTO 2100
2080     P$ = "     ##          ##.##      ###.##"
2090     LOCATE 11 + NLIN, 3
         PRINT USING P$; J1; P(J1); SO * I * 1E+13
         NLIN = NLIN + 1
2095     IF NLIN > MLIN AND J1 <> J THEN GOSUB 25000
2100     NEXT J1
2110 '
2111    LOCATE 24, 3
        PRINT "Press Q then ENTER to QUIT";
        PRINT "or    the SPACE-BAR to restart the program";
2112    ED$ = INKEY$: IF ED$ = "" THEN 2112
2113    IF ED$ <> " " AND ED$ <> "Q" AND ED$ <> "q" THEN BEEP: GOTO 2112
2114    IF ED$ = " " THEN RUN
2115    IF ED$ = "Q" OR ED$ = "q" THEN LOCATE 24, 77: PRINT "Q";
2116    LET ED$ = INKEY$
2117    IF ED$ = "" THEN GOTO 2116
2118    IF ASC(ED$) <> 13 AND ASC(ED$) <> 8 THEN BEEP: GOTO 2116
2119    IF ASC(ED$) = 8 THEN LOCATE 24, 77: PRINT " "; : GOTO 2112
2120    IF ASC(ED$) = 13 THEN CLS : SYSTEM
2130    END
2140    REM SUBROUTINE TO CALCULATE DENSITY (D), VISCOSITY (V)
2150    REM FROM TEMPERATURE (T) AND SUCROSE PERCENTAGE (P)
2160    DATA 1.0003698,3.9680504E-5,-5.8513271E-6
2170    DATA 0.38982371,-1.0578919E-3,1.2392833E-5
2180    DATA 0.17097594,4.7530081E-4,-8.9239737E-6
2190    DATA 18.027525,4.8318329E-4,7.7830857E-5
```

```
2200      DATA 342.3,18.032
2210      DATA 212.57059,0.13371672,-2.9276449E-4
2220      DATA 146.06635,25.251728,0.070674842
2230      DATA -1.5018327,9.4112153,-1.1435741E3
2240      DATA 1.0504137E5,-4.6927102E6,1.0323349E8
2250      DATA -1.1028981E9,4.5921911E9
2260      DATA -1.0803314
2270      DATA -2.0003484E1,4.6066898E2,-5.9517023E3
2280      DATA 3.5627216E4,-7.8542145E4,0,0
2290      DATA 2.1169907E2,1.6077073E3,1.6911611E5
2300      DATA -1.4184371E7,6.0654775E8,-1.2985834E10
2310      DATA 1.3532907E11,-5.4970416E11
2320      DATA 1.3975568E2
2330      DATA 6.6747329E3,-7.8716105E4,9.0967578E5
2340      DATA -5.5380830E6,1.2451219E7,0,0
2350      RESTORE 2160
2360      READ B1, B2, B3, B4, B5, B6, B7, B8, B9
2370      READ A1, A2, A3, M1, M2, C1, C2, C3
2380      READ G1, G2, G3
2390      Y1 = P / 100
2400      Y = (Y1 / M1) / (Y1 / M1 + (1 - Y1) / M2)
2410      IF T > 30 THEN 2440
2420      D = B1 + B2*T + B3*T^2 + (B4+B5*T+B6*T^2)*Y1 + (B7+B8*T+B9*T^2)*Y1^2
2430      GOTO 2460
2440      SDO = Y * M1 + (1 - Y) * M2
2450      D = SDO / (Y * (C1 + C2*T + C3*T^2) + (1 - Y) * (A1 + A2*T + A3*T^2))
2460      RESTORE 2230
2470      IF P <= 48 THEN 2490
2480      RESTORE 2260
2490      READ D0, D1, D2, D3, D4, D5, D6, D7
2500      A = D0 + D1*Y + D2*Y^2 + D3*Y^3 + D4*Y^4 + D5*Y^5 + D6*Y^6 + D7*Y^7
2510      RESTORE 2290
2520      IF P <= 48 THEN 2540
2530      RESTORE 2320
2540      READ D0, D1, D2, D3, D4, D5, D6, D7
2550      A = D0 + D1*Y + D2*Y^2 + D3*Y^3 + D4*Y^4 + D5*Y^5 + D6*Y^6 + D7*Y^7
2560      C = G1 - G2 * SQR(1 + (Y / G3) ^ 2)
2570      V = 10 ^ ((A + B) / (T + C))
2580      RETURN
2590      REM SUBROUTINE TO CALCULATE SEDIM (E) FROM PARTICLE
2600      REM DENSITY (DENS), TEMPERATURE (T) AND % SUCROSE (SUC)
2610      T = 20: P = 0: GOSUB 2140
2620      H2 = D
2630      V2 = V
2640      T = TEMP: P = SUC: GOSUB 2140
2650      H3 = D
2660      V3 = V
2670      E = ((DENS - H2) / (DENS - H3)) * (V3 / V2)
2680      RETURN
5000      RESTORE 150: GOTO 300
5001      RESTORE 160: GOTO 300
5002      RESTORE 170: GOTO 300
5003      RESTORE 180: GOTO 300
5004      RESTORE 190: GOTO 300
5005      RESTORE 200: GOTO 300
5006      RESTORE 210: GOTO 300
5007      RESTORE 220: GOTO 300
5008      RESTORE 230: GOTO 300
5009      RESTORE 240: GOTO 300
10000 ' window80
10010 PI = ATN(1) * 4
10195 IF MEN = 1 THEN LOCATE 23, 50: PRINT "Q then ENTER to QUIT"; : GOTO 10205
10200 LOCATE 24, 50: PRINT "Q then ENTER to QUIT";
```

```
10202 RETURN
10205 LOCATE 24, 46: PRINT "Press L then ENTER for last value"; :
10210 RETURN
18000 GOTO 10000
18500   ' From line 472
18510   IF J > 0 THEN 461
18520   GOSUB 10000: GOSUB 40
18550   LOCATE 7, 3: PRINT "Sample DNA", , CHR$(ASC(CED$) AND 223); "          "
18560   LOCATE 8, 3: PRINT "Alkaline gradient", CHR$(ASC(C$) AND 223); "          "
18570   LOCATE 9, 3: PRINT "Rotor", , A$(N1)
18580   LOCATE 10, 3: PRINT "Hemispherical top", CHR$(ASC(A$) AND 223); "          "
18590   LOCATE 11, 3: PRINT "Total volume", CHR$(29); V9; "          "
18600   LOCATE 12, 3: PRINT "Sample volume", CHR$(29); U2; "          "
18610   LOCATE 13, 3: PRINT "Fraction volume", CHR$(29); U3; "          "
18620   LOCATE 14, 3: PRINT "Acceleration time", CHR$(29); T1; "          "
18630   LOCATE 15, 3: PRINT "Run time", , CHR$(29); T4 * 60; "          "
18640   LOCATE 16, 3: PRINT "Deceleration time", CHR$(29); T3; "          "
18650   LOCATE 17, 3: PRINT "Average speed", CHR$(29); Q; "          "
18660   LOCATE 18, 3: PRINT "Temperature", , CHR$(29); T5; "          "
18670   LOCATE 19, 3: PRINT "Particle density", CHR$(29); P5; "          "
18680   GOTO 420
20000 '         ERROR!!!!!
20010   CLS : SCREEN 2
20015 MEN = 0
20020   GOSUB 18000
20030   GOSUB 40
20040   LOCATE 7, 28: PRINT "Your data are impossible"
20050   LOCATE 12, 29: PRINT "T r y   a g a i n ! ! !"
20055   LOCATE 13, 29: PRINT "======================="
20057   LOCATE 16, 25: PRINT "Press the space-bar to continue"
20060   E$ = INKEY$: IF E$ = "" THEN 20060
20070   IF E$ = " " THEN RUN
20080   GOTO 20060
25000   '                                   nlin>mlin
25010   LOCATE 23, 3: PRINT "Press the space-bar to continue"
25020   IF INKEY$ <> " " THEN 25020
25025   LOCATE 23, 3: PRINT "                               "
25030   FOR NLIN = 0 TO MLIN
25040   LOCATE 11 + NLIN, 3: PRINT "                                            ";
25050   NEXT NLIN
25060   NLIN = 0
25070   RETURN
```

Appendix C: A program for calculating sedimentation coefficients in zonal rotors

```
100     SCREEN 2: KEY OFF: CLS : IPO = 9: LIMA = 10
110     LOCATE 3, 26: PRINT "S-VALUES IN B-14 ZONAL ROTORS"
120     LOCATE 7, 11
130     PRINT "This  program makes it possible  to calculate  equivalent"
135     LOCATE 8, 11
140     PRINT "sedimentation coefficients by a physically correct method"
145     LOCATE 9, 11
150     PRINT "from sucrose density gradient centrifugation data."
165     LOCATE 12, 11
170     PRINT "This version is for use with the B14 zonal rotor"
200     PRINT
205     LOCATE 16, 11
210     PRINT "Two types  of inputs are required  to run  this  program,"
215     LOCATE 17, 11
220     PRINT "namely: A. GENERAL CENTRIFUGE DATA,                   and"
```

```
225   LOCATE 18, 11
230   PRINT "           B. VOLUME AND SUCROSE CONC. OF EACH FRACTION."
240   LOCATE 23, 3: PRINT "Press the space-bar to continue..."
242   GOSUB 10000
245   ED$ = INKEY$: IF ED$ = "" THEN 245
246   IF ED$ = "Q" OR ED$ = "q" THEN CLS : SYSTEM
247   IF ED$ <> " " THEN 245
249   CLS : LOCATE 3, 28: PRINT "S-values in zonal rotors"
400   MEN = 0
401   GOSUB 10000
405   LOCATE 23, 3: PRINT "ROTOR SPEED (rpm) =                                    "
406   LOCATE 23, 3: INPUT "ROTOR SPEED (rpm) = ", R1$
407   IF R1$ = "L" OR R1$ = "l" THEN 400 ELSE R1 = VAL(R1$)
408   IF R1$ = "q" OR R1$ = "Q" THEN 20000
409   LOCATE 7, 3: PRINT "Rotor speed (rpm)", , R1; "                          "
410   MEN = 1
411   GOSUB 10000
415   LOCATE 23, 3: PRINT "RUN DURATION (min.) =                                  "
416   LOCATE 23, 3: INPUT "RUN DURATION (min.) = ", D1$
417   IF D1$ = "L" OR D1$ = "l" THEN 400 ELSE D1 = VAL(D1$)
418   IF D1$ = "q" OR D1$ = "Q" THEN 20000
419   LOCATE 8, 3: PRINT "Run duration (min.)", , D1; "                         "
420   ' periods of acc.
425   LOCATE 23, 3: PRINT "PERIODS OF ACCELERATION (min.) =                       "
426   LOCATE 23, 3: INPUT "PERIODS OF ACCELERATION (min.) = ", D2$
427   IF D2$ = "L" OR D2$ = "l" THEN 410 ELSE D2 = VAL(D2$)
428   IF D2$ = "q" OR D2$ = "Q" THEN 20000
429   LOCATE 9, 3: PRINT "Periods of acceleration (min.)", D2; "                "
430   GOSUB 15000: GOSUB 10000
435   LOCATE 23, 3: PRINT "PERIODS OF DECELERATION (min.) =                       "
436   LOCATE 23, 3: INPUT "PERIODS OF DECELERATION (min.) = ", D3$
437   IF D3$ = "L" OR D3$ = "l" THEN 420 ELSE D3 = VAL(D3$)
438   IF D3$ = "q" OR D3$ = "Q" THEN 20000
439   LOCATE 10, 3: PRINT "Periods of deceleration (min.)", D3; "               "
440   GOSUB 10000
445   LOCATE 23, 3: PRINT "SAMPLE VOLUME (ml) =                                   "
446   LOCATE 23, 3: INPUT "SAMPLE VOLUME (ml) = ", V1$
447   IF V1$ = "L" OR V1$ = "l" THEN 430 ELSE V1 = VAL(V1$)
448   IF V1$ = "q" OR V1$ = "Q" THEN 20000
449   LOCATE 11, 3: PRINT "Sample volume (ml)", , V1; "                         "
450   GOSUB 2000
452   REM
453   REM     ****    MEANING OF VARIABLES    ****
454   REM     E1,E2   EXPERIMENT NUMBER,DATE
455   REM     R1,D1   ROTOR SPEED, RUN DURATION
456   REM     D2,D3   PERIODS OF ACCELERATION AND DECELERATION
457   REM     V1,V2   SAMPLE AND OVERLAY VOLUMES
458   REM     P1,T1   PARTICLE DENSITY, ROT. TEMPERATURE
459   REM
460   A2$ = "--------------------------------------------------------------------------"
480   A4$ = "ROTOR SPEED          : ##### rpm    RUN DURATION          : ##### min"
490   A5$ = "ACCELERATION PERIOD: ##### min    DECELERATION PERIOD: ##### min"
500   A6$ = "SAMPLE VOLUME        : ##### ml     OVERLAY VOLUME        : ##### ml"
510   A7$ = "PARTICLE DENSITY     : #.### g/ml   ROTOR TEMPERATURE     : ##.## Deg."
520   CLS : LOCATE 3, 24: PRINT "****      SEDCOF PROGRAM      ****"
560   LOCATE 22, 3: PRINT "Input of part B of data"
565   GOSUB 3000
570   IF DDERR = 0 THEN DIM F2(999), F3(999): DDERR = 3
575   LOCATE 9, 3: PRINT , , , " Volume", " Sucr. conc."
580   I = 1
585   LOCATE 22, 3: PRINT "FRACTION"; I; ":                                  "
590   LOCATE 23, 3: PRINT "VOLUME =                                             ";
592   LOCATE 23, 3: INPUT "VOLUME = ", F2$
```

```
593     IF F2$ = "Q" OR F2$ = "q" THEN RUN
594     IF F2$ = "L" OR F2$ = "l" THEN
            GOTO 3100
        ELSE
            F2(I) = VAL(F2$)
            GOSUB 3300
        END IF
595     LOCATE 23, 3: PRINT "SUCROSE CONCENTRATION (g/ml) =                        "
596     LOCATE 23, 3: INPUT "SUCROSE CONCENTRATION (g/ml) = ", F3$
597     IF F3$ = "L" OR F3$ = "l" THEN 585 ELSE F3(I) = VAL(F3$)
598     IF F3$ = "q" OR F3$ = "Q" THEN 20000
599     GOSUB 3350
600     REM HERE IS ROOM FOR TEST OF INPUT DATA
610     I = I + 1: IF I <= F1 THEN 585
620     REM START OF CALCULATIONS
630     REM      ****     MEANING OF VARIABLES    ****
640     REM      F1       TOTAL NO. OF FRACTIONS
650     REM      F2       VOLUME OF EACH FRACTION
660     REM      F3       SUCROSE CONC. OF EACH FRACTION
670     REM      C1       FORCE/TIME INTEGRAL
680     REM      C2       BUOYANCY CONSTANT
690     REM      M1,2,3   MDO VALUES
700     REM      VO       ACCUMULATED FRACTION VOLUME
710     REM      SO       OLD S-VALUE
720     REM      V3       START OF SAMPLE MASS CENTRE
730     REM      R2       OLD R-VALUE
740     REM
750     C1 = ((3.141593 * R1) ^ 2) * (D1 + (D2 + D3) / 3) / 15!
760     C2 = (P1 - .998) / 1.005
770     MO = (-.0000058513271# * T1 + .000039680504#) * T1 + 1.0003698#
780     M1 = (.000012392833# * T1 - .0010578919#) * T1 + .38982371#
790     M2 = (-.0000089239737# * T1 + .00047530081#) * T1 + .17097594#
800     '                                                        PRINT
845 CLS
850     PRINT
860     PRINT A2$
880     PRINT USING A4$; R1; D1
890     PRINT USING A5$; D2; D3
900     PRINT USING A6$; V1; V2
910     PRINT USING A7$; P1; T1
920     PRINT A2$
925     PRINT A1$
930     PRINT
940     PRINT " NO    VOL    SUC    AVOL    RAD     DENS    VISCOS    S-VAL"
950     PRINT
960     REM
970     VO = 0
980     SO = 0
990     V3 = V1 / 2! + V2
1000    REM NEXT LINE DEFINES RADIUS-VOLUME RELATIONSHIP
1010    DEF FNR (X) = .5947 + SQR(X * .05495 + 1.394)
1020    REM START OF MAIN CALCULATIONS
1030    REM      ****     MEANING OF VARIABLES    ****
1040    REM      R3       ACTUAL ROTOR RADIUS
1050    REM      R4       DR/R
1060    REM      Y1,Y2    WEIGHT, MOLE FRACTION SUCROSE
1070    REM      D4       PRESENT DENSITY OF THE SUCROSE GRADIENT
1080    REM      A1,2,3   VISCOSITY CONSTANTS
1090    REM      V4       PRESENT VISCOSITY
1100    REM      S3       CALCULATED SEDIMENTATION COEFFICIENT
1110    REM
1120    R2 = FNR(V3)
1130    FOR I = 1 TO F1
```

```
1140      V0 = V0 + F2(I)
1150      IF V0 > 641 THEN 1430
1160      IF V0 >= V3 THEN 1200
1170 A8$ = "###    ##.#    ##.#                                                  "
1180      PRINT USING A8$; I; F2(I); F3(I): LIED = LIED + 1
1185      IF LIED > LIMA THEN GOSUB 5000
1190      GOTO 1430
1200      R3 = FNR(V0)
1210      R4 = (R3 - R2) / ((R3 + R2) / 2!)
1220      R2 = R3
1230      Y1 = F3(I) / 100
1240      D4 = (M2 * Y1 + M1) * Y1 + M0
1250      Y1 = Y1 * 100
1260      Y2 = Y1 / (Y1 + (100 - Y1) * 18.9924)
1270      REM
1280      A1 = -1.5018327# + 9.4112153# * Y2 - 1143.5741# * Y2^2
1290      A1 = A1 + 105041.37# * Y2^3 - 4692710.2# * Y2^4 + 103233490# * Y2^5
1300      A1 = A1 - 1102898100# * Y2^6 + 4592191100# * Y2^7
1310      A2 = 211.69907# + 1607.7073# * Y2 + 169116.11# * Y2^2
1320      A2 = A2 - 14184371# * Y2^3 + 606547750# * Y2^4 - 12985834000# * Y2^5
1330      A2 = A2 + 135329070000# * Y2^6 - 549704160000# * Y2^7
1340      A3 = 146.06635# - 25.251728# * SQR(1 + (Y2 / .070674842#)^2)
1350      V4 = 10^(A1 + A2 / (T1 + A3))
1360      Z = R4 * V4 / (P1 - D4)
1370      S1 = Z * C2 / C1
1380      S0 = S0 + S1
1390      S3 = S0 * 1E+13
1400      A9$ = "###    ##.#    ##.#    ###.#    #.##    #.####    #.####    ###.#    "
1410      PRINT USING A9$; I; F2(I); F3(I); V0; R3; D4; V4; S3: LIED = LIED + 1
1415      IF LIED > LIMA THEN GOSUB 5000
1420      J = I
1430      NEXT I
1440      FOR I = J + 1 TO F1
1450      IF A8$ = "" THEN A8$ = "###    ##.#    ##.#                                  "
1455      PRINT USING A8$; I; F2(I); F3(I): LIED = LIED + 1
1457      IF LIED > LIMA THEN GOSUB 5000
1460      NEXT I
1470      LOCATE 24, 1: PRINT "Press the space-bar to restart the program";
1480 IF INKEY$ <> " " THEN 1480
1490      RUN
2000      ' overlay vol.
2005      LOCATE 23, 3: PRINT "OVERLAY VOLUME (ml) =                                              "
2006      LOCATE 23, 3: INPUT "OVERLAY VOLUME (ml) = ", V2$
2007      IF V2$ = "L" OR V2$ = "l" THEN RETURN 440:  ELSE V2 = VAL(V2$)
2008      IF V2$ = "q" OR V2$ = "Q" THEN 20000
2009      LOCATE 12, 3: PRINT "Overlay volume (ml)", , V2; "                                "
2010      ' part. dens.
2015      LOCATE 23, 3: PRINT "PARTICLE DENSITY (g/ml) =                                          "
2016      LOCATE 23, 3: INPUT "PARTICLE DENSITY (g/ml) = ", P1$
2017      IF P1$ = "L" OR P1$ = "l" THEN 2000 ELSE P1 = VAL(P1$)
2018      IF P1$ = "q" OR P1$ = "Q" THEN 20000
2019      LOCATE 13, 3: PRINT "Particle density (g/ml)", , P1; "                          "
2020      ' temp.
2025      LOCATE 23, 3: PRINT "ROTOR TEMPERATURE (DC) =                                           "
2026      LOCATE 23, 3: INPUT "ROTOR TEMPERATURE (DC) = ", T1$
2027      IF T1$ = "L" OR T1$ = "l" THEN 2010 ELSE T1 = VAL(T1$)
2028      IF T1$ = "q" OR T1$ = "Q" THEN 20000
2029      LOCATE 14, 3: PRINT "Rotor temperature (DC)", , T1; "                           "
2030      GOSUB 10000
2034      LOCATE 23, 3: PRINT "ROTOR NAME       =                                                 "
2035      LOCATE 23, 3: INPUT "ROTOR NAME =           ", A1$
```

```
2036     IF LEN(A1$) > 35 THEN
            LOCATE 22, 3
            PRINT "ROTOR NAME CANNOT BE LARGER THAN 35 SIGNS          ";
            GOTO 2035
         END IF
2037     IF A1$ = "L" OR A1$ = "l" THEN GOTO 2020
2038     IF A1$ = "q" OR A1$ = "Q" THEN 20000
2039     LOCATE 15, 3: PRINT "Rotor name          ", , " "; A1$; "                  "
2040     GOSUB 10000: LOCATE 22, 3: PRINT "                                        ";
2041     LOCATE 23, 3: PRINT "DATA OK                              Press Q to quit";
2042     LOCATE 23, 3: INPUT "DATA OK"; ED$
2045     IF ED$ = "q" OR ED$ = "Q" THEN RUN
2046     IF ED$ = "L" OR ED$ = "l" OR ED$ = "N" OR ED$ = "n" THEN 2030
2049     RETURN 452
3000     GOSUB 10000
3005     LOCATE 23, 3: PRINT "TOTAL NUMBER OF FRACTIONS =                      "
3006     LOCATE 23, 3: INPUT "TOTAL NUMBER OF FRACTIONS = ", F1$
3007     IF F1$ = "L" OR F1$ = "l" THEN
             GOTO 4000
         ELSE
         F1 = VAL(F1$)
           IF F1 = 0 THEN F1 = 1
         END IF
3008     IF F1$ = "q" OR F1$ = "Q" THEN 20000
3009     LOCATE 7, 3: PRINT "Total number of fractions", , F1; "               "
3010     RETURN
3100     '
3110     I = I - 1: IF I < 1 THEN 520
3120     LOCATE 22, 3: PRINT "FRACTION"; I; ": "
3130     GOTO 595
3300     '
3310     FOR EDDY = I - 3 - (I < 4) - (I < 3) - (I < 2) TO I
3320     LOCATE 13 - I + EDDY + (I < 4) + (I < 3) + (I < 2), 3
         PRINT "Fraction"; EDDY, , , F2(EDDY); "     "
3325     IF EDDY <> I THEN
            LOCATE 13 + EDDY - I + (I < 4) + (I < 3) + (I < 2), 57
            PRINT F3(EDDY); "              "
         END IF
3326     IF EDDY = I THEN
             LOCATE 13 + EDDY - I + (I < 4) + (I < 3) + (I < 2), 57
             PRINT "                  "
          END IF
3330     NEXT EDDY
3340     RETURN
3350     '
3360     FOR EDDY = I - 3 - (I < 4) - (I < 3) - (I < 2) TO I
3370         LOCATE 13 + EDDY - I + (I < 4) + (I < 3) + (I < 2), 57
             PRINT F3(EDDY); "                  "
3380     NEXT EDDY
3390     RETURN
4000     '                                    Return B => A
4010     CLS : LOCATE 3, 24: PRINT "****     SEDCOF PROGRAM     ****"
4020     LOCATE 7, 3: PRINT "Rotor speed (rpm)", , R1; "                    "
4030     LOCATE 8, 3: PRINT "Run duration (min.)", , D1; "                  "
4040     LOCATE 9, 3: PRINT "Periods of acceleration (min.)", D2; "          "
4050     LOCATE 10, 3: PRINT "Periods of decceleration (min.)", D3; "       "
4060     LOCATE 11, 3: PRINT "Sample volume (ml)", , V1; "                  "
4070     LOCATE 12, 3: PRINT "Overlay volume (ml)", , V2; "                 "
4080     LOCATE 13, 3: PRINT "Particle density (g/ml)", , P1; "             "
4090     LOCATE 14, 3: PRINT "Rotor temperature (DC)", , T1; "              "
4100     LOCATE 15, 3: PRINT "Rotor name          ", , " "; A1$; "          "
4150 GOTO 2030
5000     '                                          From line 1415
```

```
5010      PRINT : PRINT "Press the space-bar to continue"
5020      IF INKEY$ <> " " THEN 5020
5025      LIED = 0
5030      CLS
5040      PRINT A2$
5050      PRINT USING A4$; R1; D1
5060      PRINT USING A5$; D2; D3
5070      PRINT USING A6$; V1; V2
5080      PRINT USING A7$; P1; T1
5090      PRINT A2$
5095      PRINT A1$
5100      PRINT
5110      PRINT " NO    VOL    SUC    AVOL    RAD    DENS    VISCOS    S-VAL"
5120      PRINT
5130 RETURN
10000 ' window
10010 PI = 103.638 / 32.989
10195 IF MEN = 1 THEN LOCATE 23, 65: PRINT "Press Q to quit"; : GOTO 10205
10200 LOCATE 24, 65: PRINT "Press Q to quit";
10202 RETURN
10205 LOCATE 24, 57: PRINT " Press L for last value";
10210 RETURN
15000 RETURN
20000 RUN
```

Appendix D: Properties of sucrose solutions calculated from the polynomials devised by Barber

```
100       SCREEN 2: KEY OFF: CLS
105       LOCATE 3, 13
110       PRINT "PROPERTIES OF SUCROSE SOLUTIONS. COMPREHENSIVE PROGRAM."
115       LOCATE 7, 12
120       PRINT "This program produces a table showing W/W concentrations,"
125       LOCATE 9, 11
130       PRINT " W/V  concentrations,  mole  fraction  sucrose,  density"
135       LOCATE 11, 11
140       PRINT " and  viscosity  of  sucrose  solutions  for  a  given"
145       LOCATE 13, 11
150       PRINT " temperature.  Polynomial  coefficients  are  taken"
155       LOCATE 15, 11
160       PRINT " from   E.J. BARBER,  NATL.  CANC.  INST.  MONOGRAPH 21,"
165       LOCATE 17, 11
170       PRINT " 1966, 219-239."
175       LOCATE 24, 50: PRINT "(Press Q then ENTER to QUIT)";
180       REM
190       REM C=CONCENTRATION IN WEIGHT FRACTION (=W/W PC/100)
200       REM C1=CONCENTRATION IN W/V PERCENT
210       REM T=TEMPERATURE IN CENTIGRADE
220       REM Y=MOLE FRACTION SUCROSE
230       REM D=DENSITY (G/ML)
240       REM V=VISCOSITY IN CENTIPOISES
250       REM
255       GOSUB 10000: LOCATE 23, 3
260       INPUT "SELECT TEMPERATURE [Degrees Centigrade] ", T$
261       IF T$ = "Q" OR T$ = "q" THEN CLS : SYSTEM
262       IF VAL(T$) < 0 THEN
             BEEP
             LOCATE 23, 3
             PRINT "                                                    "
             LOCATE 23, 2
             GOTO 260
          END IF
```

```
263     LET T = VAL(T$)
280     IF T < 0 OR T > 60 THEN
            LOCATE 22, 2
            PRINT "TEMPERATURE IS OUTSIDE PROPER RANGE (0-60 DC)"
            LOCATE 23, 2
            GOTO 250
         END IF
282     CLS
283     PRINT "The selected temperature: "; T; " Degrees Centigrade."
285     LOCATE 3, 8
        PRINT "   W/W %     W/V %     MOL.FR.     DENS. g/ml     VISC. mPa.s"
288     LOCATE 4, 9
        PRINT " ---------------------------------------------------------------"
290     FOR C = 0! TO .76 STEP .05
300     Y = (C / 342.3) / (C / 342.3 + (1 - C) / 18.032)
310     IF T <= 30! THEN GOSUB 1000 ELSE GOSUB 2000
320     IF C <= .48 THEN GOSUB 3000 ELSE GOSUB 4000
330     L$ = "                ##    ###.##     #.####             #.###        #####.###"
335     C1 = D * C
340     PRINT USING L$; C * 100!; C1 * 100!; Y; D; V
350     NEXT C
360     LOCATE 24, 50: PRINT "Press Q then ENTER to QUIT";
450     LOCATE 23, 2: INPUT " Another calculation? (Y/N) ", Y$
455     IF Y$ = "Q" OR Y$ = "q" THEN RUN
460     IF Y$ = "Y" OR Y$ = "y" THEN RUN
470     IF Y$ = "N" OR Y$ = "n" THEN CLS : SYSTEM
480     GOTO 450
1000    REM DENSITY FOR 0<T<30 AND 0<C<.75
1010    D1 = 1.0003698# + .000039680504# * T - .0000058513271# * T ^ 2
1020    D2 = .38982371# - .0010578919# * T + .000012392833# * T ^ 2
1030    D3 = .17097594# + .000475300081# * T - .0000089239737# * T ^ 2
1040    D = D1 + D2 * C + D3 * C ^ 2
1050    RETURN
1150    REM
2000    REM DENSITY FOR 30<T<60 AND 0<C<.75
2010    D1 = Y * 342.3 + (1 - Y) * 18.032
2020    D2 = (212.57059# + .13371672# * T - .00029276449# * T ^ 2) * Y
        D3 = (1 - Y)
2030    D3 = D3 * (18.027525# + .00048318329# * T + .000077830857# * T ^ 2)
2040    D = D1 / (D2 + D3)
2050    RETURN
2150    REM
3000    REM VISCOSITY FOR 0<T<60 AND 0<C<.48
3010    G3 = SQR(1! + (Y / .070674842#) ^ 2)
3020    C2 = 146.06635# - 25.251728# * G3
3030    A = -1.5018327# + 9.4112153# * Y
3040    A = A - 1143.5741# * Y ^ 2 + 105041.37# * Y ^ 3
3050    A = A - 4692710.2# * Y ^ 4 + 103233490# * Y ^ 5
3060    A = A - 1102898100# * Y ^ 6 + 4592191100# * Y ^ 7
3070    B = 211.69907# + 1607.7073# * Y
3080    B = B + 169116.11# * Y ^ 2 - 14184371# * Y ^ 3
3090    B = B + 606547750# * Y ^ 4 - 12985834000# * Y ^ 5
3100    B = B + 135329070000# * Y ^ 6 - 549704160000# * Y ^ 7
3110    V = 10 ^ (A + B / (T + C2))
3120    RETURN
3220    REM
4000    REM VISCOSITY FOR 0<T<60 AND .48<C<.75
4010    G3 = SQR(1! + (Y / .070674842#) ^ 2)
4020    C2 = 146.06635# - 25.251728# * G3
4030    A = -1.0803314# - 20.003484# * Y
4040    A = A + 460.66898# * Y ^ 2 - 5951.7023# * Y ^ 3
4050    A = A + 35627.216# * Y ^ 4 - 78542.145# * Y ^ 5
4060    B = 139.75568# + 6674.73293# * Y
```

```
4070     B = B - 78716.105# * Y ^ 2 + 909675.78# * Y ^ 3
4080     B = B - 5538083# * Y ^ 4 + 12451219# * Y ^ 5
4090     V = 10 ^ (A + B / (T + C2))
4100     RETURN
4150     REM
4200     END
10000 ' window
10010 PI = 103.638 / 52.989
10200 RETURN
```

Appendix E: Interconversion of molecular parameters

```
1        ON ERROR GOTO 30000
50       SCREEN 2: KEY OFF: CLS
100      REM INTERRELATION OF MOLECULAR PARAMETERS
105      CLS : LOCATE 7, 1
110      PRINT "       The   physically   correct  relationship   between   the"
120      PRINT "       sedimentation  coefficient, S (in water at 20 D.C)  the"
125      PRINT "       molecular  weight, M ,the partial  specific volume, N ,"
130      PRINT "       and  the  frictional  ratio,  F (F/FO)  is  as  follows"
150      PRINT
160      PRINT "                         S=M^0.67*(1-N)/(N^(1/3))/K/F"
170      PRINT "                         ----------------------------"
180      PRINT "       where K is  a  natural  constant.  The  present  program"
190      PRINT "       can calculate any  one  of  these  parameters,  provided"
200      PRINT "       the remaining three variables are known."
201      PRINT
         PRINT "   (The value of N must be >0.3 and <1 .The value of F must be >=1 and
205      GOSUB 10000
210      P1 = 4 * ATN(1)
220      K = 6.02E+23 * 6 * P1 * .01002 * (.75 / (6.02E+23 * P1)) ^ (1 / 3)
230      T1$ = "Ss"
240      T2$ = "Mm"
250      T3$ = "Ff"
260      T4$ = "Nn"
262      T5$ = "Qq"
270      LOCATE 23, 2: PRINT "                                                 ";
274      LOCATE 23, 4: INPUT "INPUT PARAMETER TO BE COMPUTED (S, M, F OR N) "; T$
276      LOCATE 24, 4: PRINT "                                    ";
280      '
290      IF INSTR(T1$, T$) <> 0 THEN 1000
300      IF INSTR(T2$, T$) <> 0 THEN 2000
310      IF INSTR(T3$, T$) <> 0 THEN 3000
320      IF INSTR(T4$, T$) <> 0 THEN 4000
325      IF INSTR(T5$, T$) <> 0 THEN 20000
330     LOCATE 24, 4: PRINT "TRY AGAIN.";
340      GOTO 270
350      '
1000     REM CALCULATION OF SEDIMENTATION COEFFICIENT
1010     LOCATE 23, 3: PRINT "M =
1011     LOCATE 23, 3: INPUT "M = ", M$
1012     LOCATE 20, 13: PRINT "
1015     LOCATE 23, 3: PRINT "F =                                              "
1016     LOCATE 23, 3: INPUT "F = ", F$
1017     IF F$ = "Q" OR F$ = "q" THEN
              RUN
         ELSE
              F = VAL(F$)
              IF F = 0 THEN 1015
         END IF
1018     IF F < 1 OR F > 10 THEN GOTO 1015
```

```
1020    LOCATE 23, 3: PRINT "N =                                                      "
1021    LOCATE 23, 3: INPUT "N = ", N$
1022    IF N$ = "Q" OR N$ = "q" THEN
              RUN
         ELSE
              N = VAL(N$)
              IF N = 0 THEN 1020
        END IF
1023    IF N >= 1 OR N <= .3 THEN GOTO 1020
1030    S = M ^ (2 / 3) * (1 - N * .9982) / (N ^ (1 / 3)) * 1E+13 / K / F
1035    I1$ = "===>  THE S-VALUE (SVEDBERG UNITS) IS ####.#"
1037    LOCATE 20, 13
1040    PRINT USING I1$; S
1050    GOSUB 5000
1060    GOTO 1000
2000    REM CALCULATION OF MOLECULAR WEIGHT    Input S, F, N
2010    LOCATE 23, 3: PRINT "S =                                                      "
2011    LOCATE 23, 3: INPUT "S = ", S$
2012    LOCATE 20, 13: PRINT "                                                        "
2015    LOCATE 23, 3: PRINT "F =
2016    LOCATE 23, 3: INPUT "F = ", F$
2017    IF F$ = "Q" OR F$ = "q" THEN
              RUN
        ELSE
              F = VAL(F$)
              IF F = 0 THEN 2015
        END IF
2018    IF F < 1 OR F > 10 THEN GOTO 2015
2020    LOCATE 23, 3: PRINT "N =                                                      "
2021    LOCATE 23, 3: INPUT "N = ", N$
2022    IF N$ = "Q" OR N$ = "q" THEN
             RUN
        ELSE
             N = VAL(N$)
             IF N = 0 THEN 2020
        END IF
2023    IF N >= 1 OR N <= .3 THEN GOTO 2020
2025    I2$ = "THE MOLECULAR WEIGHT IS #######"
2030    M = (S * 1E-13 * K * F / ((1 - N * .9982) / (N ^ (1 / 3)))) ^ (3 / 2)
2035    I2$ = "===>   THE MOLECULAR WEIGHT IS #######"
2040    LOCATE 20, 13: PRINT USING I2$; M
2050    GOSUB 5000
2060    GOTO 2000
3000    REM CALCULATION OF FRICTIONAL RATIO     Input S, M, N
3010    LOCATE 23, 3: PRINT "S =                                                   "
3011    LOCATE 23, 3: INPUT "S = ", S$: LOCATE 24, 3: PRINT "                      ";
3012    LOCATE 20, 13
        PRINT "                                                   ";
        IF S$ = "Q" OR S$ = "q" THEN
             RUN
        ELSE
             S = VAL(S$)
        END IF
3015    LOCATE 23, 3: PRINT "M =                                                 "
3016    LOCATE 23, 3: INPUT "M = ", M$
3017    IF M$ = "Q" OR M$ = "q" THEN RUN:  ELSE M = VAL(M$)
3018    IF M < 1000 * S THEN
           BEEP
           LOCATE 24, 3
           PRINT "Values of M and S not compatible";
           GOTO 3010
         END IF
3020    LOCATE 23, 3: PRINT "N =                                                "
```

```
3021    LOCATE 23, 3: INPUT "N = ", N$
3022    IF N$ = "Q" OR N$ = "q" THEN
                RUN
        ELSE
                N = VAL(N$)
                IF N = 0 THEN 3020
        END IF
3023    IF N >= 1 OR N <= .3 THEN GOTO 3020
3030    F = (M ^ (2! / 3!)) * ((1 - N * .9982) / (N ^ (1 / 3))) / K / (S * 1E-13)
3035    I3$ = "===>   THE FRICTIONAL RATIO IS #.###"
3040    LOCATE 20, 13: PRINT USING I3$; F
3050    GOSUB 5000
3060    GOTO 3000
3120    STOP
4000    REM CALCULATION OF PARTIAL SPECIFIC VOLUME
4001    REM THIS IS DONE BY A BISECTIONAL METHOD      Input S, M, F
4010    LOCATE 23, 3: PRINT "S =                                                  "
4011    LOCATE 23, 3
        INPUT "S = ", S$
        LOCATE 24, 3
        PRINT "                                    ";
4012    LOCATE 20, 13: PRINT "                                                        "
4015    LOCATE 23, 3: PRINT "M =                                                  "
4016    LOCATE 23, 3: INPUT "M = ", M$
4017    IF M$ = "Q" OR M$ = "q" THEN
              RUN
        ELSE
              M = VAL(M$)
        END IF
4018    IF M < 1000 * S THEN
              BEEP
              LOCATE 24, 3
              PRINT "Values of M and S not compatible";
              GOTO 4010
        END IF
4020    LOCATE 23, 3: PRINT "F =                                              "
4021    LOCATE 23, 3: INPUT "F = ", F$
4022    IF F$ = "Q" OR F$ = "q" THEN RUN:  ELSE F = VAL(F$)
4023    IF F < 1 OR F > 10 THEN GOTO 4020
4024    N1 = 1.2: N2 = .4
4025    N = .8
4026    Q9 = S * 1E-13 * K * F / (M ^ (2 / 3))
4030    Y = (1 - N * .9982) / (N ^ (1 / 3)) - Q9
4040    IF ABS(Y) < .000001 THEN 4090
4050    IF Y > 0 THEN N2 = N
4060    IF Y < 0 THEN N1 = N
4070    N = (N1 + N2) / 2
4080    GOTO 4030
4090    I4$ = "===>   THE PARTIAL SPECIFIC VOLUME IS #.### ML/G"
4100    LOCATE 20, 13: PRINT USING I4$; N
4110    GOSUB 5000
4170    GOTO 4000
4180    END
5000    ' changing "CONTINUE WITH SAME PARAMETERS INPUT 1, OTHER 2, STOP 3"
5010    LOCATE 23, 3: PRINT "Press the space-bar to continue"
5020    EDDY$ = INKEY$: IF EDDY$ = "" THEN 5020
5030    IF EDDY$ = "q" OR EDDY$ = "Q" THEN RUN
5040    IF EDDY$ <> " " THEN 5020
5050    RETURN
10000   ' window
10005   LOCATE 3, 23: PRINT "CONVERSION OF MOLECULAR PARAMETERS"
10010   PI = 103.638 / 32.989
10200   LOCATE 24, 53: PRINT "Press Q then ENTER to quit";
```

```
10210     RETURN
20000     CLS : SYSTEM
30000     REM ERROR HANDLING ROUTINE
30010     BEEP
30020     CLS
30030     LOCATE 12, 16
          PRINT "AN ERROR WHICH WAS NOT EXPECTED BY THE WRITERS OF"
          LOCATE 13, 24
          PRINT "THIS SOFTWARE HAS OCCURRED"
          LOCATE 17, 21
          PRINT "Press ENTER to restart the program"
30040     LET A$ = INKEY$: IF A$ = "" THEN GOTO 30040
30050     IF ASC(A$) = 13 THEN RUN ELSE GOTO 30040
```

6

Isolation and characterization of membranes and cell organelles

W. HOWARD EVANS

1. Introduction

Centrifugal techniques have contributed crucially to the subcellular dissection of cells, providing important knowledge on their structure and function (1). Centrifugation is the basis of traditional methods for separating cell membranes and organelles. A number of other techniques which exploit various physical parameters (e.g. electrical charge) or biological properties (e.g. ligand affinity) have been investigated as ways of examining the complexity of organelles and membranes. However, even when using these other methods it is often the case that better results will be obtained if the material is first purified by centrifugal methods.

The subcellular fractionation of biological material at its highest development aims to provide the experimenter with reasonable amounts of membranes and organelles with known properties. The production and separation of subcellular particles is a complex procedure and only guidelines can be provided in this chapter, because detailed protocols that work well for one type of cell may prove to be totally unsuitable for another. Thus, the aim is to present a general account of the strategy and scope of centrifugation techniques in the preparation of membrane fractions and cell organelles from disrupted cells or tissues, and to point out the various experimental strategies which can be employed to enhance the purity of the desired membrane fraction, thus allowing its characterization.

This chapter is divided into three main sections. First, the relative merits of the various methods used for disrupting cells and tissues in a controlled fashion are discussed. Next, the applications of centrifugal methods for separating subcellular components are considered, with emphasis on the applications of new density-gradient materials which allow rapid and reproducible separations mainly under iso-osmotic conditions. The use of two-phase polymer systems is also described, since these offer a rapid and economical method for the preparation of membranes using low-speed centrifuges. Finally, methods, mainly enzymic, for the identification and yield assessment of the subcellular fractions will be commented upon. This chapter is concerned primarily with animal cells and tissues but whenever appropriate, information on plant systems is included. Other reviews dealing with cell fractionation of animal (2, 3) and plant (4–6) material have also been published.

2. Methods for homogenizing cells and tissues

Controlled and uniform breakage is the first and crucial step in any subcellular fractionation protocol. 'The best homogenate is the one which lends itself most successfully to fractionation' (1). The major aim is to disrupt the majority of cells uniformly and release into media of defined composition the various organelles in an intact, dispersed state, with the simultaneous conversion of the cell's endomembrane networks (e.g. endoplasmic reticulum, endocytic compartment, and Golgi apparatus) into sealed vesicles of a predominantly uniform size and of a right-side-out configuration with respect to the intrinsic bilayer asymmetry of membranes. It is important to limit the generation of polydispersed membrane components, for example, from the plasma membranes, of polarized cells, although under certain circumstances (e.g. when the separation of apical and basolateral plasma membranes is required) this can be exploited in differential centrifugation. Different general guidelines apply to hard and soft tissues of organisms. The subcellular fractionation of cells grown in tissue culture also poses some problems which are specific to cultured cells and these will also be considered. As a general rule, use a homogenization method which involves physical force to disrupt the outer cell membrane; in particular, avoid methods that use detergents to disrupt cells since their use can have very serious implications for most types of cell membranes. Having chosen a method for homogenization, it is always good practice to check routinely the degree of cell breakage obtained using one of the methods described in Section 2.5; try to use a homogenization method which will give at least 90% disruption of the cells without causing excessive damage to the subcellular membranes.

2.1 Tissues and organs

Many methods, summarized in *Table 1*, exist for disrupting tissues and organs. With 'soft tissues' (e.g. liver, brain, endocrine glands, and so on) the Potter–Elvehjem and Dounce homogenizers (*Figure 1*) which utilize liquid shear to various extents are usually the methods of choice in that they are very efficient, easy to use, and generally cause minimal damage to cell organelles. It should be noted that the shear forces generated by high-torque, motor-driven Potter–Elvehjem homogenizers can be very damaging to organelles, especially when insufficient attention is given to ensuring that the clearance between the pestle and the outer vessel is correct and that local heating does not occur. A major practical problem which often confronts the novice are the initial steps of the homogenization procedure. Before homogenization, tissues should be carefully

Table 1. Techniques for tissue and cell breakage

Apparatus	Mechanism	Comments
Dounce homogenizer	Liquid shear	Widely available in various sizes (1–30 ml). 'Tight'-fitting, useful for homogenizing cells first, swollen by suspension in hypotonic media. Hand operations lead to low shear forces. Inexpensive.
Potter–Elvehjem homogenizer	Liquid shear	Widely available and inexpensive. Hand operated or used with pestle attached to high-torque motor. Speed can be calibrated.
Polytron, Ultraturrax	Mechanical shear	Utilize circular saw at tip to disperse tissue. Commercially available in various sizes.
Nitrogen bombs	Gaseous shear	Single cycle dispersion of cells. Require calibration to minimize organelle damage.
Presses and pumps	Liquid shear	Pressure used to propel sample (mainly cells—especially bacteria) through small orifice (French) or wire or plastic mesh. Pumps manually, hydraulically or electrically operated with spring operated ball bearing. Pumps with controlled orifice through which cells are 'stripped' also available.
Sonicators	Liquid shear	Generate high shear forces which can damage macromolecules, e.g. DNA. Used for bacterial disruption. Generally used for tissue disruption when subfractionation is not required.
Glass beads	Mechanical shear	Grinding or blending by hand or by machine. Bead size used depends on the size of cell. Method used for fungal, bacterial, and plant cells.

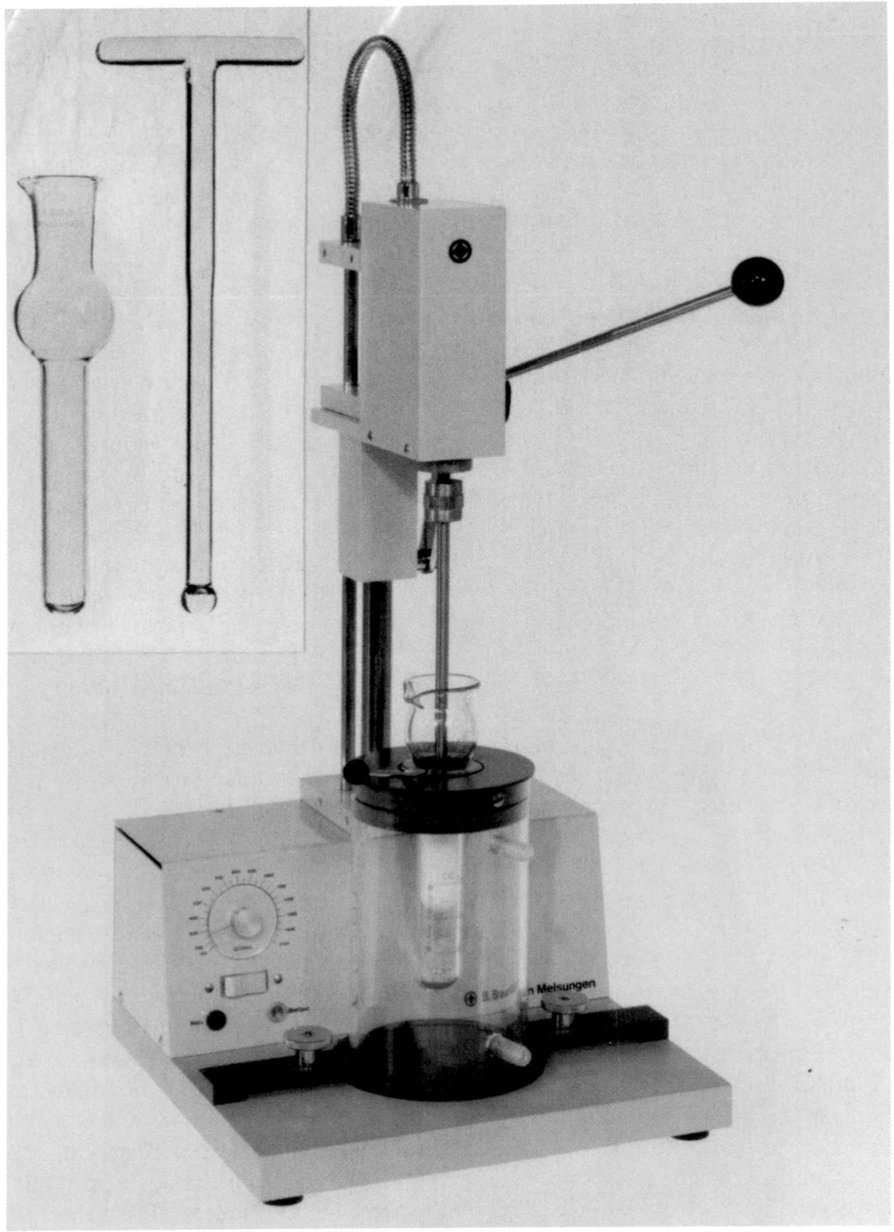

Figure 1. A Potter–Elvehjem homogenizer. In this apparatus the wall of the homogenizer vessel may be cooled by passing iced-water through the jacket. The remote control handle is a useful safety feature. The inset shows a Dounce homogenizer.

dissected to remove fibrous connective tissue and then diced into small pieces (2–3 mm cubes) using scissors or a sharp scalpel blade. If these procedures of choice are not feasible, then it is possible to subject the tissue to a short burst in a 'Polytron' or 'Ultraturrax' homogenizer (see next paragraph). If a tight-fitting homogenizer is to be used, it is useful to disrupt the tissue first (at about 5–10 g wet weight/100 ml homogenizing medium) using a loose-fitting homogenizer followed by rapid filtration through 3–4 layers of muslin cloth or fine nylon filters. When filtering homogenates avoid squeezing the cloth to get as much of the homogenate as possible since this will also force larger particulate material through the filter. The main homogenization step is carried out under standardized conditions, that is using a glass vessel and pestle of known clearance. Most commercially available Potter–Elvehjem and Dounce homogenizers have a narrow ('tight') clearance of 0.07 mm as opposed to the looser pestles where the clearance is 0.13 mm. It is important to be aware that if the clearance is too small then large organelles such as nuclei may be ruptured as they pass between the rotating pestle and the outer vessel; such damage may often prejudice subsequent fractionation and analytical procedures. The number of up and down passes should be recorded and, in addition, with the Potter–Elvehjem homogenizer, the speed of rotation should be noted. Typically, for soft tissues, 8–10 strokes of the pestle rotating at a speed of 1500 r.p.m. is sufficient to obtain good disruption. The Potter–Elvehjem homogenizer should be immersed in ice between the single up and down passages. A cooled homogenization apparatus (*Figure 1*) is available from Braun GmbH.

With 'hard tissues' (e.g. muscle) the approach varies. In the case of young tissues (e.g. neonatal hearts) then the methods described in the previous paragraph may prove effective. However, usually it is necessary to use homogenizers such as 'Polytron', 'Ultraturrax', or 'Omni' homogenizers all of which have similar modes of action involving mainly mechanical shear forces (*Figure 2*). These commercial machines utilize a rotating saw contained at the tip of the shaft (interchangeable shafts of various dimension are available). Some versions of these homogenizers also have a speed indicator and this helps the user to standardize conditions. Generally, a setting corresponding to 3–4000 r.p.m. for about 60 sec suffices to homogenize most tissues satisfactorily. For small pieces of tissue, for example, biopsy material, it is possible to use smaller probes, for example, a micro-probe of 5 mm diameter that can be used in various tubes holding 0.1–5 ml is available for some homogenizers. Alternatively, the Tissue-Tearor also appears useful for homogenizing small volumes (Biospec. Inc). For lymphoid tissue, proprietary tissue grinders, for example, from Sigma Chemical Co, are available that, in combination with a glass pestle, disrupt the cells teased out from tissues. They consist of an autoclavable, stainless steel wire mesh with screen sizes down to 0.737 mm, and a glass pestle is used to force the tissue through the holes in the mesh.

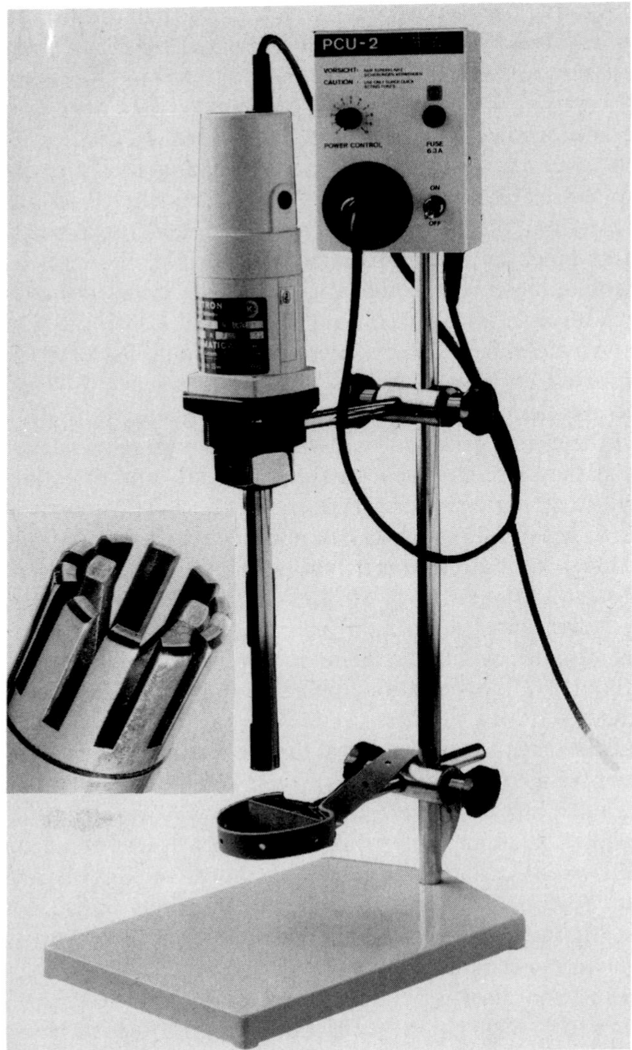

Figure 2. A Polytron homogenizer. Homogenizers of similar construction are available from other manufacturers. Probes of various sizes are available for most models. The inset shows the circular saw arrangement used to disperse the tissue.

2.2 Single-cell suspensions and cultured cells

Cultured cells are used increasingly in cell fractionation studies, but if strict adherence is kept to the general principles for homogenizing tissues just described then special problems can arise. The difficulties associated with the efficient homogenization of many types of cultured cells are often credited to the presence of a highly-developed cytoskeletal network. Thus, during cell breakage, the cytoskeletal filaments can wrap around the cell organelles in a manner similar to that encountered in the homogenization of muscle tissue. This can lead to the loss of an unacceptably large proportion of the desired subcellular component during the initial low-speed centrifugation steps. This problem can be overcome by treating the cell homogenate and derived fractions with extremely low concentrations of trypsin (see *Protocol 3* for details).

With various categories of cultured cells, the homogenization conditions need to be optimized according to the type of cell, its shape, density, growth conditions, and so on. Confluent cells are more difficult to disrupt than cells 12–36 h after passage. In general, about 1×10^7 cells or more are required to carry out a meaningful subcellular fractionation, that is one in which the subcellular component of interest will be subject so some degree of biochemical characterization; for example, its identification by marker enzymes.

Iso-osmotic media are to be preferred for cell breakage. In some cases the presence of 10 mM triethanolamine-acetic acid (pH 7.0) in the iso-osmotic medium can make cultured cells fragile enough to be susceptible to homogenization in a Potter–Elvehjem or Dounce homogenizer. Although clearly iso-osmotic media are to be preferred when disrupting cells, so as to minimize the possibility of damage to the organelles, a technique frequently employed with cultured cells involves suspending them for 5–10 min at 5 °C in a hypotonic medium (e.g. 5 mM Tris–HCl, Hepes, or imidazole, buffered to pH 7.4 and containing 0.5 mM $MgCl_2$) prior to homogenization. The swelling of cells induced by the hypotonic conditions improves the geometry for homogenization (and may detach some microfilaments from the inner surface of the plasma membrane), while the presence of magnesium ions helps to stabilize the structure of the cell nucleus. Once homogenization is completed, it is important to establish an iso-osmotic environment (0.25–0.3 M sucrose) without delay; this is usually done by adding a tenth volume of 2.5 M sucrose solution which avoids any substantial increase in volume. Low-speed centrifugation to sediment nuclei and cell debris is then carried out (*Figure 3*). It should be noted that cultured cells released from substrata by trypsin treatment should be allowed at least an hour to repair their cell surfaces, especially if surface antigens, receptors, and so on, are to be investigated.

The many different methods described for homogenizing cells have to be tailored depending on the type of cell. Repeated pipetting through a steel needle fitted with a gauze may suffice with some cells, especially after they have been subject to a hypotonic shock. With macrophages, for example, up to ten passes through the narrow orifice of two syringes connected by plastic tubing (0.25 mm inside diameter, 0.75 mm outside diameter) is effective and gives up to 80% cell disruption in 0.25 M sucrose, 0.5 mM EDTA, 20 mM Hepes–Tris (pH 7.4). However, in general, more vigorous methods have to be resorted to. The use of small, tight-fitting Dounce homogenizers (e.g. 1 ml Wheaton Scientific, Catalogue No. 357538) involving 20–30 passes may suffice to provide 70–80%

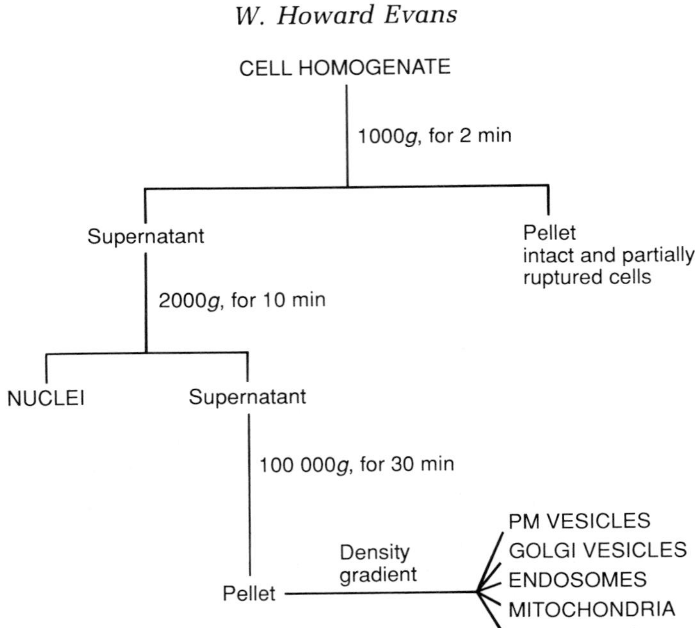

Figure 3. A typical protocol for the fractionation of subcellular organelles and membranes of cultured animal cells. It is standard procedure with cultured cells to analyse the post-nuclear supernatant directly (if volume is sufficiently small) or after pelleting in a Percoll or Nycodenz gradient. The positions of the organelles can be established by assay of the relevant markers. The resolution achieved is less than that obtained, for example, with liver tissue, as illustrated in *Figure 6* and discussed in the text. PM: plasma membrane.

breakage of a wide range of cultured cells; small plastic homogenizers (autoclavable) of 1.5 ml capacity are also manufactured by Treff, Switzerland.

2.3 Pressure cavitation

Nitrogen 'bombs' are available in various sizes and they operate on the basis of dissolving an inert gas (usually N_2) at high pressure (20–100 atmospheres) for specified periods in the cell suspension (*Figure 4*). In general, 60 atmospheres for 30 min appears adequate for the disruption of many categories of cultured cells. Mini-sized vessels (5–15 ml) are convenient for most procedures and these are available from most major manufacturers. The use of pressure cavitation, although providing quantitative disruption of cells, needs to be carefully controlled and monitored in terms of time and the pressure applied, since excessive decompression can lead to extensive damage to the organelles. The disruption of nuclei as the gas is released can result in chromatin attaching to negatively-charged macromolecules on membranes, leading to cross-contamination and the aggregation of subcellular components. The addition of DNase to the homogenate may help to minimize this problem but it is best

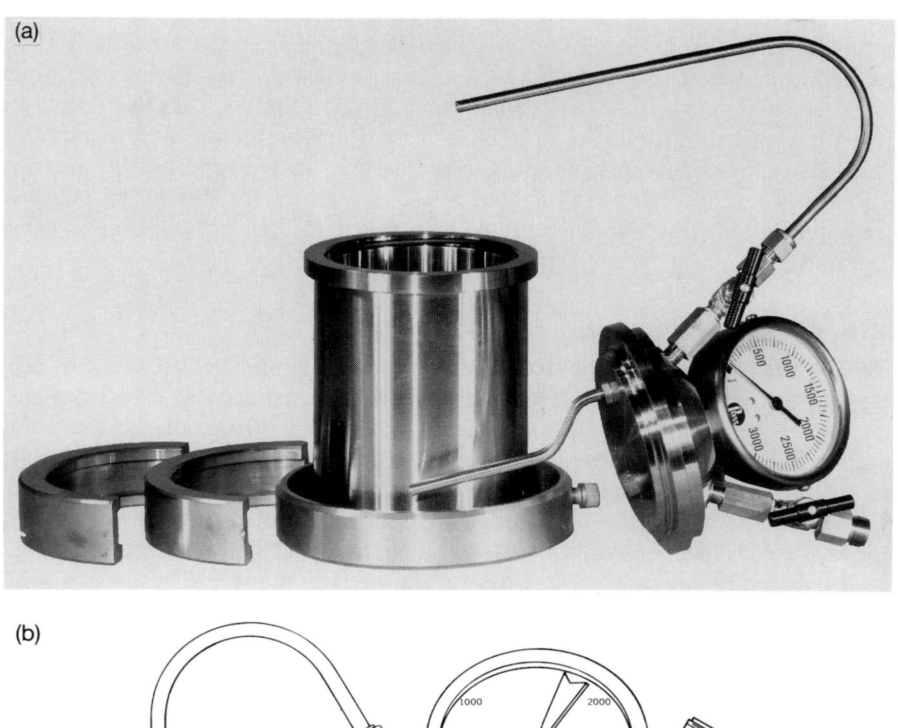

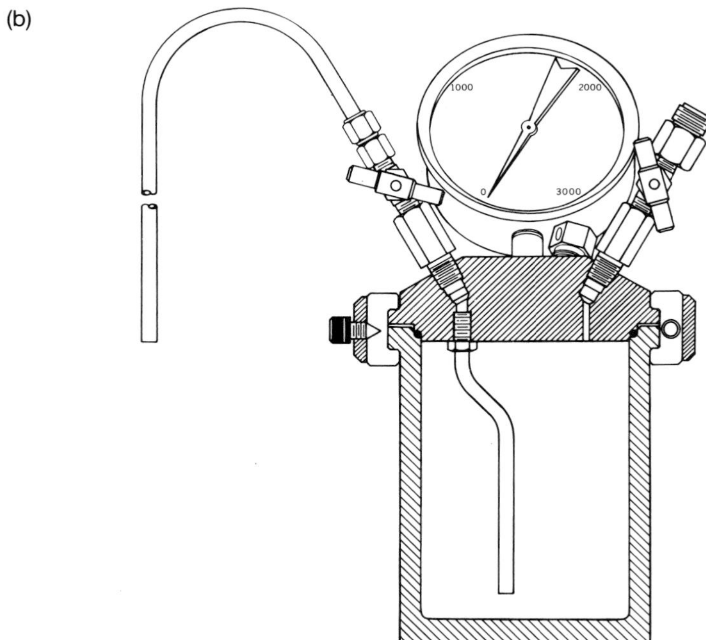

Figure 4. Pressure homogenizer. (a) View of disassembled parts of the homogenizer, (b) shown in cross-section.

to try and use conditions which completely avoid this problem. A further major problem arising from the over-zealous use of high pressures can be the conversion of endomembrane systems (and even organellar membranes such as the outer mitochondrial membrane) into small vesicles which can be difficult to separate when the membranes in the post-mitochondrial supernatant are fractionated by density gradient centrifugation. Nevertheless, used properly, the pressure cavitation method has proved to be a useful method for disrupting cultured cells such as fibroblasts which are recalcitrant to efficient disruption by other methods.

2.4 Cell crackers and pressure cells

These devices operate on the basis of effecting the controlled breakage of cells as they are pushed through an aperture often located around a ball-bearing; in some equipment the ball is flexibly attached to a spring, thereby avoiding 'jamming'. A variant of this method, used to homogenize Chinese hamster ovary (CHO) cells by Balch and Rothman (7) is similar but simpler in concept to the cell disruptors (e.g. 'Stansted') which are operated by electricity or compressed air and used to disrupt lymphocytes (8). The 'Stansted' cell disrupter (*Figure 5*) is distributed by Energy Service Co. In addition to lymphocytes, various parasitic organisms, yeast, bacteria, and plant cells can be broken open by a single passage at a pressure of 3 atmospheres (8).

The 'traditional' French pressure cells and laboratory presses in general that also operate at high hydraulic pressures (100–500 atmospheres) to push cells through a narrow orifice using a piston are used mainly with tough, cell-wall surrounded cells such as bacteria and plant cells. Equipment for small volumes (~4 ml) is available as well as higher-volume devices from Sim-Aminco, SLM Instruments Inc (UK distributors: D. G. Electronics). An alternative, relatively simple method which has proved to be very efficient for breaking open cells such as yeast and bacteria is grinding, blending, or shaking with glass beads. The glass beads are mixed with a 30% suspension of cells; 3 g of glass beads are added for each millilitre of cell suspension. The size of beads best suited for efficient breakage depends on the size of the cell. For yeast cells it is best to use 0.5–0.75 mm diameter beads since smaller beads can cause fragmentation of the cell organelles (9). Grinding with beads of 0.1 mm diameter can be used to disrupt bacteria quite effectively.

2.5 Determining the efficiency of homogenization

The efficiency of disruption of cells can be rapidly assessed by phase-contrast microscopy and by the permeability of the cells to vital stains such as trypan blue. This can be done by staining the homogenate with 0.1% toluidine blue in isotonic medium, then examination using a magnification of ×400 will reveal the extent of cell disruption with intact cells remaining unstained. As the fractionation proceeds, the morphological nature of the low-speed pellet (1000*g*

Figure 5. The 'Stansted' cell disruptor. The central unit is a pump (A) coupled to the valve in (B) where cell disruption occurs. Disrupted cells are collected in a beaker (D) via the delivery tube (C). Compressed air is supplied to the pump via the air regulator (F) which controls the pumping rate and to the adjustable valve regulator (G) which controls pressure applied to disrupt cells. An air outlet (J) which, in the case of biohazardous material, allows exhaust air to be passed through flask (H) filled with a sterilizing agent.

for 10 min) provides an immediate indication as to the effectiveness of the disruption step (*Figure 3*). Measurement of enzymic activities can also indicate the efficiency of cell breakage; for example, lactate dehydrogenase should be recovered mainly in the supernatant after centrifuging the cell homogenate at 1000*g* for 10 min. If it is intended to isolate plasma membranes, then an enzymic marker such as 5′-nucleotidase should be recovered predominantly (>75%) in the low-speed supernatant. It is acceptable if 85–95% of the cells have been broken open; attempts to achieve higher efficiency is often at the cost of organellar damage, especially of the larger organelles such as nuclei. With pressure cavitation, as described previously, different considerations apply since all cells are subject to the same disruption conditions.

Table 2. Protease inhibitors

Inhibitor	**Specificity**	**Normal concentration range**	**Comments**
Amastatin (also epiamastatin)	Amino exopeptidases	1–10 µg/ml	
Antipain	Cathepsin B, papain, trypsin	1–10 µg/ml	
Benzamidine	Serine proteases	Up to 10 mM	
Benzylmalic acid	Carboxypeptidases	1–10 µg/ml	
Bestatin (also epibestatin)	Amino exopeptidases	Up to 1 µg/ml	
Chymostatin	Cathepsin B, chymotrypsin, papain	1–10 µg/ml	
Diisopropylphosphorofluoridate (DFP)	Serine proteases	Up to 0.1 mM	Very toxic
Diprotin A and B	Dipeptidylamino peptidases	10–50 µg/ml	
Elastatinal	Elastase	10 µg/ml	
Ethylenediaminetetra-acetic acid (EDTA)	Metalloproteases	0.1–5 mM	Useful general inhibitor
Leupeptin	Cathepsin B, papain, plasmin, trypsin	1–100 µg/ml	
Pepstatin A	Carboxyl proteases (e.g. pepsin, renin)	1–10 µg/ml	Dissolve in dry ethanol or methanol rather than water

Phenylmethylsulphonyl fluoride (PMSF)	Serine proteases	Up to 0.1 mM	Limited half-life in aqueous buffers. Dissolve in dry isopropanol
Phosphoramidon	Collagenase, thermolysin	1–10 µg/ml	Not a general zinc protease inhibitor
Sodium tetrathionate	Thiol proteases	Up to 5 mM	
Tosyl-methyl ketone	Papain, trypsin		
Tosyl-phenylalanine chloromethyl ketone	Chymotrypsin		
Trypsin inhibitors Types I–IV	Chymotrypsin, trypsin		From various sources (e.g. chicken egg white, soybean) and with various activities

In addition to the above, other inhibitors to note include *p*-chloromercuribenzoate (PCMB), iodoacetamide, and heavy metal ions (for inhibition of -SH groups), α_2-macroglobulin (for inhibition of collagenases), aprotinin (from lung tissue), and other chelating agents besides EDTA (e.g. 2-phenanthroline, α,α^1-bipyridyl, sodium fluoride). Many of the inhibitors are available from Sigma Chemical Co. and Boehringer. Aprotonin is available as Trasylol® from Bayer, FRG, with subsidiaries in various countries. Nupercaine or tetracaine (0.15 mM) prevent lipid hydrolysis.

Proteolytic enzymes are classified according to type. Thus, serine proteases include trypsin, chymotrypsin, elastase, cathepsin G, and plasminogen activators active at pH 7–9 and are inhibited by DFP. Thiol proteases include various cathepsins active at pH 3–8 and inhibited by PCMB and iodoacetamide. Carboxyl proteases include cathepsin D and pepsin active at 2–7 and inhibited by pepstatin. Metallo-proteases include microvillus proteases and collagenases active at pH 7–9 and inhibited by EDTA and dithiothreitol.

A commonly used proteinase inhibitor cocktail includes A: pepstatin, 0.5 mg; leupeptin, 5.0 mg; chymostatin, 5.0 mg; antipain, 5.0 mg; aprotinin, 5.0 mg in 10 ml of water as a suspension and used at 1:100 dilution. All these reagents are available from Sigma Chemical Co. B: PMSF, 4.3 mg dissolved in 1 ml of ethanol and used at 200 µl/100 ml. C: 1 mM EDTA, DFP, and PCMB, *p*-chloromercuribenzoate.

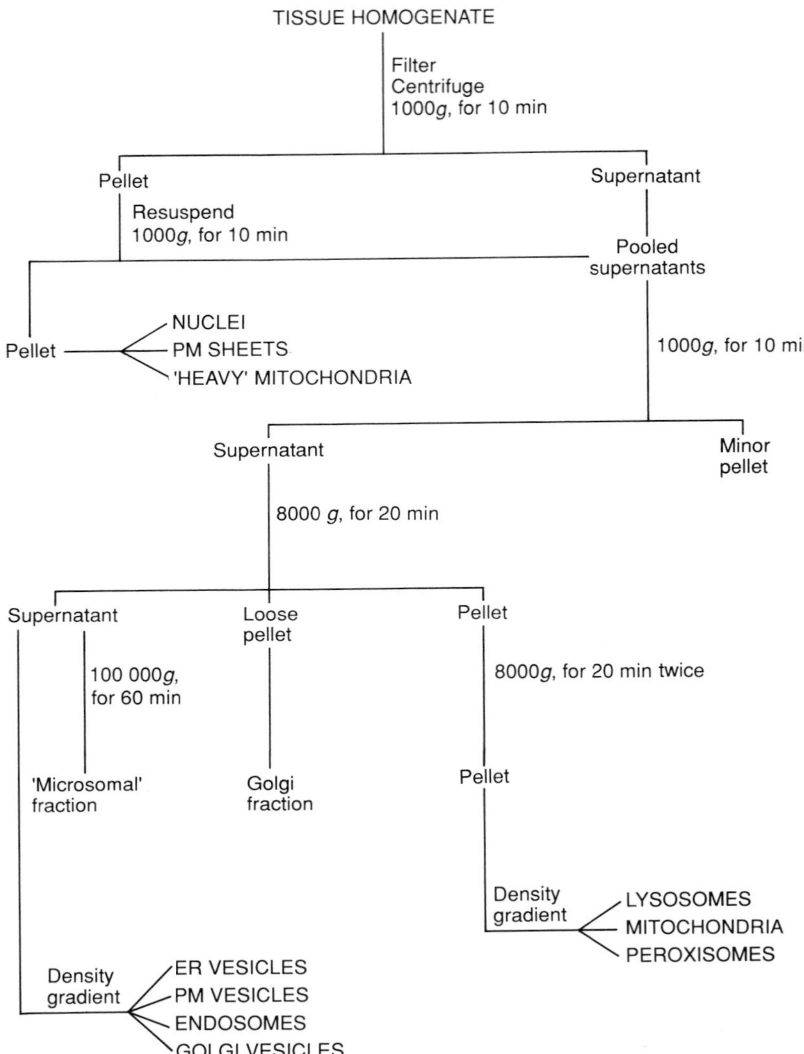

Figure 6. A generalized subcellular fractionation protocol for animal tissues based on the procedures used for liver. It is unusual for all membranes and organelles to be isolated from a single homogenate, without compromising purity, although comprehensive methods for liver (27) and intestine (47) have been reported. Plasma membrane sheets are obtained in highest yield when tissues are loosely homogenized in hypotonic alkali media and they are recovered by flotation in sucrose gradients from a 8000*g* for 10 min pellet (21). Density gradient separations involve the use of discontinuous sucrose gradients, or continuous Nycodenz or Percoll gradients (see Sections 3.3.3 and 3.3.4). With sucrose, 100 000*g* for 4 h or longer is required to resolve the fractions using swing-out or vertical rotors. Self-generated or constructed Nycodenz or Percoll gradients provide more rapid resolution. PM: plasma membrane. ER: endoplasmic reticulum.

2.6 Homogenization media and precautions to avoid membrane damage

The conversion of isolated cells or ordered arrays of different interacting cells in tissues and organs into a heterogeneous mixture of subcellular particles suspended in a diluted cell sap in readiness for fractionation by centrifugation is bound to introduce the possibility of damage to the various particles and hence the possibility of artifactual separations. Adherence to a carefully-defined homogenization protocol combined with rapid preparation at 4 °C helps to minimize damage to membranes by proteolytic enzymes which are either released from intracellular compartments (lysosomes) or are present as constitutive components of membranes. However, it may be advantageous to resort to the inclusion in the media of proteolytic inhibitors during subcellular fractionation. A commonly-used cocktail of proteinase inhibitors includes pepstatin, leupeptin, chymostatin, antipain, and aprotonin (*Table 2*). Typical concentrations used are 50 mg pepstatin and 500 mg of the other inhibitors as a stock suspension in 1 ml H_2O added to the homogenate to a final concentration of 1 µg/ml. Additionally, benzamidine (17 µg/ml) and phenylmethylsulphonyl fluoride (PMSF) (0.1–1 mM) can be added. Note that the PMSF must be prepared in dried ethanol or isopropanol as a $\times 100$ stock solution and it should be stressed that PMSF should be added drop-by-drop with stirring at 30 min intervals during the subcellular fractionation owing to its rapid hydrolysis in aqueous media at neutral pH. Calcium ions activate phospholipases and many proteolytic enzymes. Hence the addition of calcium ions should be avoided; indeed its chelation by addition of 1 mM EDTA is often standard practice for membrane fractionations. Media used in subcellular fractionation routines are normally buffered to neutral pH (e.g. by addition of 5–10 mM Tris or Hepes pH 7.5–8.0) that reinforces the intrinsic buffering capacity of subcellular components. If the pH of homogenates or derived fractions becomes more acidic, this can increase the activity of lysosomal proteolytic enzymes and so should be avoided. High ionic strength media should be avoided since these can lead to the aggregation of membranes. However, in some instances (see Section 3.2) it should be noted that selective aggregation and precipitation of membranes can be exploited to obtain membrane fractions.

A full list of the wide range of proteolytic inhibitors and their detailed properties is given in *Table 2*. However, before employing any of the enzyme inhibitors it is a wise precaution to consider whether any of them may have undesirable effects in that they may modify components of membranes or organelles that are to be studied. Particular care should be taken when using chemical inhibitors such as PMSF that, in general, are less specific in their action.

3. Centrifugal separations of subcellular components

Centrifugal techniques effectively exploit intrinsic differences in subcellular particles such as size, shape, density and, to a lesser extent, surface charges. In addition, intact, membrane-bound vesicles, whether occurring naturally in cells (e.g. secretory vesicles), or formed by vesicularization of membrane fragments during the homogenization procedure (e.g. endoplasmic reticular vesicles and endosomes), have their behaviour governed also by the contents trapped inside the vessicles. Indeed, as shown in Section 3.4, the exploitation of the content of vesicles can be used to modify selectively their density characteristics in gradients, and thus achieve a separation from other components with similar density characteristics.

3.1 Differential centrifugation or pelleting

This is the simplest method for obtaining a crude separation, on the basis of mass of the major organelles and membrane systems; the low-speed pellet (1000*g* for 10 min) contains nuclei, cell debris, and some larger subcellular components. For example, in polarized cells of tissues the baso-lateral plasma membrane sheets of cells attached to each other by intercellular junctions are also found in the low-speed pellet (*Figure 6*). It is customary to wash the nuclear pellet, by gentle resuspension in the homogenization medium using a loose-fitting (0.13 mm clearance) Dounce homogenizer, and then to repellet the material by repeating the centrifugation step.

Because nuclei are so much larger and denser than almost all other cell components they are ideal candidates for purification by differential centrifugation. However, nuclei isolated by simple low-speed centrifugation of cell homogenates are always very contaminated with membrane material. Membrane contamination can be greatly reduced by washing the crude nuclear pellet once or twice in isotonic homogenization medium containing 5 mM $MgCl_2$ (to stabilize the nuclei) and 0.5% Triton X-100 which dissolves the membranes. This treatment removes much membrane material, including the nuclear membrane, but gives nuclei which are enriched in nuclease activity, presumably as a result of contamination with lysosomal enzymes. In order to purify nuclei which appear normal in terms of morphology and most functions, it is best to avoid the use of detergents and instead to pellet them through dense (2–2.2 M) sucrose or band them in a 20–50% Nycodenz gradient (24); the latter approach is particularly useful if the cell contains other dense material such as melanin granules, and this is discussed further in Section 3.3.4. Note that nuclei isolated using any of the standard aqueous methods are likely to lose important proteins by leaching during the isolation procedure; this can be reduced by carrying out all of the isolation procedures in dense (2.2 M) sucrose. The use of non-aqueous procedures will stop all leachage of nuclear components (see Section 3.1.8 of Chapter 3) but the inconveniences associated with these techniques has discouraged most people from using this particular approach.

The pooled supernatants from the low-speed spin (post-nuclear supernatants) are next centrifuged for 20 min at 8000*g* to give a complex multilayer pellet containing mainly mitochondria, lysosomes, and peroxisomes. These steps are normally carried out using fixed-angle rotors of varying capacity; however, swing-out rotors can also be used. The pellets are usually washed two or three times by resuspension (using a loose Dounce homogenizer) followed by recentrifugation under the same conditions and the supernatants from the 8000*g* spin are then centrifuged at 100 000*g* for 40–60 min to give a 'microsomal fraction'. The microsomal pellet, which is often difficult to resuspend into an even suspension, is complex as it is composed not only of endoplasmic reticulum-derived vesicles containing either many (designated rough) or few (designated smooth) attached ribosomes but also, in the case of tissues such as liver, substantial amounts of non-membrane-bound polysomes and glycogen granules may be present. The microsomal fraction also contains endosomes (endocytic vesicles) which are derived by fragmentation of the complex vesico-tubular network (endocytic compartment/apparatus) that functions in animal cells in the uptake and intracellular processing of a variety of hormones, growth factors, and metabolites that attach initially to cell surface receptors prior to their internalization. Clathrin-coated vesicles originating mainly from coated pits on the plasma membrane tend to sediment slowly and are also recovered in this fraction. The location in such a fractionation scheme of components deriving from the Golgi apparatus is variable. The method of homogenization is clearly crucial in retaining, as far as possible, the intactness of Golgi stacks; the extent of vesicularization (in addition to the extent of the 'trafficking' vesicles that constitute the trans-Golgi networks) induced by homogenization in combination with the extent of development of the Golgi apparatus in various cells (secretory cells will contain more prominent Golgi stacks) determine the relative amounts recovered (as assessed by marker enzymes such as galactosyltransferase, see *Table 5*) in the mitochondrial-lysosomal pellet and in the microsomal fraction. In liver, Golgi components can be recovered as a loose fluffy overlay of the mitochondrial-lysosomal pellet, although the heterogeneity of the Golgi apparatus can lead to components becoming polydisperse during homogenization, resulting in variable recovery in either the mitochondrial or microsomal fractions. If Golgi vesicles are recovered mainly in the microsomal fraction, they are difficult to separate from endosomes and some plasma membrane vesicles during subsequent density gradient centrifugation. Further complicating the situation is the fact that during cell division the Golgi apparatus loses its characteristic structure, and it is possible that the Golgi vesicles will then vary in their subcellular fractionation properties.

The description of the distribution of organelles and membrane vesicles in the three major particulate subcellular fractions given here is based mainly on the fractionation characteristics of soft tissues, primarily liver tissue. The properties described, however, do apply in principle to most, if not all, animal cells. In plant cells, a similar subcellular fractionation routine can be used but other subcellular organelles such as tonoplasts, chloroplasts, and chromoplasts are present (4, 6). In the fractionation of lower eukaryotic cells such as yeast and ciliates some organelles, for example, nuclei, exhibit unusual characteristics and so these have to be borne in mind when deciding on the isolation procedures to be used (see, for example, 9). The key to success for all differential pelleting fractionation procedures is the choice of homogenization method and the nature of the homogenization medium as described in Section 2. For cells with cell walls (e.g. plant and microbial cells) the first step in fractionation involves breaking or the removal of the cell wall. In the latter case this is usually carried by mixed enzyme digestion; for yeast an extract of snail-gut containing a range of cellulases is often used. In cultured animal cells (*Figure 3*) the fractionation of well-disrupted cells is less complex although the ease of breakage is very dependent on the type of cell.

3.2 Cation-catalysed pelleting

An adjunct to differential centrifugation is the technique of specific precipitation of plasma membranes (especially discrete functional areas) by the addition of divalent cations, usually to a post-nuclear supernatant. Divalent cations (Ca^{2+}, Mg^{2+}) bind to sialic acid groups present on plasma membrane sheets and membrane vesicles (of right-side-out orientation), especially vesicles originating from the apical surface regions in polarized epithelial cells.

An example of the use of this relatively simple method which has the advantages of speed and simplicity is given in *Protocol 1*.

Protocol 1. Preparation of bile canalicular plasma membranes by calcium-ion induced aggregation (10).

1. Dice the liver into 2–3 mm cubes using fine scissors or a sharp scalpel blade and place in a nitrogen bomb.
2. Add 10 volumes of 0.25 M sucrose, 10 mM Hepes–Tris (pH 7.4), 14 mM $CaCl_2$, and homogenize by nitrogen cavitation (12 atmospheres for 15 min) as described in Section 2.3.
3. Centrifuge the homogenate at 7500*g* for 20 min to pellet nuclei, cell debris, mitochondria, lysosomes, and peroxisomes; carefully pour off the supernatant.
4. Centrifuge the supernatant at 47 000*g* for 20 min to pellet the bile canalicular plasma membrane fraction of acceptable purity; the endoplasmic reticular and endosomal vesicles remain in the supernatant.
5. Further purification, if required, can be achieved by fractionating the pelleted material on a Nycodenz gradient (see Section 3.3).

A similar procedure, using 10 mM $MgCl_2$ instead of calcium, has been used to precipitate microvilli differentially from a kidney homogenate post-nuclear fraction; this procedure involved only four short differential centrifugation steps (11).

As indicated in Section 2.6, the addition to homogenates and subcellular fractions of high amounts of calcium ions is not without its problems, because it can activate membrane-bound proteases and phospholipases as well as nucleases. Also, high concentrations of cations may interfere with the transport properties of apical vesicles.

3.3 Density gradient centrifugation

To obtain subcellular fractions of high purity it is usually necessary to fractionate further the components present in the crude fractions obtained by simple differential pelleting. The diversity of sizes of organelles and membrane fractions means that rate-zonal gradient separations are normally of little use. In contrast, isopycnic centrifugation in various types of density gradient media is a very powerful method of purifying subcellular fractions, because it exploits differences in density that exist between subcellular membranes and organellar systems. The major determinant of the position of membrane particles in density gradients is their ratio of lipid to protein; for example, mitochondrial inner membranes are protein-rich and thus have a high density (1.22 g/ml in sucrose gradients), whereas endosomal membranes are lipid-rich and are of low-density (1.10–1.13 g/ml in sucrose gradients) (*Table 3*). Other parameters determining density include the contents of vesicles; for example, secretory low-density-lipoproteins contained within Golgi vesicles make them more buoyant, whereas the protein content of secretory granules makes them denser (e.g. pituitary secretory vesicles, 45). The presence of attached components (e.g. ribosomes on endoplasmic reticulum membranes and clathrin on coated vesicles) also affects the density of membranes. While differences in composition do determine relative densities of subcellular components the degree of separation obtained also depends on the nature of the gradient medium (see Section 2.2 of Chapter 3).

Once the density characteristics of the subcellular components of interest have been established, and the overall complexity of the banding patterns in the gradient medium ascertained using continuous gradients, then it is often advantageous to do separations on discontinuous gradients in order to concentrate the particles of interest at an interface, thus allowing their rapid collection by using a pipette or syringe. It is important to ensure that discontinuous gradients are not overloaded in order to avoid the formation of 'mats' of material at interfaces, since these can limit the movement of particles to their isopycnic density. As mentioned previously, similar isopycnic separations can be achieved by using vertical rotors. The short pathlength and higher centrifugal forces generated by these rotors can save time or allow separations to be carried out using a high-speed centrifuge. However, care should be taken not to overload

Table 3. Properties of cell organelles and membranes

Organelle	% of homogenate protein	Particle dimension (μm)	Sedimentation force required to pellet	Density in sucrose (g/ml)
Plasma membrane[a]	0.4–2.5	3–20 (large sheets)[b]	1500*g*, 15 min	1.15–1.18
		0.05–3 (vesicles)	100 000*g*, 60 min	1.12–1.14
Nuclei	13	3–12	600*g*, 15 min	>1.30
Nuclear membrane	—	3–12	1500*g*, 15 min	1.18–1.22
Golgi apparatus[c]	1	1 (large)	2000*g*, 20 min	1.12–1.16
		0.05–0.5 (vesicles)	150 000*g*, 20 min	1.12–1.16
Mitochondria[d]	16	0.5–2	10 000*g*, 25 min	1.17–1.21
Lysosomes	2	0.5–0.8	10 000*g*, 25 min	1.19–1.22
Peroxisomes	3	0.5–0.8	10 000*g*, 25 min	1.18–1.23
Endoplasmic reticulum	24	0.05–0.30	150 000*g*, 40 min	1.06–1.23 (smooth)
				1.18–1.23 (rough)
Endocytic vesicles	1	0.1–0.5	150 000*g*, 40 min	1.10–1.13
Secretory vesicles[e]	—	0.1–0.5	150 000*g*, 40 min	

[a]Although in theory 5.4 mg/g liver should be recovered from rat liver, in practice values of 1 mg/litre (18–20%) are obtained. With cultured cells (e.g. RLT-28 hepatoma, 2.5 mg should be recovered from 10^8 cells, although in practice 0.3 mg (= 12%) is an acceptable value (43).

[b]Large sheets derive mainly from basolateral region and produced under low shear conditions. Vesicles derived from microvilli at apical or secretory aspect of the basolateral membrane. Note that under carefully defined conditions of tissue homogenization, large areas of the apical plasma membrane (e.g. bile canaliculi of liver parenchymal cells) when attached to lateral membranes sediment at low centrifugal forces. Tight-homogenization of these structures can then release the apical plasma membrane as vesicles (21).

[c]Golgi apparatus can be isolated as a single fraction in which the stacks or cisternae are intact or as those subfractions designated light, intermediate, and heavy (37–39). These are claimed to correspond roughly to the trans (light) and cis (heavy) aspects of the Golgi apparatus. Since Golgi fractions are cross-contaminated with endocytic vesicles, a method that isolates Golgi fractions and endocytic vesicles from the same homogenate offers advantages. (2, 42).

[d]Methods for the separation of mitochondria into inner and outer membranes exist (43). Note that the density of released outer membranes is 1.12–1.14 g/ml.

[e]Density of secretory vesicles and amount recovered varies according to the nature of the content. In liver, secretory vesicles are recovered mainly in Golgi-light fraction (see above).

gradients resulting in membrane components streaking along the walls of the centrifuge tubes and to avoid pelleting of the sample (see Section 4.3.3 of Chapter 2).

Density gradient separations, carried out mainly with the post-nuclear supernatants of cell or tissue homogenates, often aim to be comprehensive in the sense that a wide range of enzymic markers are measured across the gradient (unloaded into 10–20 fractions) and their distributions across the density gradient are analysed. Probably the best examples of the comprehensive analytical approach to identify the subcellular distribution and properties of organelles and membranes have been carried out using non-commercially-available zonal rotors as described by de Duve and his colleagues (12, 13; see also *Figure 7*). However, as stated previously, most isopycnic separations, whether applied to whole cell or tissue homogenates or to specific fractions (e.g. post-nuclear or post-mitochondrial supernatants), can also be carried out using conventional swing-out or vertical rotors with the additional advantage that up to six or more variables in the cells or tissues being studied can be examined simultaneously. The analytical density gradient technique is especially applicable to diseased tissues obtained by biopsy (15). Many density gradient separations are now carried out using gradient media other than sucrose and these are described in the following sections.

3.3.1 Use of sucrose gradients

Some subcellular fractions such as nuclei and free polysomes can be purified by centrifugation through a 'cushion' of 2 M sucrose; both of these fractions are very much denser than cell membranes and other organelles and so they will pellet through dense sucrose solutions leaving the membrane contaminants floating in the upper part of the tube. However, for other cell fractions it is usual to use a density gradient. As described in Section 5.3 of Chapter 3, density gradients are routinely prepared by mixing equal volumes of 'heavy' and 'light' solution in a mixing chamber as the gradient (dense solution first) is pumped via a tube into a centrifuge tube held at an angle to reduce disturbance. A number of proprietary Perspex density gradient generators are available. The most common density gradient material used is sucrose, since it is freely soluble in water, innocuous, chemically inert, and transparent to visible and UV light. Most importantly, it is cheap and available in high purity. Sucrose gradients extending linearly from 30–60% (w/v) are often employed to purify subcellular particles; it is a frequent practice to underlayer linear gradients with a cushion of 70% (w/v) sucrose occupying about 10% of gradient volume. Resolution and analysis of subcellular fractions, previously prepared by differential centrifugation, is usually carried out using linear sucrose gradients in a swing-out rotor; vertical rotors can also be used but fixed-angle rotors tend to give poorer separations because membrane fractions tend to pellet on to the wall of the tube (see Section 4.3.2 of Chapter 2). Typically, one can use the 17 ml or 34 ml gradients in a Beckman SW28 rotor or equivalent. The subcellular fraction to be fractionated (usually about 10–15% of the gradient volume) is suspended in 20% sucrose and layered carefully on the top of the gradient using a pipette. As one of the largest and densest of cell organelles, nuclei can be readily separated by centrifugation at 60–80 000*g* for an hour through dense sucrose. Most membrane fractions will equilibrate within 2–5 h at the maximum speeds of these rotors (about 80–100 000*g*). However, it should be noted that, because it is an isopycnic separation, the sample to be fractionated can also be 'sandwiched' into an appropriate part of the gradient after appropriate adjustment of the density of the sucrose. This allows the low density membrane components to 'float' to lighter parts of the gradient, whereas others sediment and equilibrate in the denser parts of the gradient. At the other extreme, samples may be suspended in a dense (~70% w/v) sucrose solution (a loose-fitting Dounce homogenizer is usually needed to obtain a well-dispersed suspension) and underlayered using a syringe fitted with a long metal cannula into the bottom of the tube below the gradient. After centrifugation for 4 h at 100 000*g*, particles should reach their isopycnic points in the density gradient located above the cushion.

3.3.2 An introduction to the use of other gradient media

The separations of subcellular particles afforded by use of other types of density gradient media are generally similar to those achieved in sucrose gradients but these media have been selected on the basis of their ease of use and low viscosity. Hence, gradients prepared from these materials provide the important advantages of speed and more especially of resolution (owing to lower viscosity), as well as the ability to carry out separations under iso-osmotic conditions, a highly desirable property when fragile organelles or membrane vesicles for transport studies are required. Iso-osmotic media limit damage induced by hypertonic shock of secretory vesicles (e.g. from the adrenal medulla or pituitary) that can lead to loss of contents and polydispersity of the vesicles. Percoll gradients also have the added advantage that they can be formed *in situ* by high-speed centrifugation.

3.3.3 Fractionation using Percoll gradients

Percoll is a sterile colloidal suspension of silica particles coated with polyvinylpyrrolidone (PVP) available from Pharmacia-Biosystems AB; a description of the properties of Percoll are given in Section 3.1.7 of Chapter 3. Separations are usually carried out in iso-osmotic (0.25 M) sucrose, buffered to pH 7.4–7.6 with 5–10 mM Tris–HCl or Hepes. Percoll gradients have been used extensively for the fractionation of plasma membranes, mitochondria, peroxisomes, synaptosomes, and lysosomes. In a generalized protocol, post-nuclear supernatants (suspended in 0.25 M sucrose) are mixed with 50% Percoll (in 0.25 M sucrose, pH 7.4) to give a final Percoll concentration of 30% (v/v) (this is equivalent to about 7% w/v silica). Centrifugation of this suspension at 25 000*g* for 20 min both generates the gradient and sediments the different membrane components to their isopycnic positions in the gradient. For this type of separation it is best to use a fixed-angle rotor since this will give the best gradient profile. Separations can be done in swing-out or vertical rotors, but often less effectively, and there is also more of a problem of pelleting of the Percoll in these rotors. Sometimes, when the sample is present while the gradient is forming, artifactual bands can be generated; if this occurs then form the gradient before loading on the sample and this usually solves the problem. If the separations are carried out in gradients made iso-osmotic with salt (0.14 M) then the gradient forms much faster and so the centrifugation conditions will need to be modified accordingly. Most of the major subcellular components can be separated on self-forming Percoll gradients (2).

The centrifugation conditions required for different tissues and cells vary greatly, but Percoll gradients have been applied successfully; for example, for the preparation of liver basolateral plasma membranes (15), apical and basolateral plasma membranes of kidney (16), endosomes and lysosomes of macrophages (17), and kidney lysosomes (18). More detailed practical information on the use of Percoll is given in Section 3.1.7 of Chapter 3 and has been published elsewhere (ref. 2; booklets provided by Pharmacia-Biosystems AB). It is claimed that, in the case of A431 human carcinoma cells, fractionation in Percoll gradients at pH 9.6 gives a better separation of endoplasmic reticulum, lysosomes, mitochondria, and plasma membranes than at neutral pH, suggesting that the use of alkaline media may prove advantageous (19). However, there are risks that working at such a non-physiological pH may

Table 4. Organelle and membrane densities in Percoll gradients

Fraction	Source	Density (g/ml)[a]
Chromaffin granules	Bovine adrenal	1.067–1.081
Endocytic vesicles	K562 cells	1.03
	Rabbit alveolar macrophages	1.05
Glycosomes	*Trypanosoma brucei*	1.087
Golgi	Rat liver	1.028–1.057
Lysosomes	Rat liver	1.087
	Rat kidney cortex	1.15
	Human lymphoblasts	1.085
	Pig thyroid	1.14
Mitochondria	Rat liver	1.085–1.100
	Bovine skeletal muscle	1.035–1.070
Peroxisomes	Rat liver	1.075
Plasma membranes	Rat liver sinusoidal	1.02–1.04
	Rabbit kidney cortex basolateral	1.037
	Rabbit kidney cortex brush-border	1.042
Synaptosomes	Rat brain	1.04–1.05

[a]Density applies to Percoll in 0.25 M sucrose.

alter the native conformation of subcellular components. *Table 4* summarizes the densities of various organelles and membrane fractions from different sources in Percoll gradients.

The density of fractions collected in Percoll gradients can be determined by simultaneously running a blank gradient containing coloured density marker beads obtainable from Pharmacia-Biosystems AB. A drawback of using Percoll is the probable need to remove the colloidal silica particles from the fractionated material. This can be achieved, to some extent, by differential centrifugation exploiting the differences in the rates of sedimentation of the fraction constituents and the colloidal silica particles; gel filtration in Sephacryl S-1000 superfine beads is an alternative method which appears to work. However, the colloidal silica particles do tend to stick to membranes and so neither of these methods can remove particles which are stuck to membranes. Percoll does interfere with protein determination by the Lowry procedure, although after precipitation with Triton X-100 the Bradford method can be used (see Appendix 4). Percoll is inert and has little or no effect when present in low amounts on enzyme marker assays, but it does absorb UV light and so may also interfere in some assays.

3.3.4 Fractionations using Nycodenz and metrizamide gradients

Nycodenz and metrizamide share with Percoll many similar advantages over sucrose gradients in terms of their low viscosity and osmolarity. However, both Nycodenz and metrizamide are completely different to Percoll in that they are not colloidal media. Nycodenz and metrizamide are non-ionic derivatives of triiodobenzoic acid linked to three aliphatic hydrophilic side chains, and they form true solutions. The structures and detailed properties of these media are described in Section 3.1.5 of Chapter 3. Information on the physico-chemical properties and use of Nycodenz is also available in booklets provided by the manufacturers, Nycomed A/S. Like Percoll, Nycodenz gradients can be generated *in situ* by centrifugation (self-forming gradients) generating continuous gradients to allow the separation of most subcellular components of densities between the range of 1.10 g/ml ('light' endosomes) and 1.24 g/ml (nuclei). However, because gradients form fairly slowly, in practice, it is most convenient to use preformed gradients using one of the methods given in Section 5 of Chapter 3. In practice, Nycodenz gradients have been applied to isolate a wide range of membrane fractions and organelles in high purity including liver endosomal fractions (density 1.090–1.110 g/ml) (20), liver apical (bile-canalicular) membranes (density 1.05 g/ml), basolateral plasma membranes (density 1.12–1.16 g/ml) (21), and liver lysosomes (1.11–1.13 g/ml) by a modification of the Wattiaux method (22). Indeed, lysosomes can be separated more effectively from mitochondria in Nycodenz gradients than in Percoll gradients (23), although such separations of organelles take 1–2 h at 100 000*g* in continuous density gradients involving the use of a Beckman SW28 swing-out or equivalent ultracentrifuge rotors. It is also possible, because of the low viscosity of Nycodenz solutions, to use lower centrifugal forces without radically increasing the time of centrifugation. Thus, using a post-nuclear rat-liver supernatant suspended in 0.25 M sucrose, 5 mM Tris–HCl (pH 7.6), 1 mM EDTA, good resolution of nuclei, mitochondria, lysosomes, and Golgi membranes (assessed by enzyme marker distribution) can be obtained using 20–50% (w/v) discontinuous Nycodenz gradients centrifuged in a swing-out rotor at 20 000 r.p.m. (80 000*g*) using a high-performance high-speed centrifuge (24). One of the advantages of purifying nuclei by banding them in a gradient of Nycodenz or metrizamide is that nuclei can be readily separated from other dense cellular components; this has proved particularly useful for the isolation of nuclei from the liver of *Xenopus laevis* which is rich in dense granules of melanin (48).

The majority of Nycodenz or metrizamide can be simply removed from the purified fractions by pelleting the particulate material after dilution with a suitable buffer, and the remainder can be removed by washing the pelleted material. Metrizamide and Nycodenz can also be removed from fractions by ultrafiltration or dialysis. Neither of these media has any significant affect on most enzymic activities. Protein can be assayed in the presence of Nycodenz or metrizamide using the Bradford method. The density of Nycodenz and metrizamide gradients can be determined from the refractive index of the solution (Appendix 3). For Nycodenz containing 0.25 M sucrose, as is often used for the separation of organelles, the relationship between density and refractive index (RI) is calculated as follows:

$$\text{Density (g/ml)} = 3.410 \times \text{RI} - 3.555$$

An alternative method, applicable to standard solutions, for determining density is to measure optical density at 350 or 360 nm, since the relationship between density and absorbance is linear and is given by:

$$\text{Density (g/ml)} = 0.0815 \times \text{OD}_{350} + 1.0 \text{ or}$$

$$0.1325 \times \text{OD}_{360} + 1.0$$

3.3.5 Fractionations using other gradient media

Ficoll (Pharmacia-Biosystems AB) is a neutral, highly-branched polymer of sucrose with an average molecular weight of 400 000. It dissolves readily in water and gradients up to 50% (w/v), equivalent to 1.2 g/ml, can be obtained, although such concentrated solutions are very viscous and can have a high osmolarity. Ficoll has been used in the isolation of plant vacuoles (4) and for rat-liver endosomes (25). However, its use has been largely superseded by the density gradient materials described in the preceding sections.

3.4 Enhancement of isopycnic gradient separations by density perturbation methods

A variety of approaches have been developed for modifying selectively the density of subcellular components. These density perturbants may be classified generally according to whether they interact directly with membrane components or whether they modify the density of vesicles by gaining entry to the vacuolar space and, in some instances, change the density of vesicles by forming precipitates within the vesicles, enabling selective fractionation of specific populations of membranes.

A commonly-used membrane perturbant is digitonin; this interacts with cholesterol increasing the density of the membrane and so its use is limited to those classes of membranes enriched in cholesterol such as plasma membranes, Golgi membranes, and endosomes. Although digitonin treatment has been carried out at 0.1–3.0 mg/mg of membrane protein, depending on the source and nature of the subcellular fractions examined, digitonin should strictly be used at an equimolar ratio with that of cholesterol determined in the membrane (2). Digitonin treatment of membranes at higher concentrations causes leakiness of membrane vesicles and can also result in solubilization of membrane constituents and cause damage to the functional integrity of membranes (2).

A wide range of vesicle content perturbants have been applied to the purification of subcellular membranes. For the subfractionation of muscle reticular and sarcolemmal vesicles, the differential distribution of Ca^{2+} pumping capacity, that is Ca^{2+}-ATPase activity, can be exploited. The separation of muscle membrane domains containing high Ca^{2+}-ATPase activity, for example, T-tubules, can be achieved by increasing the levels of calcium phosphate inside microsomal vesicles in the presence of ATP (26) or, alternatively, by inducing the precipitation of calcium oxalate inside the vesicle so that the sedimentation properties of the vesicles of interest are radically modified.

A further density perturbation method in current use involves the separation of liver endocytic vesicles from other vesicles of similar density (e.g. Golgi membranes, apical plasma membranes) by incorporation of horse-radish peroxidase which has first to be chemically conjugated to ligands to which cells are exposed. After receptor-mediated endocytosis there is a time-dependent concentration of the labelled ligands in the membranes that comprise the endocytic compartment. Incubation of the liver microsomal fraction with diaminobenzidine leads to an increase in the density of endocytic vesicles containing the chemically-modified ligand owing to the formation, in the presence of H_2O_2, of peroxidase-ligand-diaminobenzidine polymers (28). This method is described in *Protocol 2* and *Figure 7*.

Protocol 2. Diaminobenzidine density perturbation method for endosomes

1. Conjugate the appropriate ligand (e.g. galactosylated BSA) to horse-radish peroxidase (HRP) and separate this HRPL from the unconjugated ligand and HRP.
2. Define the experimental conditions for endocytosis so that the HRPL can be concentrated in the endocytic compartment of interest.
3. If possible, remove external and membrane-adsorbed HRPL by washing the cells or perfusing the tissue as appropriate.
4. Isolate the subcellular fraction enriched in the vesicles of interest by differential pelleting. Resuspend the sample gently in isotonic sucrose to avoid damaging the membrane vesicles and so releasing HRPL.
5. Purify the vesicles by isopycnic centrifugation in a suitable density gradient medium, typically 30–60% (w/v) sucrose or 20–50% (w/v) Nycodenz. This also removes any free HRPL which remains in the loading zone.
6. After centrifugation unload the gradient into fractions and locate those containing the vesicles of interest.
7. Incubate the fractions containing the vesicles with the HRPL in 5.5 mM 3,3′diaminobenzidine (DAB), 11 mM H_2O_2 in the dark at 25 °C for 15–30 min.
8. After incubation, load the samples on to a second density gradient which has a density at the top that is the same as the density at which the vesicles banded in the first gradient.
9. Centrifuge again to band the membrane vesicles, at the same time banding control samples which have been incubated with DAB in the absence of H_2O_2. At the same time also do controls to ensure that the density of vesicles not exposed to HRPL is not affected by incubation with DAB and H_2O_2 (*Figure 7*).
10. Those membrane vesicles containing the HRPL will be found banding at a significantly higher density than before.

However, after density labelling with DAB these endocytic vesicles, although enzymically characterized, cannot be analysed by SDS–polyacrylamide gel electrophoresis because many of the membrane proteins become cross-linked following this density-perturbing procedure.

A further example, involving the use of biological affinity of ligand-membrane receptor complexes to modify the subcellular behaviour of particles, is provided by the use of colloidal gold to isolate endosomal vesicles from the cultured cells. The density of gold–colloid–transferrin receptor antibody complexes localized inside the endosomal vesicles can allow the isolation, by centrifugation, of

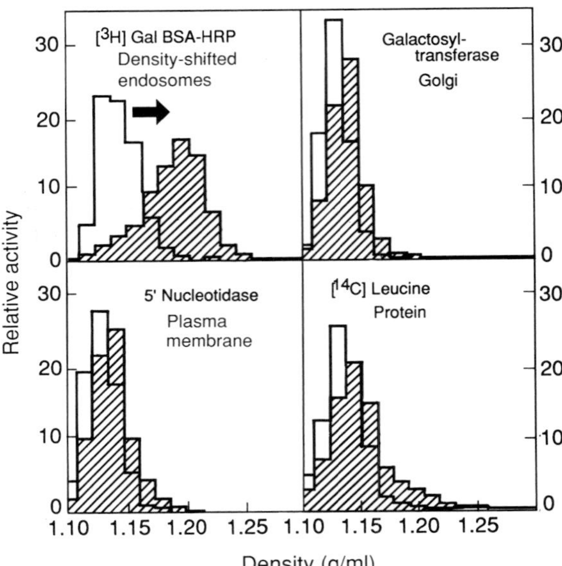

Figure 7. Density perturbation of liver endosomes. The marker used is endocytosed ^{3}H-galactosylated bovine serum albumin attached to horse-radish peroxidase. Ten minutes after uptake of the marker perfused through the liver, homogenates were prepared and the post-mitochondrial supernatant fraction collected. The figure shows that the banding density of the endosomes containing ^{3}H-labelled marker in a sucrose gradient is increased from 1.12 g/ml to 1.20 g/ml (hatched area) after the addition of diaminobenzidine. Little change in the density of Golgi or plasma membranes, as shown by marker enzymes, occurs since the control (no H_2O_2 added) (—) and the experimental (hatched areas) coincide. Protein distributions within the gradients were assessed after labelling with ^{14}C-leucine for 40 h *in vivo*. Details of the density shift method is given in *Protocol 2*. (Figure derived from reference 28.)

endosomes from A431 human carcinoma cells. The density of the colloidal gold complexes can modify the normal density of endosomes (1.11–1.13 g/ml) such that they become much denser and sediment to the bottom of a sucrose density gradient (>1.23 g/ml) (29) as shown in *Figure 8* and described in *Protocol 3*.

Protocol 3. Density perturbation of endosomes using colloidal gold labelled antibodies

1. Incubate confluent cells grown in culture in phosphate-buffered saline with ^{125}I-labelled gold antibody complex (~50 kBq) at 20 °C for 60 min.
2. Wash the cells three times with 0.25 M sucrose, 1 mM $MgCl_2$, 10 mM triethanolamine-acetic acid (pH 7.4) to remove excess gold-labelled antibodies. Scrape the cells from the dish using a rubber policeman and lyse the cells by repeated pipetting through an automatic pipettor. This should give about 80% breakage; if the cells are more resistant try one of the more rigorous methods listed in Section 2.2.

3. Treat the homogenate with 0.5 μg/ml of trypsin for 3 min at 37 °C followed by 0.5 μg/ml of soybean trypsin inhibitor and cool the homogenate to 0 °C.
4. Pellet unbroken cells and nuclei by centrifugation at 1000*g* for 10 min at 5 °C. Centrifuge the post-nuclear supernatant again at the same speed.
5. Take the post-nuclear supernatant and repeat Step 3; this helps to enhance the separation by dispersing the cytoplasmic filaments.
6. Pour 0.7 ml of 4% agar solution containing 52% sucrose in to a 14-ml centrifuge tube. When it has solidified cover the pellet of agar with a 13-mm disc of Millipore filter (0.1 μm) and load on top of the pellet a 26–52% sucrose gradient containing 1 mM EDTA and 10 mM triethanolamine-acetic acid (pH 7.4).
7. Load the post-nuclear supernatant from Step 5 derived from about 10^7 cells (in <1 ml) on to the gradient and centrifuge the gradients in a 6×14 ml swing-out rotor at 200 000*g* for 18 h at 5 °C.
8. Unload the gradients into 1-ml fractions; the gold-labelled endosomes will be found pelleted on the Millipore filter at the bottom of the tube (*Figure 8*).

3.5 Centrifugation in aqueous two-phase systems

The technique of aqueous two-phase partitioning has been widely used to isolate membrane fractions, especially plasma membranes from a number of sources, plant and animal (30). The procedure is relatively simple, rapid, and reproducible, and avoids the time-consuming aspects of some density-gradient methods that are accompanied often by the need for expensive centrifuges and rotors. The method also has the advantage of being easily scaled-up.

This method, which exploits membrane properties such as surface charge, density, weight, and hydrophobicity, is based on the differential distribution of subcellular particles present in the homogenate or subcellular fraction between the two compartments and an interface generated when dextran and polyethyleneglycol (PEG) aqueous systems are mixed. Low-speed centrifugation separates subcellular components mainly into the upper or lower phases generated (*Figure 9*). The order of affinity of animal cell membranes for the upper phase is: endoplasmic reticulum < mitochondria < lysosomes < Golgi < plasma membranes. In plant cells, the order is: inside-out thylakoids < intact chloroplasts < rough endoplasmic reticulum < mitochondria, peroxisomes, tonoplasts < smooth endoplasmic reticulum < thylakoids, Golgi, lysosomes < plasma membranes.

In general, tissue or cell homogenates are first centrifuged at low speed (3000*g* for 15 min in a swing-out rotor preferably) to yield a pellet that is then dispersed in the solutions as described in *Protocol 4*.

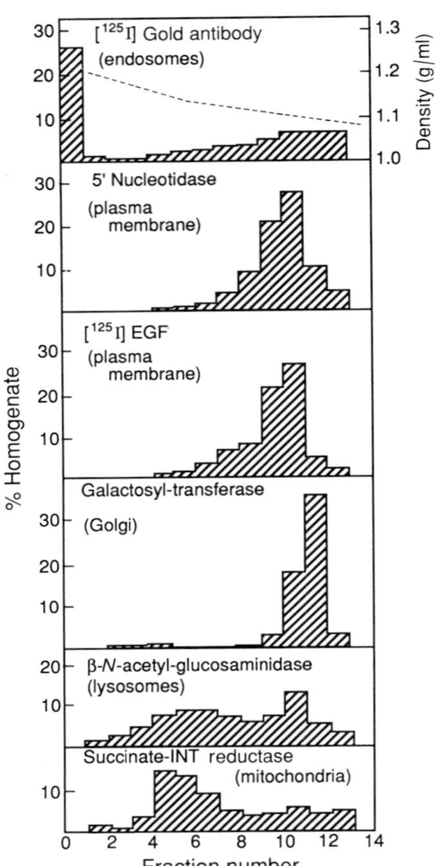

Figure 8. Density perturbation of endosomes from a cultured cell line (A431). The marker used is a colloidal-gold labelled antibody against the epidermal growth factor receptor that is endocytosed by the cells. The analysis of a post-nuclear supernatant fraction on a 26–52% sucrose gradient is shown. Details of the procedure are given in *Protocol 3*.

Protocol 4. Two-phase separation of membranes and organelles

1. Prepare stock solutions: Dextran (20% w/w). Dissolve 220 g Dextran T-500 (Pharmacia-Biosystems AB) in 780 g H_2O; stir and heat gently. Check and adjust, if necessary, to 20% using a polarimeter. Polyethyleneglycol (PEG) 3350 (Union Carbide) Dissolve 300 g in 1 litre of H_2O.
2. Prepare top and bottom phases as follows. Mix 200 g 20% dextran and 103 g PEG stock solutions with 33 ml of 0.2 M sodium phosphate (pH 6.5) and 179 ml H_2O. Allow the phases to separate (overnight) and collect the two phases—PEG is the top one and dextran the lower phase.

3. Homogenize the cells (10^7–10^8 cells) or tissue as described in *Table 1*. Centrifuge at 500*g* for 3 min to remove non-disrupted cells/tissue, and centrifuge supernatant to produce a pellet (3000*g* for 15 min). For tissue subfractions, various pelleted fractions may be used. Add particulate material to two-phase mixture prepared as described in Step 2.

4. Mix phases well (30–40 inversions) and then centrifuge at 2000*g* for 10 min. Keep the temperature at 4 °C. Collect the resolved phases and repartition against fresh upper or lower phases as shown in *Figure 9*.

5. Collect the relevant phases. Dilute upper phase 3–5-fold before collecting desired membranes by centrifugation; dilute lower phase 8–10-fold to collect relevant constituents.

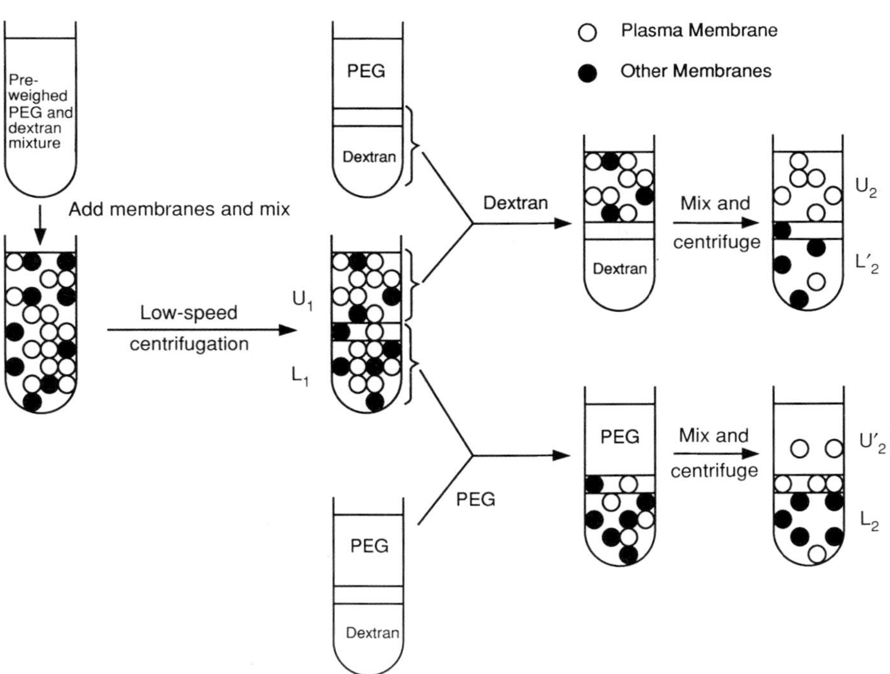

Figure 9. Purification of plasma membranes by two-phase separation. The subcellular fraction containing the plasma membranes and other membranes and organelles is added to a pre-weighed mixture of polyethyleneglycol, dextran, and a buffer (pH 6.5–7.0) to give a phase system (see *Protocol 4*). The centrifuge tube is inverted thoroughly 30–40 times and the phases are then separated by a low-speed centrifugation step. The upper phase (U_1) is removed and then repartitioned against fresh lower phase. Plasma membranes remaining in the lower phase after the first partitioning step are recovered by re-extraction with fresh upper phase. (Figure derived from reference 32.)

The procedure described in *Figure 9* requires modification according to the source of the membrane material, and variations in the centrifugation speed, polymer concentrations, pH and ionic concentrations have been used to obtain optimal separations. The dextran/polyethyleneglycol two-phase system has also been applied for the affinity partitioning of membranes, a purification procedure which exploits biological rather than general physical properties; for example, the covalent attachment of the lectin wheat germ agglutinin to activated dextran (33) allows liver plasma membranes containing a lectin binding site (*N*-acetylglucosamine) to partition into the lower (dextran) phase in contrast to their usual location in the upper phase shown in *Figure 9*. Such affinity-based procedures widen the scope and utility of these two-phase separation procedures.

4. Subcellular markers

The detection and quantification of the yield of subcellular membranes and organelles is carried out by the measurement of selected markers—usually biochemical (enzymic) but increasingly by immunological markers (*Table 5*). Previously, the assignment of markers in specific organelles and membrane systems was based on the concept advanced initially by de Duve (34), which stated that markers are located predominantly in one subcellular component or a specific region of that component and that they are retained during subcellular fractionation (34). This was amply illustrated by studies involving comprehensive analytical fractionation of liver tissue (13, 14, 35, 36). As fractionation procedures of cells, especially animal cells, become more refined with the subfractionation of the Golgi apparatus (37–39), the identification of 'early' and 'late' endosomes, clathrin-coated membranes, and the dissection of the discrete plasma membrane domains of polarized cells, the specific and exclusive assignment of markers has become compromised, but the general principles are still valid.

A list of the classical enzymic markers normally employed in subcellular fractionation studies is provided in *Table 5*. The present lack of ideal functional markers for the nuclear membrane, the cis-aspect of the Golgi apparatus, and the 'early' and 'late' vesicles isolated from the endocytic compartment has led to a search for immunological markers which can be applied in subcellular fractionation studies. As an example, the generation of antibodies to specific membrane receptors (e.g. mannose-6-phosphate receptors) in endosomes, mannosidase II in Golgi membranes, or to synthetic peptides of the deduced sequence of putative immunolocalized proteins of unassigned function can be used to assay, by Western blotting of fractions analysed by SDS–PAGE, the position of unknown membrane components in fractions resolved in density gradients (*Table 6*). These methods (often combined with immuno-cytochemical approaches) have shown, for example, that the mannose-6-phosphate receptor may be a useful marker for 'late' endosomal membranes present in polarized cells (e.g., MDCK cells and liver parenchymal cells).

Table 5. Markers for subcellular fractionation

Membrane type	**Enzymic or chemical marker**	**Membrane subtype**
Plasma membrane	Na^+K^+ ATPase	Basolateral
	Adenylate cyclase	Basolateral
	Specific cell surface receptors	Basolateral
	5′-nucleotidase	Apical
	Leucine aminopeptidase-glutamyltranspeptidase	Apical
	Alkaline phosphatase	Apical
Endoplasmic reticulum	Glucose-6-phosphatase	
	NADPH-cytochrome *c* reductase	
	Choline phosphotransferase	
	Cytochrome b_5	
Mitochondria	Succinate dehydrogenase	inner membrane
	Cytochrome oxidase	inner membrane
	Monoamine oxidase	outer membrane
	Kynurenine-3-hydrolase	outer membrane
Golgi apparatus	Galactosyltransferase	trans and middle regions
	Sialyltransferase	trans and middle regions
	NADP-phosphatase	trans and middle regions
	Mannosidase II	cis region
Lyosomes	Acid phosphatase	
	Glucuronidase	
	Aryl sulphatase	
Endosomes	Monensin-activated Mg^{2+}-ATPase in addition to un-degraded internalized ligands	
	Specific GTP binding proteins *rab* 7 'late' endosomes	
	Specific GTP binding proteins *rab* 4 and 5 'early' endosomes	
Peroxisomes	Catalase	
	Carnitine palmitoyl trans-ferase	
Cytosol	Lactate dehydrogenase	
Plant Cells (in addition to above)		
Plasma membrane	K^+ stimulated, Vanadate inhibited, ATPase, glucan synthetase II	
Tonoplast	C^1-stimulated, NO_3-inhibited ATPase	
Chloroplasts	Chlorophyll	
Amyloplasts	Monogalactosyldiglyceride synthetase	
Golgi apparatus	Inosine diphosphatase	
Bacterial membranes (specific)		
Inner membranes	D-lactate dehydrogenase	
Outer membranes (gram negative)	Phospholipase A1	

Table 6. Advantages and disadvantages of various subcellular markers

Method	Advantages	Disadvantages
A. Endogenous markers		
Enzymic	Standard and convenient to measure Organellar profiles established Allow quantification of subcellular fractionation	Some enzymes (e.g. glycosyl transferase) need to be measured immediately Latency in sealed vesicles (opened up by addition of saponin) Nuclear membranes, endosomes lack clear markers Absolute confinement to organelle not always tenable
Chemical	Mainly supportive	Often tedious to measure, e.g. sialic acid for plasma membrane, cytochrome b_5 for endoplasmic reticulum
Morphological	Essentially supportive, e.g. brush borders, intercellular junctions	Normally highly selective, although morphometry possible
Immunological	Antibodies to specific organelle proteins and receptors in sub-compartments, e.g. Golgi or endosomes used in Western blots of PAGE	Antibodies not always commercially available Receptors undergo recycling from cell surface to intracellular membranes

B. Exogenous markers		
Surface labelling		
(a) Proteins	Covalent, highly-sensitive	I_2-permeability
Lactoperoxidase	Labelled proteins can be examined by PAGE directly (with ^{125}I) or indirectly (with ^{125}I-streptavidin)	Quantification
Bolton + Hunter reagent		Internalization of plasma membrane proteins (if temperature $> 20\ °C$)
Diazotized di-iodosulphanilic acid		
Biotinylation		
(b) Carbohydrates	Exclusive to cell surface carbohydrates	Low sensitivity
Galactose-oxidase borohydride tritiation		
C. Internalized markers		
(a) Labelled ligands	Can identify endocytic vesicles when administered to intact cells or perfused tissues	Ligand dissociation during subcellular fractionation
	Peroxidase sensitivity can be exploited to density label endosomes	Membrane proteins cross-linked preventing PAGE
(b) Viral markers	Viral VSV G-proteins identify endomembrane vesicles	High infectivity may modify trafficking pathways

VSV: vesicular stomatitis virus

A further method for identifying endosomal subcellular membrane vesicles lacking a clear biochemical marker entails their isolation by following the location of radiolabelled endocytosed ligands in fractions separated in density gradients (e.g. see *Figures 7* and *8*). Since membrane migration in cells is a temperature-dependent process, the manipulation of the temperature (e.g. to 21 °C) at which the cells are incubated prior to homogenization can arrest intracellular traffic and thus internalized marker ligands at appropriate intracellular positions. As an example, 'early' and 'late' endocytic vesicles were isolated from liver and other sources on the basis of their high concentration (relative to cell/tissue homogenates) of receptor-bound radioactive ligands in microsomal fractions analysed by density gradient centrifugation, and purified further on Nycodenz gradients (20, 42). In a similar way, the intermediate or 'salvage' vesicles which migrate between the endoplasmic reticulum and the 'cis' aspect of the Golgi apparatus were identified on the basis of the location of viral proteins inside the cell (46).

5. Conclusions

Centrifugal methods continue to be the main procedure for the isolation of subcellular components from animal and plant sources. Indeed, the increasing availability of centrifuges and rotors, especially compact, cooled, bench-top centrifuges which allow either low, intermediate, or even very high speeds to be attained, make centrifugation a useful, if not exceptional, method for organelle/membrane isolation. Nevertheless, it is important to bear in mind that the care exercised in homogenizing tissues or cells is amply repaid when reproducible separations are required. The availability of Percoll and Nycodenz also allow the rapid analysis of primary fractions prepared first by differential centrifugation, with their low osmolality and iso-osmotic properties ensuring the preparation of subcellular organelles and membranes useful for functional studies (e.g. membrane transport). The number of characterized membrane components gradually increases as organellar networks are subject to finer dissection using new subcellular markers, often the products of immunological/immunocytochemical approaches. The application of centrifugal separations is frequently a necessary prelude to the provision of material for further analysis according to other physical (e.g. electrophoretic), or biological (e.g. affinity) properties.

References

1. De Duve, C. and Beaufay, H. (1981). *J. Cell Biol.* **91**, 293.
2. Evans, W. H. (1988). In *Biological membranes: a practical approach* (ed. J. B. C. Findlay and W. H. Evans) p. 1. Oxford University Press, Oxford.
3. Kinne-Saffran, E. and Kinne, R. K. H. (1989). *Methods Enzymol.* **172**, 3.
4. Morré, D. J., Brightman, A. O., and Sandelius, A. S. (1988). In *Biological membranes: a practical approach* (ed. J. B. C. Findlay and W. H. Evans) p. 37. Oxford University Press, Oxford.

5. Wagner, G. J. (1987). *Methods Enzymol.* **148**, 55.
6. Briskin, D. P., Leonard, R. T., and Hodges, T. K. (1987). *Methods Enzymol.* **148**, 542.
7. Balch, W. E. and Rothman, J. E. (1985). *Arch. Biochem. Biophys.* **240**, 413.
8. Wright, B. M., Edwards, A. J., and Jones, V. E. (1974); *J. Immunol. Methods* **4**, 281.
9. Walworth, N. C., Goud, B., Ruohala, H., and Novick, P. J. (1989). In *Methods in cell biol.* 31A (ed. A. M. Tartakoff) p. 355. Academic Press, New York.
10. Inoue, M., Kinne, R., Tran, T., Biempica, L., and Arias, I. M. (1983). *J. Biol. Chem.* **258**, 5183.
11. Booth, A. G. and Kenny, A. J. (1974). *Biochem. J.* **142**, 575.
12. De-Duve, C., Berthet, J., and Beaufay, H. (1959). *Progress in Biophys. Biochem.* **9**, 325.
13. Smith, G. D. and Peters, T. J. (1980). *Europ. J. Biochem.* **104**, 305.
14. Peters, T. J. (1976). *Clin. Sci. Mol. Med.* **51**, 557.
15. Epping, R. J. and Bygrave, F. L. (1984). *Biochem. J.* **223**, 733.
16. Hilden, S. A., Johns, C. A., Guggino, W. B., and Madias, N. E. (1989). *Biochim. Biophys. Acta* **983**, 77.
17. Wileman, T., Boshans, R. J., Schlesinger, P., and Stahl, P. (1984). *Biochem. J.* **220**, 665.
18. Harikumar, P. and Reeves, J. P. (1983). *J. Biol. Chem.* **258**, 10403.
19. Payrastre, B., Plantavid, M., Etievan, C., Ribbes, G., Carratero, C., Chap, H., and Douste-Blazy, L. (1988). *Biochim. Biophys. Acta* **939**, 355.
20. Evans, W. H. and Flint, N. (1985). *Biochem. J.* **232**, 25.
21. Ali, N., Aligue, R., and Evans, W. H. (1990). *Biochem. J.* **271**, 185.
22. Wattiaux, R., Wattiaux-de Coninck, Ronveaux-Dupal, M-F, and Dubois, F. (1978). *J. Cell Biol.* **78**, 349.
23. Holzman, E. (1989). In *Lysosomes* p. 8. Plenum Press, New York.
24. Graham, J., Ford, T., and Rickwood, D. (1990). *Anal. Biochem.* **187**, 318.
25. Branch, W. J., Mullock, B. M., and Luzio, J. P. (1989). *Biochem. J.* **244**, 311.
26. Rosemblatt, M., Hildago, C., Vergava, C., and Ikemoto, N. (1981). *J. Biol. Chem.* **256**, 8140.
27. Croze, E. M. and Morré, D. J. (1984). *J. Cell Physiol.* **119**, 46.
28. Courtoy, P. J., Quintart, J., and Baudhuin, P. (1984). *J. Cell Biol.* **98**, 870.
29. Beardmore, J., Howell, K. E., Miller, K., and Hopkins, C. R. (1986). *J. Cell Sci.* **87**, 495.
30. Fisher, D. and Sutherland, I. A. (ed.) (1989). *Separations using aqueous phase systems. Applications in cell biology and biotechnology*. Plenum Press, New York.
31. Morré, D. J. and Morré, D. M. (1989). *Biotechniques* **7**, 946.
32. Larsson, C. (1985). In *Modern methods of plant analysis*, Vol 1, (ed. H. F. Linskins and J. F. Jackson) p. 85. Springer, Berlin.
33. Persson, A., Johansson, B., Olsson, H. and Jergil, B. (1991). *Biochem. J.* **273**, 173.
34. De Duve, C. (1971). *J. Cell Biol.* **50**, 20D.
35. Amar-Costesec, A., Wilson, M., Thines-Sempoux, D., Beaufay, H., and Berthet, J. (1974). *J. Cell Biol.* **62**, 717.
36. Mircheff, A. K. (1983). *Am. J. Physiol.* G-347.
37. Futerman, A. H., Stieger, B., Hubbard, A. L., and Pagano, R. E. (1990). *J. Biol. Chem.* **265**, 8650.

38. Trinchera, M. and Ghidoni, R. (1989). *J. Biol. Chem.* **264**, 15766.
39. Trinchera, M., Pirovano, B., and Ghidoni, R. (1990). *J. Biol. Chem.* **265**, 18424.
40. Sun, I., Morré, D. J., Crane, F. L., Safranski, K., and Croze, E. M. (1984). *Biochim. Biophys. Acta* **797**, 266.
41. Beckers, C. J., Plutner, H., Davidson, H. W., and Bakh, W. E. (1990). *J. Biol. Chem.* **265**, 18293.
42. Evans, W. H. (1985). *Methods Enzymol.* **109**, 246.
43. Navas, P., Nowack, D., and Morré, D. J. (1989). *Cancer Res.* **49**, 2147.
44. Keesey, J. (1987). *Biochemica information, a revised biochemical reference source.* p. 106. Boehringer Mannheim Biochemicals, PO Box 50816, Indianapolis, IN 46250, USA.
45. Devi, L., Gupta, P., and Fricker, L. D. (1991). *J. Neurochem.* **56**, 320.
46. Brandli, A. W. (1991). *Biochem. J.*, **276**, 1.
47. Moktari, S., Feracci, H., Gorvel, J. P., Mishal, Z., Rigal, A., and Maroux, S. (1986). *J. Membrane Biol.* **89**, 53.
48. Risley, M. S., Gambino, J., and Eckhardt, R. A. (1979). *Develop. Biol.* **68**, 299.

7

Centrifugation separations of mammalian cells

ADRIAAN BROUWER, HENK F. J. HENDRIKS, TERRY FORD, and DICK L. KNOOK

1. Introduction

Detailed analysis of the biochemistry, function, and behaviour of a specific cell population can be ascertained only on relatively homogeneous cell populations. However, cell preparations obtained from blood or tissues represent a mixture of different cell types. Many techniques are currently available for fractionating heterogeneous cell populations and for isolating fractions which are enriched for a specific type of cell, even if it constitutes only a minor component of the original population.

The cellular properties used as the basis for cell separation or purification include cell size, density, electrical charge, (antigenic) surface properties, and light-scattering characteristics. Since cell separations on the basis of parameters other than cell size and density cannot be accomplished by centrifugation techniques, they will not be considered in this chapter.

In practice, centrifugation proves to be a very powerful and flexible tool for the separation of cell populations. Examples of separations of cells by centrifugation for both analytical and preparative purposes will be described in this chapter. The major techniques available, namely, differential centrifugation, isopycnic density centrifugation, rate-zonal centrifugation (velocity sedimentation), and centrifugal elutriation, will be described in detail with a brief introduction outlining their theoretical background. For a more extensive introduction to the basic principles of centrifugation and sedimentation of particles in general, the reader is referred to other chapters in this volume and to other reviews (1–4).

2. Use of centrifugation for cell separations

Centrifugation can be applied to separate cells solely on the basis of differences in cell density as well as differences in sedimentation rate, which reflects both cell size and density. The methods available are summarized and compared with

Table 1. General characteristics of the various methods for the separation of cells by centrifugation

Property used	Nomenclature	Centrifugal force	Gradient shape	Specific equipment	Limitations	Major advantages
Density (ρ)	Isopycnic equilibrium density centrifugation	Relatively high (100–30 000g)	Continuous/discontinuous (ρ gradient $\cong \rho$ cells)	Swing-out, fixed angle, or zonal rotor	Overlapping density profiles; high centrifugal forces; capacity	Least sensitive to cell aggregation
Sedimentation rate (D and ρ)	Differential pelleting	1–300g	No gradient	—	Low resolution	Rapid; easy
	Unit gravity velocity sedimentation	1g	Continuous gradient	Special separation chamber	Capacity $\pm 50 \times 10^6$ cells; special equipment; long duration	Simple; inexpensive
	Velocity sedimentation or rate-zonal centrifugation	20–1000g	Continuous gradient: (linear/isokinetic (ρ gradient $\ll \rho$ cells)	Swing-out or zonal rotor	Low resolution or special equipment; capacity	Large capacity (zonal); simple (swing-out)
	Centrifugal elutriation	100–1000g	No gradient	Elutriator rotor and Beckman J-21 or J-6B centrifuge	Special equipment	Rapid; high viability; high resolution and capacity

D: cell diameter; ρ: density.

respect to their essential characteristics in *Table 1*. More detailed descriptions of the methods summarized in *Table 1* are given in the following sections (Sections 2.1 and 2.2).

2.1 Separation on the basis of density

For density separations, cells are included in, or layered on top of, a continuous or discontinuous density gradient. The density range of the gradient should cover the densities of the cells to be separated. After centrifugation to equilibrium, each cell type will be found at a position in the gradient which corresponds to its own density, irrespective of its size. Equilibrium is generally achieved only if the cells are subjected to centrifugal forces much higher than those used for velocity sedimentation or centrifugal elutriation. Isopycnic density centrifugation can be used for both preparative and analytical purposes. For preparative separations, the applications are limited, since, even within a homogeneous population of one type of cell, there is heterogeneity in cell density (3). This often results in a serious overlap in the density profiles of different cell types (see Section 5.2.1). However, in some cases changing the osmotic environment from isotonicity (~ 300 mOsm) can modify the density of cells by causing differential swelling or shrinkage of cells (31); cells can tolerate osmolarities as high as 500 mOsm without any loss of viability. Isopycnic density centrifugation has proved to be very efficient for separating cells, especially when combined with other techniques.

Cells can be introduced into the centrifuge tube in two ways, either by layering a (concentrated) suspension of cells on top of a preformed continuous or discontinuous density gradient or by distributing the cells throughout the gradient. In the latter case, cells are initially suspended throughout the density gradient and some (lighter) cells will find their equilibrium position by moving centripetally (flotation), while others sediment.

The choice of using continuous or discontinuous gradients depends on the aims of the experiment. Discontinuous gradients are useful for routine preparative purposes, for example, for the separation of red blood cells from white blood cells by a single-step density separation in Ficoll-metrizoate (see Section 5.1) or from non-parenchymal liver cells using Nycodenz or metrizamide (see Section 5.2). Discontinuous gradients are generally not recommended because of the higher risks of artefacts (selection) at interfaces and the generally poorer results obtained (3). Continuous gradients are often used analytically to correlate differences in cell density within populations of cells with differences in functional or biochemical parameters (see Section 5.2.1).

2.2 Separation according to sedimentation rate

For separations on the basis of sedimentation rate, cell size is the major determinant, but cell density is also important. This is reflected by the equation describing the ideal behaviour of cells in suspension subjected to a centrifugal field. The sedimentation velocity of a cell (dr/dt) can be expressed as:

$$\frac{dr}{dt} = \frac{r_p^2(\rho_p - \rho_m)\omega^2 r}{K_\eta}$$

where r_p is the radius of the cell, ρ_p is the density of the cell, ρ_m is the density of the medium, ω is the angular velocity of centrifugation, r is the distance to the centre of rotation, η is the local viscosity of the medium, and K is a constant. Several techniques can be employed to separate cells on the basis of sedimentation rate and these are described in the following sections.

2.2.1 Differential pelleting

This term refers to a procedure by which cells can be separated in the absence of a density gradient. The method involves allowing a homogeneous suspension of cells to sediment either under the Earth's gravity or using low-speed centrifugation. The larger, more rapidly sedimenting cells pellet to the bottom of the tube leaving the smaller, more slowly sedimenting cells in suspension. For all its simplicity and rapidity, this method is handicapped by the poor resolution of this approach for selecting pure populations of cells; for example, it is not easy to separate parenchymal liver cells from the much smaller non-parenchymal cells. However, this method can be used to isolate functional groups of cells and it is widely used for preparing the buffy coat fraction of blood and for preparing leukocyte-rich plasma from dextran-treated blood. Hence differential pelleting is used mainly for preparing enriched fractions of cells prior to further purification on a gradient. The resolution of differential pelleting may be enhanced in some instances by use of a supporting column of a Percoll solution (4); alternatively, if the cells are pelleted on to a pad of isotonic metrizamide or Nycodenz then all the non-viable cells pellet to the bottom of the tube leaving the viable cells on top of the cushion (32).

2.2.2 Unit gravity velocity sedimentation

This technique does not involve centrifugation. A thin layer of cell suspension is placed on the top of a preformed density gradient and left to sediment in to the gradient for one to several hours depending on the cells being separated. The shallow gradient has a maximum density less than that of the cells to be separated and is used only to stabilize the position of bands in the column against convection currents. The collection of fractions is started before the first cells reach the bottom of the gradient. Because the cells are sedimented only by the Earth's gravitational field, this technique is relatively slow. The application of unit gravity velocity sedimentation is discussed in Section 5.3.

2.2.3 Velocity sedimentation by rate-zonal density gradient centrifugation

This method is essentially the same as that for unit gravity separations, but employs centrifugation to decrease the time required to achieve the separation. Velocity sedimentation can be performed either in a normal swing-out rotor or in a reorienting zonal rotor. For small numbers of cells ($1-50 \times 10^6$) and for the separation of cells with large differences in sedimentation rate, centrifugation in swing-out rotors may give sufficiently good separations. For high cell numbers ($\geqslant 10^9$ cells), large capacity reorienting zonal rotors can be used. Zonal rotors require more skill and care, especially when working under sterile conditions.

A rather specialized method for the velocity sedimentation of cells, with a high resolution, has been developed by Tulp *et al.* (5). This method employs a special separation chamber which can be used in an adapted swing-out rotor of a low-speed centrifuge (see *Figure 2*). However, the capacity of the chamber ($< 50 \times 10^6$ cells) is rather low for preparative purposes.

In summary, velocity sedimentation can be a very simple technique when working with cells which are of very different sizes. However, for cells with relatively small differences in sedimentation rate and for larger numbers of cells, special equipment is necessary. The use of velocity sedimentation is discussed further in Section 5.3.

2.2.4 Centrifugal elutriation or countercurrent centrifugation

The term centrifugal elutriation describes a technique which involves the balance between a centrifugal force and a centripetal flow of liquid within a separation chamber (*Figure 1*). Because of the conical shape of the separation chamber, a gradient of liquid flow rate which opposes the centrifugal force is generated. Cells present in the separation chamber will be found in a position at which the two forces acting on them are at equilibrium. The position of each cell is determined by its size, shape, and density. No pelleting occurs and the fractions can be harvested by increasing the flow rate or decreasing the centrifugal force, allowing each fraction to be eluted from the chamber separately (*Figure 1*). The advantages of centrifugal elutriation include:

(a) one can use any medium in which cells sediment
(b) the viability of cells remains high
(c) the rotor has a large capacity range
(d) the rotor is capable of high resolution and
(e) only a short time is needed for separation.

On the other hand, the equipment is relatively expensive and requires an experienced operator. Application of this technique is described in detail in Section 5.4.

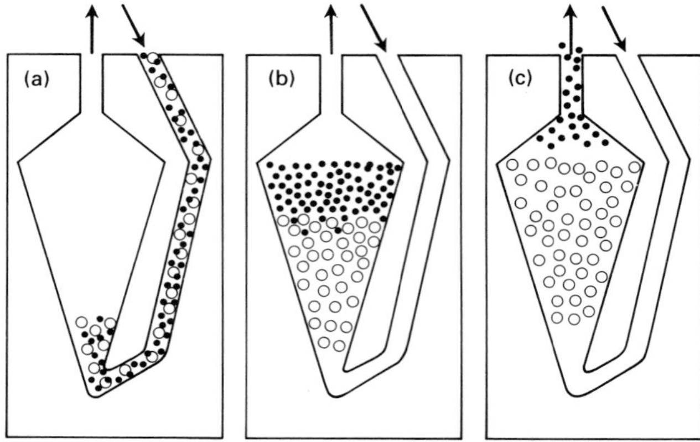

Figure 1. A diagramatic illustration of the behaviour of cells in the standard Beckman separation chamber during centrifugal elutriation. (Reproduced from the Beckman manual for the JE-6B elutriator rotor with permission.)

2.3 Gradients for cell separations

The range of gradient media that can be used for preparing density gradients for cell separations is rather limited. As described in Section 3 of Chapter 3, there are several criteria which must be fulfilled for a material to be an ideal gradient medium. Particularly relevant for cells are that the medium:

(a) is non-toxic to cells
(b) forms isotonic gradients of sufficient density
(c) does not permeate cells
(d) is easy to remove from the cells after fractionation and
(e) forms gradients of low viscosity.

The list of gradient media used for separating cells includes Percoll, Nycodenz, metrizamide, Ficoll, and serum albumin. Details of the properties of these various gradient media have been given in Section 3 of Chapter 3. In this chapter protocols are given for the separation of cells using either Percoll (Section 5.4.2) or Nycodenz (Sections 5.2.1–5.2.3) which are the most widely used gradient media for separating mammalian cells. Nycodenz has similar properties to metrizamide, widely used previously, but it has some practical advantages (e.g.

autoclavability and metabolic inertness) that now make it the iodinated gradient medium of choice for cell separations (7).

The shape of density gradients (which can be determined by plotting the density of each fraction of the gradient) can be varied. The shape of a gradient can be discontinuous, or continuous; in the latter case either linear or non-linear gradients can be generated. Linear and non-linear gradients can be generated easily using simple gradient makers (see Section 5.2.1 of this Chapter and Section 5 of Chapter 3). In addition, one can use isokinetic gradients which are designed in such a way that the increase in the density and viscosity of the gradient solution balances the increase in centrifugal force along the length of the centrifuge tube. Isokinetic gradients were specifically developed for rate-zonal separations.

When separating cells on density gradients special care must be taken to make sure that the gradients are isotonic and that the osmolarity varies by less than 10% throughout the gradient. Otherwise a cell travelling through the gradient may shrink or swell during the separation, altering its sedimentation behaviour and ending up at a position reflecting a density and size other than that of the original cell. With gradients formed by a gradient maker, homogeneous osmolarity is easily achieved by starting with two isotonic solutions. In the case of discontinuous gradients and continuous ones generated by the diffusion of discontinuous gradients, there is a risk of local changes in osmolarity due to differences in the diffusion properties of the solute forming the density gradient and that used as the osmotic balancer, usually NaCl. One advantage of using colloidal media (e.g., Ludox or Percoll) is that they are osmotically inert and hence stable iso-osmotic gradients can be prepared simply by adding NaCl to a final concentration of 0.9% prior to forming the gradients by centrifugation.

2.4 Problems and artefacts of cell separations

Various problems and artefacts may result in suboptimal or misleading separations. The significance of these artefacts and the means available to circumvent them are briefly discussed below. The major artefacts include the following.

2.4.1 Cell clumping

A major problem encountered in cell separation is cell clumping, since clumped cells will behave differently during centrifugation. Serious clumping will prevent sufficient purification and result in a reduction of the cell yield obtained. The usual measures which can be taken to reduce cell clumping include working at low cell concentrations and the addition of components such as albumin, serum, DNase (to reduce gelation), protease, or EDTA to the medium (6). In some instances, working at 4 °C may reduce clumping (3), although in the authors' experience with sinusoidal liver cells, the opposite has been found, in that the clumping of cells during the initial steps of the isolation procedure can be largely prevented by keeping the temperature above 20 °C.

2.4.2 Overloading of gradients with cells

For the optimal separation of cells by centrifugation in a density gradient, the number of cells applied should not exceed the capacity of the gradient. The gradient capacity can be defined as the number of specific cells that can be sedimented in a given experimental situation without observing any deviation from the ideal sedimentation pattern (3). Overloading a gradient with cells will result in the broadening of bands and an apparently faster sedimentation of cells. The gradient capacity must be estimated experimentally for any given situation, since no accurate theoretical descriptions are available (3).

2.4.3 Wall effects

Wall effects are the consequence of the fact that the direction of sedimentation within a normal centrifuge tube is not parallel to the wall of the centrifuge tube, even in swing-out rotors (see Section 3.2.1. of Chapter 1). This leads to concentration of cells at the periphery of the tube, thus increasing the risk of clumping and the loss of cells sedimented on to the wall of the centrifuge tube. In addition, the local density of the gradient may increase, resulting in a disturbance in the gradient stability. The loss of cells due to wall effects can be minimized by reducing the distance which cells have to travel, through a gradient and by increasing the distance of the tube from the axis of rotation. Centrifugation of cells in a fixed-angle rotor, as is sometimes used for isopycnic separations, generally results in serious wall effects.

2.4.4 Swirling

Swirling or vortexing of the tube contents during acceleration and deceleration can induce serious disturbances in the gradient and the separated bands. The changes in the angular velocity of the rotor are accompanied by an angular momentum termed Coriolis forces which cause rotational movement of the fluid. Swirling can be reduced by increasing the distance from the axis of rotation and by slow acceleration and deceleration. Sometimes it is possible to introduce a special anti-vortex baffle into the separation chamber (5). The effects of swirling can also be partially counteracted by using steeper density gradients.

2.4.5 Streaming (droplet formation)

When a suspension of cells is layered on top of a density gradient and left to stand without centrifugation, small droplets which fall into the gradient may be formed at the interface; this can occur only after a few minutes. This phenomenon appears to result from the diffusion of the gradient medium into the sample layer (30). It may be circumvented by reducing the cell concentration in the sample, by increasing the steepness of the gradient at the sample/gradient interface, or by centrifuging the gradients immediately after loading the sample. (See Section 2.1.3 of Chapter 3).

2.4.6 Centrifugal force

Depending on the type of cell, all cells are affected to some extent by centrifugation; typically, changes are seen in the viability of cells, internal organization of cells, disruption of the cell cycle, and changes in the patterns of transcription. Most of the disruptive effects of centrifugation on most types of cells appear to be readily reversible within a few hours. Cells can be damaged both by fast acceleration of the rotor and by using high centrifugal fields; if the centrifugal fields are very high then the hydrostatic pressure generated can damage the cells. In general, isopycnic centrifugation tends to be carried out using higher centrifugal forces than is used for rate-zonal sedimentation in order to ensure that the cells actually reach their isopycnic positions. The centrifugal forces can be reduced by decreasing the distance that the cells have to travel through the gradient. In addition, the use of low viscosity media (see Section 2.3) is recommended to ensure optimal sedimentation velocities of the cells. Centrifugal elutriation has the advantage that it can be carried out using relatively low centrifugal forces which have a minimal effect on the functioning of the cells.

2.4.7 Osmolarity effects

Because mammalian cells have no rigid cell wall, any change in the osmolarity of the medium surrounding the cell will induce a change in cell volume. Swelling or shrinkage of cells due to hypo- or hyper-osmolarity, respectively, will affect their density and sedimentation properties. For almost all types of cell separations, the gradients must have the same osmolarity throughout; usually the osmolarity reflects that of the medium in which the cells live *in vivo*, typically about 290 mOsm. Cells are much more resistant to hypertonic conditions and although hypertonic conditions up to 450 mOsm cause cells to shrink, their viability does not appear to be affected during the time of a typical separation procedure. Hypo-osmolarity, leading to swelling, appears to be more critical in causing changes to ultrastructural morphology and reduced viability; as a general rule, the exposure of cells to hypotonic conditions should be avoided.

Most of the problems described in this section are not encountered when using centrifugal elutriation, which is described in detail in Section 5.4. A rather sophisticated alternative is offered by special separation chambers which resemble those used for unit gravity sedimentation but that can be inserted into a centrifuge (*Figure 2*). This chamber combines a relatively large effective surface and a short sedimentation distance for cells with special devices to prevent swirling and disturbances during layering and collecting of samples. The chamber is used in a large rotor of a low-speed centrifuge which maximizes the distance from the centre of rotation. In this arrangement, artifacts due to swirling, wall effects, and centrifugal forces are minimized. The equipment can be used for both isopycnic and rate-zonal (see Section 5.3) separations. Disadvantages are the relatively high price and the limited capacity ($< 50 \times 10^6$ cells).

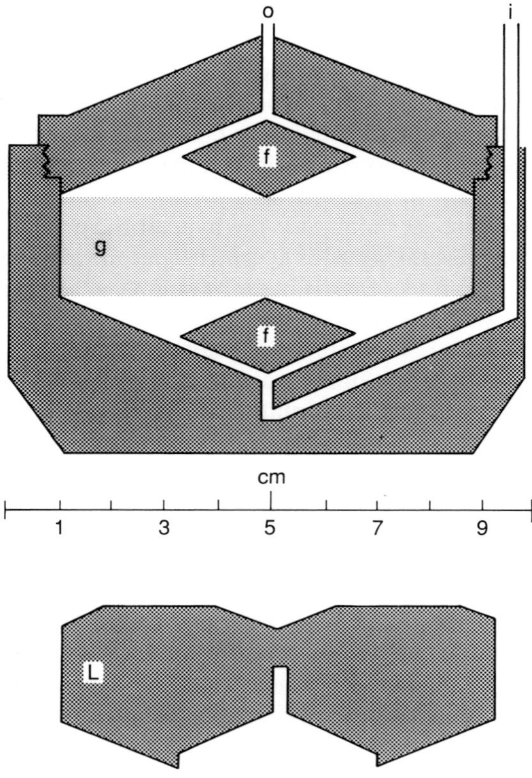

Figure 2. Special separation chamber for cell separations by centrifugation. This separation chamber, developed by Tulp *et al.* (5), can be installed in a modified centrifuge swing-out bucket, and it allows the separation of cells with high resolution using moderate centrifugal forces. The Perspex chamber (darkly shaded) can be loaded and unloaded through the inlet (i) and outlet (o), respectively, without disturbing the gradient because of the presence of flow deflectors (f). The lightly shaded area within the chamber (g) indicates the position of the gradient during centrifugation. One of the vanes (L) present to prevent swirling in the chamber is also shown.

3. Characterization of cells and analysis of results

Before the separation of a heterogeneous population of cells can be attempted, methods must be devised for distinguishing between the various cell types that are present. Therefore, the selection of appropriate markers for the characterization of the cells is very critical. The markers which can be selected depend not only on the cells to be purified, but also on the nature of the contaminating cells and on the preference of the investigator.

Generally, microscopy forms the basis of most methods for the identification of cells. Routine light microscopic examination is sometimes sufficient to identify cell types; for example, in the separation of large parenchymal and the smaller non-parenchymal liver cells. In most other instances, the cells must be stained before they can be identified. This can be done by routine histological staining of cell smears (e.g. May–Grünwald, Giemsa) such as are used for the morphological identification of various types of white blood cells, as well as by immunochemical or cytochemical staining, by which the presence of a certain antigenic property or enzymatic activity can be visualized. These microscopic methods allow an assessment of the actual percentage of cells in the sample which are associated with the chosen marker. A different type of information can be obtained by biochemical analysis of purified cells. A positive marker for the desired cell type is of less value here, since variations in the levels of a marker from one cell to another can prevent its accurate use for the determination of cell purity. A negative marker which is absent from the target cell but present in all cell types that are possible contaminants is clearly more useful.

For the different types of liver cells that are used for most of the experimental examples given in this chapter, there are several independent markers, both biochemical and morphological, that can be used to distinguish each liver cell type from the other ones (6). These methods include light and electron microscopy and biochemical analysis (8–10). In this chapter the discussion will be limited to light-microscopy techniques combined with enzyme cytochemistry. Parenchymal cells can be recognized directly using a light microscope because of their large size (diameter $\simeq 20\,\mu m$). Fat-storing cells also have a very characteristic morphology associated with the presence of numerous large lipid droplets, which also allows direct recognition by light microscopy. Kupffer cells, endothelial cells, and lymphocytes can be distinguished from each other by light microscopy only after they have been stained histologically or cytochemically. Here, peroxidase and non-specific esterase will be described as examples of cytochemical markers. In the rat, peroxidase is found exclusively in Kupffer cells while, although esterase activity is present in all liver cells, lymphocytes do not exhibit this activity (11). The value of these markers has been confirmed in previous studies, by comparison with other, independent, markers such as ultrastructural morphology (8–13), histochemical staining, other enzyme markers (12), and specific vitamin A fluorescence (fat-storing cells) (13, 14).

Apart from purity, the viability of a cell preparation must also be assessed. A rapid method for assessing cell 'viability' which is frequently used is based on the capacity of cells to exclude trypan blue. This gives an indication only of the integrity of the cellular membrane. In practice, cells which exclude trypan blue may in fact prove to be non-viable when assessed for functional or morphological properties (6). Therefore, each method for cell separation should be checked for its influence on cell morphology and function. The methods presented in this chapter have been extensively studied for their effects on these parameters and were found to be satisfactory with regard to ultrastructural morphology (8, 9, 13), biochemistry (10, 12, 14), and the ability of cells to survive in culture (15–17). Most of these studies employed rather laborious techniques which will not be included in the experimental protocols presented here.

4. Guidelines for devising a method for cell separations

This section is designed to give guidelines for deciding on the best procedure to be used for the purification of a particular type of cell ('*A*' cells) from a heterogeneous suspension.

(a) As a prerequisite, a method to distinguish *A* cells from all other cell types present in the suspension should be selected. This marker will be used to determine the percentage of *A* cells present in the suspension and in any cell sample obtained after fractionation. In addition, criteria for cell viability should be defined to evaluate the effect of the separation procedure on the cells.

(b) Determine the density profiles of *A* cells and contaminating cells, by isopycnic density centrifugation in a continuous gradient using various density media (see Section 2.3), in isotonic or hypertonic solutions. Compare the density profiles obtained using different media, check the recovery and viability of the cells. Select the best gradient medium. Prepare a density gradient with a density range which includes all cells present in the suspension and determine the density profile of the cells. Any of the following results might be obtained: complete purification of *A* cells; considerable enrichment of *A* cells although not sufficiently pure (in this case, density separation in combination with other techniques may prove useful); or there may be no enrichment of *A* cells.

(c) Determine the size distribution of *A* cells and contaminating cells by light microscopy. The extent of overlap will give a first indication as to whether a separation on the basis of sedimentation rate is likely to be successful.

A combination of the results from steps (b) and (c) will indicate whether there is a good possibility of purifying *A* cells on the basis of sedimentation rate by rate-zonal or elutriation techniques. A theoretical estimation of the sedimentation behaviour of *A* cells can be calculated from the average size and density of the cells. The homogeneity of *A* cells with respect to these parameters and the extent of overlap with contaminating cells will indicate whether high resolution methods such as centrifugal elutriation or those employing special separation chambers are necessary. The estimated sedimentation rate of *A* cells can be used to predict the experimental conditions necessary for purification by rate-zonal centrifugation or centrifugal elutriation. A trial separation can be carried out; in the first instance try a rate-zonal separation, and alternatively, if you are familiar with the procedure and equipment, try centrifugal elutriation. The experimental conditions for the centrifugal elutriation can be selected using the rotor speed and flow-rate nomogram that is included in the Beckman instruction manual (see Appendix A) of this chapter. A fixed rotor speed can be selected and the suspension of cells fractionated by applying a series of flow rates centred around the predicted flow rate necessary for the elution of *A* cells. The results can be used to establish the conditions necessary for the preparative purification of *A* cells.

If neither of these methods alone leads to complete separation from the contaminating cells, sequential application of two methods which select for differences in density and sedimentation velocity may give better results. If this approach is also unsuccessful, attempts can be made to apply methods of cell separation that select for other criteria, such as mentioned in Section 1.

5. Experimental protocols for cell separations

The experimental examples to be described in this chapter will demonstrate the various techniques which can be used to achieve the separation and purification of the different types of animal cells contained in mixed-cell suspensions. The starting material can be quite variable, cells may occur as a mixed-cell suspension; for example, blood or semen. Alternatively, the cells may be present as an organized tissue; for example, liver. In the latter case it is necessary to prepare a single-cell suspension, by dissociation of the cells of the tissue (for example, see Appendix B) of this chapter.

5.1 The isolation of blood cells

Blood cells exist as a single-cell suspension of a number of different cell types, each type with its own characteristics and functions. Research investigations into the properties of the cells require the various fractions to be available in a purified form, but blood is also widely used medically for diagnostic purposes. For this reason, given the number of blood samples that are taken for testing, much effort has been devoted to designing simplified, reliable methods for blood-cell purifications.

The most widely used of these methods is that developed by Bøyum (31) in which a single solution, composed of a mixture of sodium metrizoate and Ficoll, is used to separate the mononuclear leukocytes (lymphocytes and monocytes) from the erythrocytes and polymorphonuclear cells. This solution, under various trade names such as Lymphoprep (Nycomed Pharma, A/S) and Ficoll-paque (Pharmacia-Biosystems, AB), has been, and still is, widely used for this purpose. Subsequently, media for the same purpose, but replacing sodium metrizoate with Nycodenz which is non-ionic and omitting the polysaccharide component (32), have been developed.

Almost all of the separation media developed for fractionating blood cells involve layering the blood or a blood cell fraction over a separation medium with a defined density and osmolarity and the discontinuous gradient so formed is centrifuged at low speed at room temperature. A typical experimental protocol for this type of separation is given in *Protocol 1*.

Protocol 1. Isolation of human mononuclear cells

1. Take the blood sample in an anticoagulant suitable for subsequent work; usually 0.32% sodium citrate works well for most purposes.
2. Dilute the blood with an equal volume of isotonic saline.
3. Place 3 ml of metrizoate-Ficoll separating medium into a 10–15 ml centrifuge tube and carefully overlayer it with 6 ml of the diluted blood.
4. Centrifuge at 600*g* for 20 min in a swing-out rotor at about 20 °C; do not centrifuge at 5 °C or the separation will be poorer.
5. After centrifugation, collect the band of cells found at the sample/medium interface using a Pasteur pipette, and this consists of the mononuclear cells and some platelets.
6. Add a few millilitres of physiological saline to the cells to reduce the density and wash by centrifugation at 250*g* for 5 min at 20 °C. Resuspend the cells in saline or homologous plasma.

The buoyant densities of monocytes and lymphocytes are very similar, making them difficult to separate under iso-osmotic conditions. Work by Bøyum has shown that in slightly hypertonic conditions lymphocytes tend to lose water, and therefore become denser more readily than monocytes (32), allowing a solution to be designed to separate the two species by a single centrifugation step. This solution is currently marketed under the trade name 'Nycoprep 1.068' by Nycomed Pharma A/S.

The polymorphonuclear cells, predominantly neutrophils, have buoyant densities which, in iso-osmotic conditions, overlap with those of the erythrocytes. Solutions have been developed (Mono-Poly Resolving Medium, Flow Laboratories, and Polymorphprep, Nycomed Pharma A/S) which allow the mononuclear cells, polymorphonuclear cells, and erythrocytes to be banded separately in a single centrifugation step (*Figure 3*). Again, this separation method exploits the changes in buoyant density brought about by exposure to the high osmolality and density of the gradient medium.

These routine methods all depend on a single-step discontinuous gradient, or density barrier, retaining some cell species at the interface and allowing others to penetrate the density barrier. More recently a method has been developed for the separation of mononuclear cells which involves mixing the whole blood sample in equal volumes with a separating medium composed of Nycodenz and NaCl. When the mixture is centrifuged, the mononuclear cells float to the top of the solution while the erythrocytes and polymorphonuclear cells sediment (33). In this case, as with some of the other media described, no polysaccharide

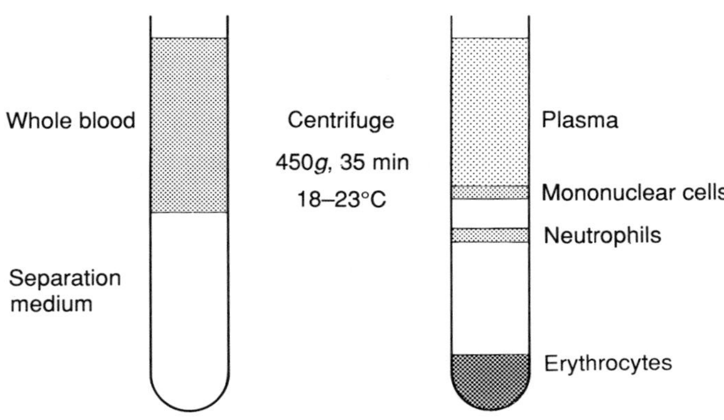

Figure 3. A one-step separation of mononuclear and polymorphonuclear cells from human blood. Freshly drawn blood (3–5 ml) mixed with anticoagulant is carefully layered over an equal volume of Polymorphprep (Nycomed Pharma A/S) to give a liquid column of 6–10 cm and centrifuged at 450*g* for 35 min at 20 °C. The mononuclear cells band isopycnically at the interface between the blood and the separation medium while the polymorphonuclear cells, mainly neutrophils, concentrate at the interface and then sediment through the separation medium as a sharp band. The polymorphonuclear cells will pellet to the bottom of the tube if the length of centrifugation is excessive or if the temperature is too high.

Table 2. Commercially-available media for the isolation different types of human blood cells

Medium	Composition	Target cell population	Supplier
Lymphoprep	9.6% Sodium metrizoate 5.6% Ficoll	Mononuclear cells	Nycomed-Pharma
Ficoll-paque	9.6% Sodium diatrizoate 5.6% Ficoll	Mononuclear cells	Pharmacia-Biosystems
Nycoprep 1.077	14.1% Nycodenz 0.44% NaCl	Mononuclear cells	Nycomed-Pharma
Nycoprep mixer	19.0% Nycodenz 0.2% NaCl	Mononuclear cells	Nycomed-Pharma
Nycoprep 1.068	13.0% Nycodenz 0.58% NaCl	Monocytes	Nycomed-Pharma
Nycoprep 1.063	12.0% Nycodenz 0.56% NaCl	Platelets	Nycomed-Pharma
Mono-Poly resolving medium	15.5% Sodium diatrizoate 8.18% Ficoll	Polymorphonuclear cells	Flow Laboratories
Polymorphprep	13.8% Sodium metrizoate 8.0% Dextran 500	Polymorphonuclear cells	Nycomed-Pharma

All percentages are given as percent weight/volume.

is used to aggregate and increase the sedimentation rate of the erythrocytes, thus obviating the possibility of the polysaccharide masking the receptor sites on the cell surfaces.

With minor modifications, the methods described for the purification of blood cells can be adapted for other mixed-cell suspensions, such as lavages of the lungs or peritoneal cavity. However, it must be appreciated that the ready-made solutions are designed for separating normal human blood cells. Hence, commercially-available blood separation media designed for human blood often do not give good results with blood of other animals. It is therefore usually necessary to modify the composition of the medium for each type of animal (e.g. for *Xenopus laevis* see reference 34). In addition, some human diseases can modify the characteristics of human blood cells in terms of density, size, and the number of cells. In such cases the media that work with normal blood may not give such good results. *Table 2* summarizes some of the media that are available for the isolation of various blood cell fractions.

5.2 Fractionation of liver cells by isopycnic centrifugation

Although the liver appears to be a fairly homogeneous tissue, it consists of a variety of cell types (11). The predominant type of cell are the parenchymal cells which account for over 90% of the total liver volume and about 60% of the total cells of the liver (11). The other cells (non-parenchymal cells) consist primarily of sinusoidal cells, of which the Kupffer, endothelial and fat-storing cells are the most abundant (11). In total, these fairly small sinusoidal cells account for about 35% of all the liver cells but together they occupy only about 6.5% of the total liver volume (11). Rat liver contains about 110×10^6 parenchymal cells, 10×10^6 Kupffer cells, 20×10^6 endothelial cells, and 16×10^6 fat-storing cells per gram of liver.

There are several methods for the preparation of cell suspensions from the rat liver (6). They fall into two categories. The first category includes methods which employ perfusion and incubation of the liver with collagenase to degrade extracellular matrix, combined with a Ca^{2+}-free perfusion which releases the junctions between parenchymal cells. These methods allow the isolation of parenchymal as well as non-parenchymal cells from the same liver. The methods from the other category employ pronase, or another reagent that selectively destroys parenchymal cells (6, 8). Such methods are especially suited for isolating sinusoidal liver cells in high yield and purity.

In order to optimize the digestion of the extracellular matrix and obtain a higher yield of cells, especially fat-storing cells, collagenase (500 µg/ml) can also be included in the incubation medium during pronase treatment (*Figure 4*) (13).

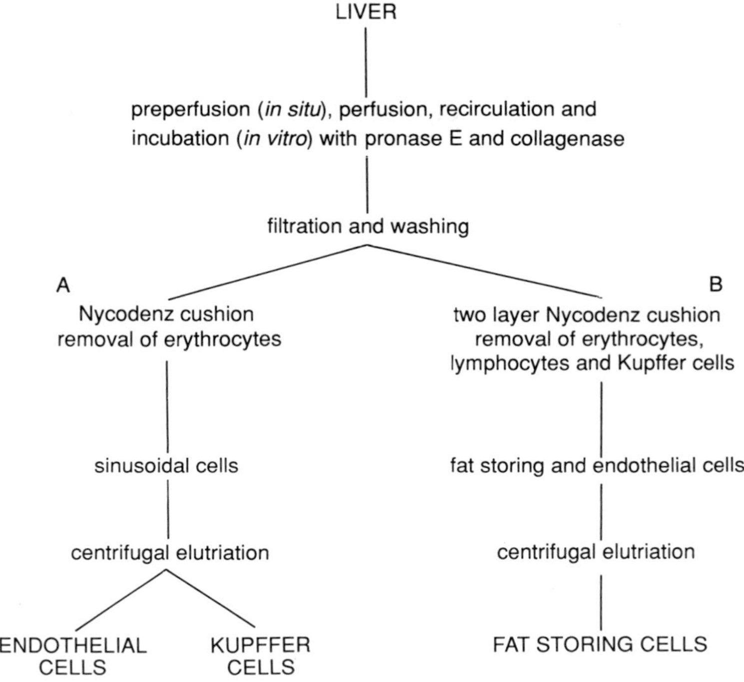

Figure 4. A schematic illustration of the methods for the isolation of sinusoidal liver cells (Section 5.2) and the subsequent purification of endothelial, Kupffer, and fat-storing cells. The preparative isopycnic density gradients are described in detail in Sections 5.2.2 and 5.2.3 for A and B, respectively. Procedures for purification by centrifugal elutriation are given in Sections 5.4.3 and 5.4.4 for A and B, respectively.

There is much variety in details of the experimental procedures between and within laboratories. Some of these differences are trivial, but some have important consequences for the usefulness of the cell preparation in a particular experiment. Conditions for the isolation of viable Kupffer and endothelial cells are generally less critical than those required for the isolation of parenchymal and fat-storing cells. For the experiments described below various methods for the isolation of liver cells have been used. Detailed descriptions of the major techniques employed are given in Appendix B of this chapter.

5.2.1 Continuous gradient fractionation of sinusoidal cells

Freshly isolated sinusoidal cells are separated according to differences in density and the density profiles of both Kupffer and endothelial cells are analysed. The following equipment, chemicals, and solutions are required:

- a simple gradient maker with magnetic spinbar and stirrer
- a peristaltic pump, silicon tubing (i.d. 1.3 mm; o.d. 3.0 mm), and glass capillary extensions (15 cm)
- an Abbé refractometer
- a low-speed centrifuge with a swing-out rotor
- centrifuge tubes (Corex, 15 ml)
- Gey's balanced salt solution (GBSS): NaCl (8000 mg/litre); KCl (370 mg/litre); $MgSO_4 \cdot 7H_2O$ (70 mg/litre); $NaH_2PO_4 \cdot 2H_2O$ (150 mg/litre); $CaCl_2 \cdot 2H_2O$ (220 mg/litre); $NaHCO_3$ (227 mg/litre); KH_2PO_4 (30 mg/litre); $MgCl_2 \cdot 6H_2O$ (210 mg/litre); glucose (1000 mg/litre); adjusted to pH 7.4; osmolarity: 275–285 mOsm. Sterilize by filtration and store at 4 °C.
- 28.7% (w/v) Nycodenz or 30% (w/v) metrizamide (Nycomed Pharma A/S) in GBSS without NaCl; osmolarity, 285 mOsm; this solution is pH 7.4
- sinusoidal liver cells (see Appendix B)
- 0.5% w/v trypan blue in physiological saline (0.9% NaCl in water)
- items required for peroxidase staining (8):

  conical tubes (15 ml)

  a Bürker counting chamber

  a microscope

  a water bath (37 °C)

  incubation medium: prepare the following mixture; dissolve 15 mg of 3,3-diaminobenzidine tetrahydrochloride·$2H_2O$ (DAB) (Merck) in 15 ml of 0.05 M Tris–HCl (pH 7.4) containing 7% (w/v) sucrose (300 mOsm). Add 10 µl of 30% H_2O_2. Use within 30 min.
  Caution: DAB is suspected of being carcinogenic

Protocol 2. Fractionation of rat liver sinusoidal cells

Sample preparation

1. Use freshly-isolated sinusoidal cells ($2–3 \times 10^6$ cells) from one rat liver, isolated as described in Appendix B of this chapter.
2. Wash the cells three times by suspension in 15 ml of GBSS and centrifuge at 300*g* for 5 min at room temperature. This reduces the risk of cell clumping and removes most of the contaminating parenchymal cell debris.
3. Resuspend the final pellet in about 6 ml of GBSS.

Gradient preparation and centrifugation

1. Dilute the stock solutions of 30% (w/v) metrizamide or 28.7% (w/v) Nycodenz to concentrations of 20% and 19.1%, respectively, with GBSS in

the ratio of 6.6 vol. of stock solution to 3.4 vol. of GBSS to give Solution *A* (note that, in most cases, Nycodenz and metrizamide can be used interchangeably, but Nycodenz has the advantage that its solutions can be autoclaved).

2. Dilute Solution *A* in the ratio of 3.1 vol. of Solution *A* to 5.4 vol. of cell suspension, to obtain concentrations of 8% (w/v) metrizamide or 7.7% (w/v) Nycodenz to use as the light solution for the gradient.
3. Prepare continuous gradients (either 8–20% metrizamide or 7.7–19.1% Nycodenz), using a simple two-chamber gradient mixer, or a more sophisticated device, the Gradient Master (see Chapter 3, Section 5.3).
4. Centrifuge the gradients at 2000*g* for 30 min at 4 °C in a swing-out rotor, using slow acceleration and deceleration. To avoid cell clumping the time of centrifugation should not exceed 45 min.

Analysis of gradients

1. Fractionate the gradients into about 25 equal fractions (approx 0.4 ml) preferably using the upward displacement method (Chapter 3, Section 7.5).
2. Determine the refractive index (RI) of each fraction (Abbé refractometer) from which the density of the fraction can be calculated using the following equations:

$$\text{Nycodenz: Density (g/ml)} = \text{RI} \times 3.242 - 3.323$$

$$\text{Metrizamide: Density (g/ml)} = \text{RI} \times 3.453 - 3.601$$

3. Determine the cell number and viability for each fraction. Viability may be assessed by mixing equal volumes of the gradient fraction and 0.5% trypan blue dissolved in physiological saline. Cells taking up the stain are non-viable. Determine the percentage of viable and non-viable cells.
4. Determine the distribution of Kupffer, endothelial, and other cells by peroxidase staining.

Peroxidase staining reaction

1. Add a few drops of each gradient fraction (0.5×10^6 cells) to about 2 ml of incubation medium and incubate for 30 min at 37 °C.
2. Centrifuge the mixture for 3 min at 300*g*, remove the supernatant, and resuspend the cells in 200 µl of GBSS.
3. Determine the percentage of positively staining cells using a haemocytometer. Erythrocytes will be stained almost homogeneously black and the Kupffer cells will have a brown-to-black colour. Erythrocytes and Kupffer cells can be distinguished on the basis of cell size. The endothelial cells and lymphocytes remain unstained.

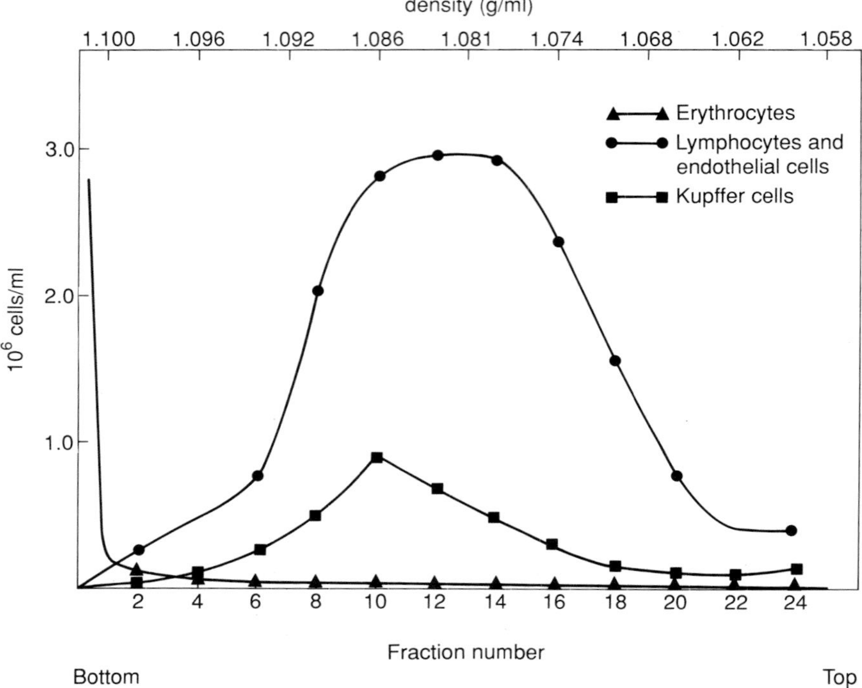

Figure 5. Fractionation of cells on a continuous metrizamide gradient. Liver sinusoidal cells were centrifuged, fractionated, and analysed as described in Section 5.2.1. The viability of cells in all fractions was greater than 90%.

The results obtained in a typical experiment done with sinusoidal isolated cells using the pronase procedure but without collagenase are shown in *Figure 5*. Almost all of the erythrocytes present in the original preparation are recovered in the pellet ($\sim 30 \times 10^6$ cells). The peroxidase-positive Kupffer cells, the negative-staining endothelial cells, and lymphocytes all form broad bands ranging between 1.060–1.098 g/ml. There is a large overlap between the different cell types.

As a result, the maximum purity of Kupffer cells that can be obtained using this technique amounts to 25% in fraction 10. This fraction represents only a small percentage of the total number of Kupffer cells present. For endothelial cells, the maximum yield and purity is much higher. Pooling of fractions 14–24 would result in an endothelial cell preparation of 90% purity which contains about 50% of the endothelial cells in the gradient. Such results show that Kupffer and endothelial cells of the normal rat liver cannot be completely purified by isopycnic density centrifugation alone. Previous results from the authors' laboratory have shown that the density separation of these cell types can be improved by preloading of the Kupffer cell population with Jectofer® *in vivo*.

To do this rats are injected intramuscularly (biceps femoris) with 0.2 ml of Jectofer® per 150 g of body weight 4 days and 1 day before isolation of the cells as described in Appendix B of this chapter. Jectofer® is an iron–sorbitol–citric acid complex (Astra, Södertälje, Sweden) containing 50 mg iron per ml. After this treatment the density distribution of Kupffer cells is shifted towards higher densities and relatively pure fractions of each cell type can be obtained (9).

5.2.2 Single-step discontinuous gradients for the preparative purification of sinusoidal liver cells

This procedure is routinely used in the authors' laboratory to eliminate red blood cells and parenchymal cell debris from the sinusoidal cell suspension obtained after treatment of the liver with pronase (Appendix B of this chapter) (10, 16). A single-step density separation is achieved between denser erythrocytes, which sediment to the bottom, and the floating sinusoidal cells. The following equipment, chemicals, and solutions are required:

- a centrifuge with a swing-out rotor
- Gey's balanced salt solution (GBSS) (see Section 5.2.1)
- 28.7% Nycodenz or 30% metrizamide in GBSS without NaCl (see Section 5.2.1)
- 15 ml plastic centrifuge tubes
- sinusoidal liver cells (see Section 5.2.1)
- trypan blue staining (see Section 5.2.1)
- peroxidase staining (see Section 5.2.1)
- items required for esterase staining:

  equipment as for peroxidase staining (see Section 5.2.1)

  physiological saline (0.9% NaCl)

  fixative: 2.5% glutaraldehyde in 0.1 M sodium cacodylate buffer (pH 7.4). *Caution: Cacodylate is highly toxic* and may be replaced by 0.1 M phosphate.

  incubation medium: mix (in a fume cupboard) the following solutions; 9.5 ml of 0.1 M potassium phosphate buffer (pH 7.4), 0.5 ml ethylglycolmonomethylether containing 11 mg of 1-naphthylacetate (keep under a nitrogen atmosphere), 0.5 ml 1.0 M NaOH, 0.5 ml of a mixture of equal parts of 4% *p*-rosanilin in 2.0 M HCl and 4% sodium nitrite in distilled water. Adjust the mixture to pH 7.8; filter through cellulose paper and use within 30 min.

Protocol 3. Single-step gradients for the isolation of rat-liver sinusoidal cells

1. Prepare a 10 ml suspension of sinusoidal cells in GBSS ($100–400 \times 10^6$ cells).
2. Add 14 ml of the stock 28.7% (w/v) Nycodenz or 30% (w/v) metrizamide solution to the suspension and mix thoroughly but gently.

Protocol 3. *Continued*

3. Divide the mixture between two centrifuge tubes and carefully overlayer with 1–2 ml of GBSS.
4. Centrifuge at room temperature for 15 min at 400*g* and allow the rotor to coast to a halt.
5. Collect the sinusoidal cells by slowly aspirating the band at the interface between the GBSS gradient medium using a Pasteur pipette.

The sinusoidal cells can then be analysed directly or fractionated further into Kupffer and endothelial cells by centrifugal elutriation (see *Protocol 6* in Section 5.4.3).

Analysis of sinusoidal cells

Determine the cell number, viability and percentage of peroxidase-positive cells as described in *Protocol 2*.

Esterase staining reaction

1. Incubate a few drops of cell suspension ($\sim 0.5 \times 10^6$ cells), with 1–2 ml of fixative, for 7 min at 4 °C.
2. Centrifuge for 3 min at 700*g*, and wash the pellet twice with 0.9% NaCl.
3. Resuspend the final pellet in 200 µl of 0.9% NaCl, add an equal volume of incubation medium, and incubate for 10 min at 37 °C.
4. Determine the percentage of positive-staining cells in a haemocytometer. Kupffer, endothelial, and parenchymal cells will stain red, erythrocytes and leukocytes will be unstained, and fat-storing cells will be faintly stained and can be recognized by their typical appearance.

The results of a typical experiment are given in *Table 3*. The viability of the sinusoidal cell preparation, as judged by trypan blue exclusion, is 99%. This first indication of cellular integrity can be confirmed by electron microscopical studies (8–13), biochemical analysis (10, 12), and the behaviour of cells in maintenance culture (15, 16, 19). Kupffer and endothelial cells represent the majority (~70%) of cells in the preparation. The esterase-negative lymphocytes are the major contaminant while there are very few if any erythrocytes present.

5.2.3 Two-step discontinuous gradients for the preparative partial purification of fat-storing cells.

This procedure is used as the initial step in the purification of fat-storing cells from a sinusoidal cell suspension (13).

The equipment and chemicals are the same as described for *Protocol 3* in Section 5.2.2.

Table 3. Preparation of sinusoidal liver cells by use of single-step discontinuous gradients

Parameter	Nycodenz	Metrizamide
Cell yield (cells/g liver)	43×10^6	44×10^6
Viability (trypan blue)	99%	99%
Cytochemical staining:		
peroxidase-positive cells	17%	27%
esterase-positive cells	71%	80%
Composition:		
Kupffer cells	16%	26%
endothelial cells	54%	54%
fat-storing cells	10%	1%
erythrocytes	1%	1%
parenchymal cells	1%	1%
other cells (e.g. lymphocytes)	18%	18%

Sinusoidal liver cells isolated from a 6-month-old female *BN/BiRij* rat by pronase–collagenase treatment of the liver (Appendix B of this chapter) were centrifuged and fractionated as described in Section 5.2.2. In separate experiments the cells were separated either on Nycodenz or metrizamide gradients. The cells were stained and analysed by light microscopy for peroxidase and esterase activity.

Protocol 4. Isolation of fat-storing cells on a two-step gradient

1. Prepare a suspension of sinusoidal cells from an adult female *BN/BiRij* rat ($100–400 \times 10^6$ cells in 12 ml GBSS).
2. Dilute either the stock 28.7% (w/v) Nycodenz or 30% (w/v) metrizamide solutions with GBSS in the ratio of 6 vol of stock solution to 4 vol of GBSS. Place 5 ml of the diluted solution (17.2% Nycodenz or 18% metrizamide) into each of two centrifuge tubes.
3. Add 8 ml of the appropriate original stock solution to 12 ml of the cell suspension and mix gently, but well to give solutions of either 11.5% Nycodenz or 12% metrizamide.
4. Carefully overlayer half of the cell suspension on to the solution in the two centrifuge tubes and overlayer this with 1–2 ml of GBSS to form a two-step gradient.
5. Centrifuge the gradients at 1400*g* for 17 min at 20 °C and allow to coast to a halt without braking.
6. Collect the cell layers found at the two interfaces in each tube using a Pasteur pipette.

The cells at the two interfaces can be analysed and characterized with respect to yield, viability and composition as described in *Protocol 3* and the results

Table 4. Isopycnic separation of sinusoidal liver cells on a two-step discontinuous Nycodenz gradient

	Composition			
	Low-density fraction		**High-density fraction**	
	Cell number $\times 10^6$	**% of total**	**Cell number $\times 10^6$**	**% of total**
Fat-storing cells	52.5	79.6	0.9	0.4
Kupffer cells	0.7	1.0	45.5	20.3
Endothelial cells	10.9	16.5	129.7	57.9
Other cells (lymphocytes)	1.9	2.8	47.9	21.4
Total	66.0	100	224.0	100

Sinusoidal cells isolated from a 12-month-old female *BN/BiRij* rats (pronase/collagenase) were centrifuged on a Nycodenz gradient and fractionated as described in Section 5.2.3. The low density (< 1.067 g/ml) on top of 11.5% Nycodenz and the high density (1.096 g/ml, on top of 7.2% Nycodenz) fractions were stained for esterase and peroxidase activity.

of a typical analysis is shown in *Table 4*. The low density fraction contains about 45% of the total cells recovered. This fraction is relatively rich in fat-storing cells containing about 90% of the recoverable fat-storing cells. The major contaminants of this fraction are the endothelial cells, while the contamination by lymphocytes which are difficult to remove by subsequent centrifugal elutriation is insignificant. As a result, this fraction of fat-storing cells can be further purified by centrifugal elutriation if this is required (see Section 5.4.4). The high density fraction contains the majority of the Kupffer cells, endothelial cells, and lymphocytes.

5.2.4 Other applications of isopycnic centrifugation for the separation of mammalian cells

In the previous sections, representative examples of isopycnic and density-barrier separations have been described. These examples essentially reflect the major variations of density gradient centrifugation and the protocols can be easily modified for the separation of other cell types. Additional information on the densities and other details of specific cell types can be extracted from the literature. A useful overview by Pertoft and Laurent of the density ranges of a large variety of cells has been published (4). This list includes all types of blood cells and cells from a wide range of tissues.

Next to, and often in combination with, velocity sedimentation, separation on the basis of density has played an important role in the separation and characterization of bone marrow-derived cells (20). This extensive field, which involves many different cell types representing all stages of differentiation between stem cells and various types of mature differentiated cells in blood and tissues, has been reviewed by Williams (20).

Table 5. Examples of cell separation by velocity sedimentation

Cell sample	Cell type(s) separated	Gradient	Run time/centrifugal forces	Equipment	Results			Ref.
					Purity	Yield[a]	Viability	
White blood cells	Lymphocytes (a); monocytes (b)	BSA (1.5–6.5%)	9 min/ 20*g*	Special separation chamber (*Figure 2*)	(a) > 90% (b) 90%	(a) > 90% (b) 67%	99%	5
White blood cells	Lymphocytes (a); monocytes (b); basophils (c)	Ficoll (2–4%)	2 h/ 1*g*	Unit gravity separation chamber	(a) — (b) 69–77% (c) —	(a) — (b) 28% (c) —	> 98%	22
White blood cells	Lymphocytes (a); monocytes (b)	Ficoll (3.4–16.9%)	45 min/ 2000 r.p.m.	Zonal rotor	—	—	—	22
Pre-fractionated white blood cells	Lymphocytes (a); basophils (b)	Ficoll (4.5–9.5%)	15 min/ 85*g*	100-ml tubes; swing-out rotor	(a) > 95% (b) 62–72%	(a) ± 90% (b) 30–50%	—	21
Human bone marrow cells	CFU-C	Ficoll (2–4%)	2 h/ 1*g*	Unit gravity separation chamber	—	—	—	
Pre-fractionated canine gastric cells	Parietal cells (a); chief cells (b)	Ficoll (2–4%)	50 min/ 1*g*	Unit gravity separation chamber	(a) < 60% (b) 85%	(a) ± 50% (b) ± 30%	—	22
Sinusoidal liver cells	Kupffer cells (a); endothelial cells	Percoll (3.5–18%)	8 min/ 16*g*	Special separation chamber (*Figure 2*)	(a) 98% (b) 97%	(a) 68% (b) 86%	> 95%	5

[a]Yield represents the percentage of cells of one type that is recovered in the purified fraction.

5.3 Velocity sedimentation of cells

The principles of velocity sedimentation have been described in Sections 2.2.2 and 2.2.3. Several examples of cell separation by velocity sedimentation and the major experimental conditions are given in *Table 5*. Unless one uses rather specialized equipment, rate-zonal sedimentation is often limited in both its degree of resolution and the number of cells that can be separated in a single step. The specialized equipment which is used includes both large capacity reorienting zonal rotors (1) and special separation chambers (see Section 2.2.3); the use of zonal rotors for separations is described in Chapter 8. With regard to the separation of sinusoidal liver cells, the only method using velocity sedimentation that has been reported to give satisfactory results employs a special separation chamber and centrifuge (5). With this arrangement, the separation of sinusoidal cells into Kupffer and endothelial cells can be accomplished (5) with a resolution which is comparable with centrifugal elutriation (cf. Section 5.4.3). It has also been applied to the separation of dividing leukaemia cells and white blood cells (5). Other examples of velocity sedimentation can be found in studies on the fractionation of various types of colony-forming cells in the bone marrow (20, 22), which is often performed at unit gravity (22). No detailed experimental protocols using rate-zonal sedimentation will be presented here.

5.4 Centrifugal elutriation of cells

The concept of applying centrifugal elutriation to the separation of cells was described by Lindahl as early as 1948 under the name of 'counterstreaming centrifugation' (23). Lindahl's idea was taken up and developed by Beckman in the mid 1960s, resulting in a relatively simple elutriator rotor which has become commercially available. The method uses the principle of counterflow centrifugation in which the movement of particles of different sedimentation rate is opposed by a controlled centripetal flow of liquid (elutriation) (*Figure 1*). This method does not need a density gradient, and pelleting, which can damage cells, does not occur (see also Section 2.2.4). The elutriator rotor is designed to separate and/or purify suspensions of cells or particles over the size range 5–50 μm in diameter, mainly according to size (see Appendix A of this chapter). It can also be used to provide cell-free serum or media.

5.4.1 Description of the elutriator rotor system

The original model of the elutriation rotor, the JE-6B, consists of a single separation chamber plus a by-pass chamber as counter balance. The JE-6B elutriator rotor is used either in J-21 series or in modified J-6B series centrifuges. A larger elutriator rotor, the JE-10x is also available. The JE-10x rotor can be used for the same applications as the standard JE-6B rotor. The separation chamber of the JE-10x has a 10 times greater volume (40 ml) than the JE-6B rotor chamber. Between 10^7 and 10^{12} cells can be separated in a single run (cf., Appendix A of this chapter for the standard rotor). The JE-10x rotor can only

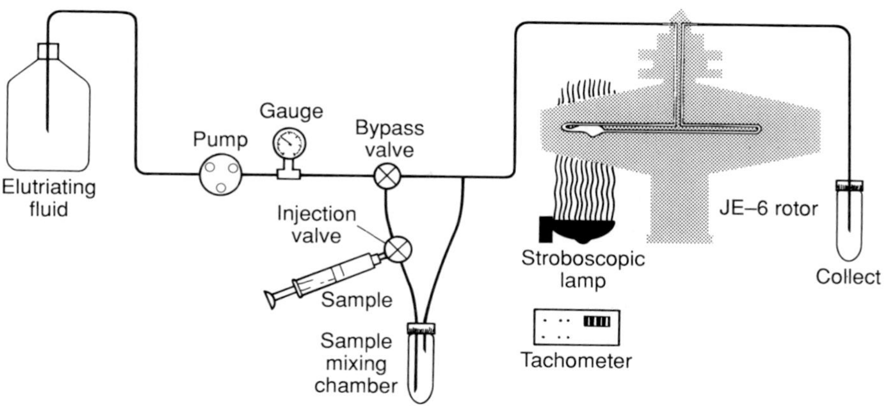

Figure 6. Scheme of the Beckman elutriator system. The figure shows the positioning of the various components of the elutriator systems within and outside of the centrifuge. (Reproduced with permission from the Beckman elutriator instruction manual.)

be used in the J6 series centrifuges. Elutriator rotors require several accessories, namely a stroboscope unit, a peristaltic pump for liquid delivery and removal (*Figure 6*), as well as a modified centrifuge door with a window.

In this section, only the smaller JE-6B rotor with its accessories, in combination with a J-21 centrifuge will be described. The later metal JE-6B elutriator rotor has a slightly different design from the original plastic rotor and has different caps on the rotor chambers. In our experience, both types of this rotor fractionate cells equally well. However, the metal JE-6B rotor is much more practical and can be assembled more easily without risks of leakage, or wrong connections.

i. Rotor chambers

The elutriator rotor has two epoxy-resin chambers. One of these chambers, the so-called bypass chamber, is always installed in the bypass position. One of the two types of separation chamber, the standard or Sanderson chamber (*Figure 7*), is always installed in the elute position of the rotor. The Sanderson chamber has a slightly higher resolving power (see Appendix A of this chapter), which makes it especially useful for the separation of cells with only small differences in sedimentation rate. In cases where there is a great tendency for the cells to form clumps, the Sanderson chamber inlet (*Figure 7*) may have an advantage over the standard chamber in reducing the formation of clumps. A third shape of chamber which can fractionate larger numbers of cells has also been described (35).

ii. Installation of the centrifugal elutriation system

Setting up the elutriator system is done in two stages involving first the assembly of the elutriator rotor, followed by the installation of the stroboscope and

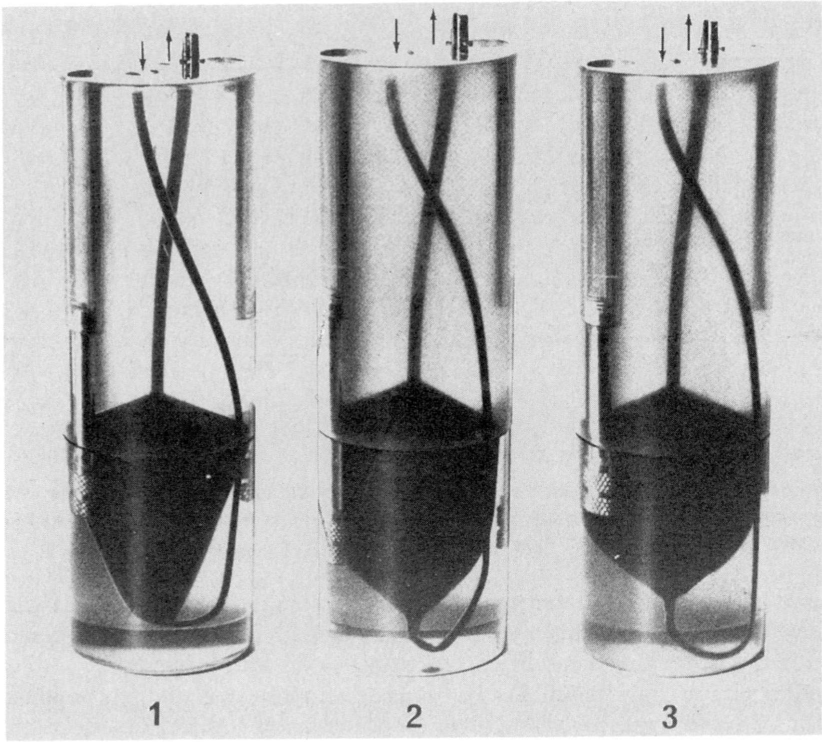

Figure 7. Separation chambers for centrifugal elutriation using the Beckman JE-6B elutriator rotor. (1) Standard Beckman chamber; (2) and (3) Sanderson chambers. Note the difference in internal chamber volume and height between the original Sanderson chamber (2) and the later one (3). The cap of the later model will only fit the original Sanderson chamber if it is modified.

rotor in the centrifuge and connection of the rotor inlet and outlet to the flow system. An accurate and complete description of all the necessary steps and preparations is included in the Beckman instruction manual. In addition, it contains an extensive section on troubleshooting. Therefore, the installation of the system will not be described in further detail here. However, it is important to emphasize the lubrication of the O-rings, in particular the two which are in contact with the rotating seal. Insufficient lubrication and/or wearing of these O-rings will cause cross-leakage between the inflow and outflow channels within the rotor. In this case, the effective flow rate within the separation chamber is reduced and does not correspond to the measured flow rate.

iii. Operation

The collection of separated cell fractions of increasing diameter during centrifugal elutriation of a heterogeneous cell sample can be accomplished by either stepwise increases in the flow rate or stepwise decreases in the rotor speed. The protocols described below all involve variation in flow rates at a fixed rotor speed, since, in our experience, it is much easier to predict and control the course of the separation. The desired flow rates must be checked in advance and can even be corrected during the run. The variations in rotor speed are more difficult to control, especially with the standard J-21 centrifuge and the original type of stroboscope unit which gives a reading of the rotor speed only after 1 min; with the later types of stroboscope units (Kit numbers 354700 and 354679 for the J2-21 and J-6B centrifuges, respectively), continuous monitoring of the rotor speed is possible. If, in addition, the relatively inaccurate standard rotor speed control is replaced by one that gives a finer regulation of speed then the system is better-suited for achieving separations by variation in rotor speed. Even then, one has to take into account that there will always be an overshoot in speed of the rotor when changing from one speed to another.

5.4.2 Separation of parenchymal liver cells into ploidy classes

Rat-liver parenchymal cells consist of a heterogeneous population with regard to their ploidy (24). The relative proportions of the different ploidy classes within the liver are dependent on the strain of the rat. In addition, there is an age-related shift towards higher ploidy classes (24). *Table 6* shows the distribution pattern of ploidy classes in livers of 3-month-old female *WAG/Rij* rats. At this age, mononuclear tetraploids (4*n*) represent the majority of the parenchymal cell population. The following section describes the separation of diploid, tetraploid, and octaploid cells by means of centrifugal elutriation (25).

The following equipment, chemicals, and solutions are required

- J-21 (or J-6B) modified centrifuge (Beckman)
- JE-6B elutriator rotor with a stroboscope unit (Beckman)
- Sanderson chamber in combination with the bypass chamber, silicone tubing (i.d. 3.0 mm; o.d. 5.0 mm, Pharmacia-Biosystems, silicone tubing (i.d. 3.18 mm; o.d. 6.35 mm, Beckman) (for construction of the flow system), centrifuge with a swing-our rotor
- Bürker counting chamber
- microscope
- peristaltic pump type 2115 (Pharmacia-Biosystems AB) or an equivalent low-pulse pump
- 50-ml centrifuge tubes
- elutriation medium; any medium suitable for the isolation of parenchymal cells can be used (Appendix B of this chapter). For the method described in *Protocol 5*, the results of which are shown in *Table 7*, the elutriation medium used was L-glutamine (131 mg/litre), L-aspartic acid (13.3 mg/litre), L-threonine (23.8 mg/litre), L-serine (31.5 mg/litre), glycine (37.6 mg/litre), L-alanine (53.5 mg/litre), L-glutamic acid (132.4 mg/litre), KCl (223.7 mg/litre),

Table 6. Proportions of diploid and polyploid parenchymal cells in the liver of 3-month-old Female *WAG/Rij* rats

Ploidy class nuclei	Diploid (2n)	Tetraploid (4n)		Octaploid (8n)		Decahexaploid (16n)
	MD	BD	MT	BT	MO	BO
% of total cell number	9	18	53	16	4	2

MD, mononuclear diploid; BD, binuclear diploid; MT mononuclear tetraploid; BT, binuclear tetraploid; MO, mononuclear octaploid; BO, binuclear octaploid (24).

Table 7. Centrifugal elutriation of parenchymal liver cells: separation into ploidy classes

Fraction	Flow rate (ml/min)	Cell number (10^6 cells)	Viability (%)	Composition
Original suspension		137.8	84	2n, 4n, 8n, 16n, aggregates, debris
I[a]	19	39.8	72	2n, 4n, debris
II	32	54.7	85	4n (90%)
III	41	15.5	83	4n, 8n, aggregates
Pellet		17.1	82	As original suspension
Mixing chamber		5.2	85	As original suspension
Total		127.0 (92%)	—	—

Parenchymal cells were isolated from 3-month-old female *WAG/Rij* rats (cf., Section 5.2) and subjected to centrifugal elutriation as described in Section 5.4.2. Total recovery of cells after elutriation was 92%. Cell fractions were analysed for their proportions of ploidy classes by TGZ-3 analysis (Section 5.4.2).

[a]Fraction I was further purified by centrifugation in a Percoll density medium (Section 5.4.2) resulting in a suspension of 32×10^6 cells with a viability of 90%. This preparation consisted of 50% diploid (2n) cells, contaminated mainly with tetraploid cells.

$NaH_2PO_4 \cdot H_2O$ (96.6 mg/litre), $MgCl_2 \cdot 6H_2O$ (101.7 mg/litre), $NaHCO_3$ (2.01 g/litre), glucose (3.60 g/litre), fructose (3.60 g/litre), sucrose (67.4 g/litre), the solution is adjusted to pH 7.4, osmolarity: 308 mOsm, store at −20 °C

- parenchymal liver cell suspension (see Section 5.2.1)
- trypan blue staining (see Section 5.2.1)

Protocol 5. Separation of parenchymal cells of different ploidy by elutriation

1. Prepare the centrifuge ready for the rotor, installing the stroboscope and flow harness.
2. Place the rotor into the centrifuge chamber, connect the silastic tubing to the rotor, and fill the entire system with the elutriation medium. Be careful to remove all the air bubbles.

3. Accelerate the rotor to 1350 r.p.m. (210*g*) and check the pump speed settings.
4. Load the cells in medium kept at 4 °C using an initial flow rate of 12 ml/min, wait until all the cells are in the chamber, this usually takes about a minute.
5. Collect three elutriation fractions I, II, and III by using successively faster flow rates of 19, 32, and 41 ml/min; collect fractions of 100 ml, 150 ml, and 100 ml for fractions I, II, and III, respectively. The larger volume for fraction II reflects the slower elutriation of the cells. Store the fractions at 4 °C.
6. Stop the pump and then the centrifuge. Collect the pelleted cells from the chamber and any cells remaining in the mixing chamber to allow the yield of cells to be determined.
7. Pellet the cells by centrifugation in a swing-out rotor at 1000*g* for 10 min; resuspend the cells from each of the fractions in 3 ml of elutriation medium.

In the analysis, determine the cell concentration, sample volume, and hence calculate the cell yield of each fraction. Determine the viability of cells using trypan blue. Calculate the overall recovery. Prepare dried slide preparations of each fraction using a sedimentation chamber. Stain the cells with Feulgen to stain the nucleus clearly. The ploidy classes in each fraction are determined from photomicrographs (× 1000) by measurement of the nuclear diameter (e.g. with a TGZ-3 particle size analyser) and determining the number of nuclei per cell (24).

Typical results are shown in *Table 7*. Fraction I contains most of the diploid cells in addition to tetraploid cells, non-viable cells, debris, and some sinusoidal cells. When fraction I is subjected to centrifugation (275*g* for 10 min at 4 °C) in an isotonic 27% (v/v) Percoll solution, a pellet consisting of 50% diploid cells with high viability is obtained (*Table 7*). Fraction II consists of a relatively homogeneous (90%) population of viable tetraploid cells. The octaploid cells cannot be separated from tetraploid cells and small aggregates (fraction III). The overall recovery is usually over 90%, with only small losses of cells in the pellet fraction and the mixing chamber.

5.4.3 Purification of Kupffer and endothelial cells

A mixed sinusoidal cells preparation isolated using a single-step discontinuous gradient (*Protocol 3*), consisting of a mixture of Kupffer and endothelial cells with variable numbers of fat-storing cells, and with lymphocytes as the main contaminants, can be used as the starting material. Alternatively, it is possible to use the bottom layer of the two-step discontinuous gradient prepared as described in Section 5.2.3. From either of these fractions, purified Kupffer and endothelial cells can be prepared using centrifugal elutriation.

The following equipment, chemicals, and solutions are required

- Equipment as in Section 5.4.2
- elutriation medium; GBSS (see Section 5.2.1)

- medium for peroxidase staining (see Section 5.2.1)
- medium for esterase staining (see Section 5.2.2)
- trypan blue staining (see Section 5.2.1)
- sinusoidal liver cells (see Section 5.2.1)

Protocol 6. Purification of Kupffer cells and endothelial cells by centrifugal elutriation

1. Prepare rat liver sinusoidal cells on a discontinuous Nycodenz gradient as described in Section 5.2.2.
2. Install the elutriation system as described in *Protocol 5*.
3. Adjust the rotor speed to 3250 r.p.m. if you are using the Sanderson chamber or 2550 r.p.m. if using the standard chamber and equilibrate the rotor temperature to 4 °C. In the authors' experience, the Sanderson chamber tends to give purer cell fractions than the standard chamber.
4. Load $100–500 \times 10^6$ cells into the elutriation system using an initial flow rate of 18 ml/min (Sanderson chamber) or 13.5 ml/min (Standard chamber).
5. Collect three different fractions, the first of 100 ml and the next two of 150 ml each, using successive flow rates of 18, 32, and 48 ml/min for the Sanderson chamber (13.5, 22.5, and 45 ml/min for the standard chamber. The three fractions contain lymphocytes, endothelial, and Kupffer cells, respectively. The cell fractions should be kept at 4 °C.
6. Collect the cells remaining in the chamber after elutriation.
7. Pellet the cells in each of fractions by centrifugation at 450*g* for 10 min and resuspend each pellet in about 3 ml of GBSS.

Determine the cell yield, and viability of cells as described in Section 5.2.1, and thence the overall recovery of cells. Estimate the percentages of Kupffer and endothelial cells in the various cell fractions on the basis of the number of cells reacting positively or negatively after staining for peroxidase (see *Protocol 2*) and esterase (see *Protocol 3*) activities. Typical results obtained with sinusoidal cells isolated from a 3-month-old *BN/BiRiJ* rat, which contain few fat-storing cells, are shown in *Table 8*. They show that both endothelial and Kupffer cells are obtained in separate fractions with a purity of 80% and 84%, respectively. Most (~70%) of the cells present in the original suspension are recovered in the purified fractions, with only a minor proportion of each cell type being lost in other fractions; the overall recovery of cells is usually about 90%. The viability of all cell fractions is at least 90%.

5.4.4 Purification of rat-liver fat-storing cells

The following protocol can be applied to partially purified fat-storing cell preparations that require further purification (13).

Table 8. Purification of rat-liver Kupffer and endothelial cells by centrifugal elutriation

Fraction	Main type of cell	Total cells ($\times 10^6$)	Viability (%)	% of cells staining for		Composition (%)[a]			
				Peroxidase	Esterase	P	L	E	K
Starting sample	Sinusoidal cells	167.4	87	24.7	86.8	0.6	13.2	61.5	24.7
L	Lymphocytes	19.9	80	3.0	28.6	—	71.4	25.6	3
E	Endothelial cells	82.5	95	9.0	89.2	—	10.8	80.2	9
K	Kupffer cells	34.7	97	83.5	95.1	—	4.9	11.6	83.5
Pellet	Cell aggregates	9.8	95	35.4	96.0	16.0	4.0	44.6	35.4

[a]P: parenchymal cells; L: lymphocytes; E: endothelial cells; K: Kupffer cells.
Sinusoidal liver cells were isolated from 3-month-old rats and separated from erythrocytes by single-step density centrifugation (see Section 5.2.2). The sinusoidal cell suspension was subjected to centrifugal elutriation using the Sanderson chamber as described in Section 5.4.3.

Table 9. Purification of rat-liver fat-storing cells by centrifugal elutriation

Fraction	Main type of cell	Cells ($\times 10^6$)	Viability (%)	% of cells staining for		Composition (%)[a]			
				Peroxidase	Esterase	FSC	L	E	K
Starting sample	Endothelial and fat-storing cells	133	94	8	98	16	2	75	8
F1	Fat-storing cells	14	80	1	88	78	12	9	1
F2	Fat-storing cells	6	85	1	94	53	6	40	1
Pellet	Endothelial cells	108	93	11	99	7	1	81	11

[a] FSC: fat-storing cells; L: lymphocytes; E: endothelial cells; K: Kupffer cells.
Sinusoidal liver cells isolated from 12-month-old rats were initally separated by a two-step density centrifugation (see Section 5.2.3). The top layer fraction, designated as starting sample, was subjected to centrifugal elutriation as described in Section 5.4.4.

The equipment is as described in Section 5.4.1, and the chemicals and solutions in Section 5.4.2. The preparation of the sinusoidal cell fraction using a two-step discontinuous gradient is described in *Protocol 4*; use the top (i.e. low density) cell fraction. Wash the cells by resuspension in GBSS and pellet them by centrifugation at 450*g* for 10 min.

Protocol 7. Purification of fat-storing cells

1. Install the elutriator system with the Sanderson chamber as described in *Protocol 5* and equilibrate it at 4 °C.
2. Set the rotor speed at 3250 r.p.m. and the flow rate to 16 ml/min.
3. Load 50–500 × 10^6 cells into the mixing chamber.
4. Collect two fractions of cells using flow rates of 16 ml/min and 18 ml/min; collect the cells remaining in the chamber after use.
5. Pellet the cells in each of the three fractions by centrifugation at 450*g* for 10 min and resuspend each cell pellet in about 2 ml of GBSS.

Determine the cell yield, and viability of cells as described in Section 5.2.1, and thence the overall recovery of cells. Identify the cells on the basis of the staining with peroxidase or esterase (*Protocols 2* and *3*). A set of typical results from this type of separation is given in *Table 9*. The fat-storing cells which constitute only a minor component of the starting cell suspension are recovered mainly in the first eluting fraction with a purity of about 80%. The main contaminants of this fraction are lymphocytes and endothelial cells which each represent about 10% of this fraction; Kupffer cells are practically absent.

The second fraction, eluted at 18 ml/min, contains some fat-storing cells, but they have a purity of only about 50% mainly as a result of contamination with endothelial cells. Overall, about 60% of the fat-storing cells are recovered in the two eluted fractions, the remainder being lost mainly as a result of cell clumping.

5.4.5 Separation of mononuclear leukocytes from human blood

During the past 15 years, several groups have reported the separation of functionally distinct types of white blood cells by centrifugal elutriation (27). Various protocols which can be used for the purification of lymphocytes (27, 28), monocytes (27, 28) and granulocytes (27, 29) have been developed. The separation of mononuclear leukocytes by centrifugal elutriation as described here is the same as that originally devised by Fogelman *et al.* (28).

The following equipment, chemicals, and solutions are required:

- equipment for centrifugal elutriation (see Section 5.4.1) using the standard chamber
- centrifuge with a swing-out rotor and tubes (50 ml)

- elutriation medium: Krebs–Ringer phosphate buffer containing 25 mM glucose and 1% (w/v) bovine serum albumin (BSA)
- equipment and chemicals for esterase staining (Section 5.2.2)
- heparinized human blood (50 ml)
- Ficoll-Paque (Pharmacia-Biosystems) or Lymphoprep (Nycomed Pharma, A/S)
- phosphate buffered saline (PBS) (50 mM; pH 7.4)

Protocol 8. Fractionation of a human mononuclear fraction by centrifugal elutriation

1. Isolate the mononuclear fraction from 50 ml of heparinized human blood as described in *Protocol 1* and dilute the cells harvested from the interface up to a final volume of 5 ml using Krebs–Ringer as the diluent.
2. Install the elutriator system, fill it with Krebs–Ringer solution, and equilibrate it to 5 °C.
3. Set the rotor speed at 2030 r.p.m. and the flow rate at 4.7 ml/min.
4. Load about 100×10^6 cells into the rotor and collect nine 50-ml fractions using flow rates of 4.7, 8.0, 10.0, 11.0, 12.0, 12.7, 13.5, 14.5, and 16 ml/min.
5. Pellet the cells in each of the elutriated fractions by centrifugation at 450*g* for 10 min.
6. Suspend each of the pellets together with the pellet from the elutriation chamber in about 3 ml of Krebs–Ringer solution.

In the analysis, each fraction from the elutriator rotor is analysed for cell concentration and the percentage of cells positive for non-specific esterase activity (monocytes). Typical results are given in *Table 10*. The majority of cells are eluted in fractions 3–6. These fractions contain primarily lymphocytes, and in particular, fractions 3–5 contain populations of highly purified lymphocytes. The larger esterase-positive monocytes are eluted only in fraction 9 or remain in the separation chamber. The pellet fraction represents a purified monocyte preparation. The esterase-negative cells which contaminate this fraction are mainly granulocytes. The platelets present in the original sample are recovered mostly in fractions 1 and 2.

5.4.6 Separation of other types of cells

Centrifugal elutriation has been extensively applied to separate a large variety of cells from different tissues and species. A complete list of all major publications employing this technique can be obtained from Dr Sussmann of Beckman Instruments International SA (Geneva). This extensive list includes a brief abstract of each publication with the essential experimental details necessary for carrying out the separations. Another possibility for obtaining specialized information on elutriation techniques is from the Elutriator Users Group (Contact address: Dr A Lodola, Biological Laboratory, University of

Table 10. Separation of mononuclear leukocytes by centrifugal elutriation

Fraction	**Flow rate (ml/min)**	**Cell number ($\times 10^6$)**	**Esterase staining**[a] **Negative (%)**	**Positive (%)**
Original sample	—	104	85.6	14.4
1	4.7	0	—	—
2	8.0	3.0	100	0
3	10.0	17.8	99.7	0.3
4	11.0	25.0	99.3	0.7
5	12.0	25.0	98.0	1.9
6	12.7	10.8	95.2	4.8
7	13.5	1.2	95.3	4.7
8	14.5	0.8	84.5	15.5
9	16.0	2.8	58.7	41.3
Pellet	—	10.0	28.4	71.6

White blood cells were obtained by Ficoll–Paque centrifugation of human blood and subjected to centrifugal elutriation as described in Section 5.4.5. The cellular composition of fractions was determined by non-specific esterase staining and light microscopic examination.
[a] Esterase-positive cells represent monocytes; esterase-negative cells represent lymphocytes (mainly fractions 1–8) and granulocytes (fraction 9 and pellet).

Canterbury, Kent CT2 7NY, UK). The Users Club issues a newsletter containing articles on problems and new applications of centrifugal elutriation.

Acknowledgements

The authors wish to thank R. J. Barelds, G. C. F. de Ruiter, E. Ch. Sleyster, A. M. Seffelaar, and S. J. Bukvic for their participation in the experiments. A. C. Ford and L. Vermeer-Greven are thanked for their help in the preparation of the manuscript.

References

1. Rickwood, D. (1978). In *Centrifugal separations in molecular and cell biology*, (ed. G. D. Birnie and D. Rickwood), p. 219. Butterworths, London.
2. Price, C. A. (ed.) (1982). *Centrifugation in density gradients.* Academic Press, New York.
3. Pretlow, T. G. and Pretlow, T. P. (1982). In *Cell separation: methods and selected applications*, Vol. 1, (ed. T. G. Pretlow and T. P. Pretlow), p. 41. Academic Press, New York.
4. Pertoft, H. and Laurent, T. C. (1982). In *Cell separation: methods and selected applications*, Vol. 1, (ed. T. G. Pretlow and T. P. Pretlow), p. 115. Academic Press, New York.
5. Tulp, A., Kooi, W., Kipp, J. B. A., Barnhoorn, M. G., and Polak, F. (1981). *Anal. Biochem.*, **117**, 354.
6. Brouwer, A., Leeuw, A. M. de, Praaning-van Dalen, D. P., and Knook, D. L. (1982). In *Sinusoidal liver cells*, (ed. D. L. Knook and E. Wisse), p. 509. Elsevier Biomedical Press, Amsterdam.
7. Ford, T. C. and Rickwood, D. (1982). *Anal. Biochem.*, **124**, 293.

8. Knook, D. L., Sleyster, E. Ch. (1976). *Exp. Cell Res.*, **99**, 444.
9. Knook, D. L., Blansjaar, N., and Sleyster, E. Ch. (1977). *Exp. Cell Res.*, **109**, 317.
10. Knook, D. L. and Sleyster, E. Ch. (1980). *Biochem. Biophys. Res. Commun.*, **96**, 250.
11. Wisse, E. and Knook, D. L. (1979). In *Progress in liver diseases*, Vol. VI, (ed. H. Popper and F. Schaffner), p. 153. Grune and Stratton, Inc.
12. Wilson, P. D., Watson, R., and Knook, D. L. (1982). *Gerontology*, **28**, 32.
13. Knook, D. L., Seffelaar, A. M. and Leeuw, A. M. de (1982). *Exp. Cell Res.*, **139**, 468.
14. Knook, D. L. and Leeuw, A. M. de (1982). In *Sinusoidal liver cells*, (ed. D. L. Knook and E. Wisse), p. 45. Elsevier Biomedical Press, Amsterdam.
15. Leeuw, A. M. de, Brouwer, A., Barelds, R. J., and Knook, D. L. (1983). *Hepatology*, **3**, 497.
16. Leeuw, A. M. de, Barelds, R. J., Zanger, R. de, and Knook, D. L. (1982). *Cell Tissue Res.*, **223**, 201.
17. Leeuw, A. M. de, Martindale, J. E., and Knook, D. L. (1982). In *Sinusoidal liver cells*, (ed. D. L. Knook and E. Wisse), p. 139. Elsevier Biomedical Press, Amsterdam.
18. Bezooijen, C. F. A. van, Grell, T., and Knook, D. L. (1977). *Mech. Ageing Dev.*, **6**, 293.
19. Barelds, R. J., Brouwer, A., and Knook, D. L. (1982). In *Sinusoidal liver cells*, (ed. D. L. Knook and E. Wisse), p. 449. Elsevier Biomedical Press, Amsterdam.
20. Williams, N. (1982). In *Cell separation: methods and selected applications*, Vol. 1, (ed. T. G. Pretlow and T. P. Pretlow), p. 85. Academic Press, New York.
21. MacGlashan, D. W., Lichtenstein, L. M., Galli, S. J., Dvorak, A. M., and Dvorak, H. F. (1982). In *Cell separation: methods and selected applications*, Vol. 1, (ed. T. G. Pretlow and T. P. Pretlow), p. 301. Academic Press, New York.
22. Wells, J. R. (1982). In *Cell separation: methods and selected applications*, Vol. 1, (ed. T. G. Pretlow and T. P. Pretlow), p. 169. Academic Press, New York.
23. Lindahl, P. E. (1948). *Nature* **161**, 648.
24. Bezooijen, C. F. A. van, Noord, M. J. van, and Knook, D. L. (1974). *Mech. Ageing Dev.* **3**, 107.
25. Bezooijen, C. F. A. van, Bukvic, S. J., Sleyster, E. Ch., and Knook, D. L. (1984). In *Pharmacological, morphological and physiological aspects of liver aging*, Vol. 1, (ed. C. F. A. van Bezooijen), p. 115.
26. Sleyster, E. Ch. and Knook, D. L. (1982). *Lab. Invest.* **47**, 484.
27. Sanderson, R. J. (1982). In *Cell separation: methods and selected applications*, Vol. 1, (ed. T. G. Pretlow and T. P. Pretlow), p. 153. Academic Press, New York.
28. Fogelman, A. M., Seager, J., Edwards, P. A., Kokom, M., and Popjak, G. (1977). *Biochem. Biophys. Res. Commun.* **76**, 167.
29. Lionetti, F. J., Hunt, S. M., and Valeri, C. R. (1980). In *Methods of cell separation*, Vol. 3, (ed. N. Catsimpoolas), p. 141. Plenum Publishing Co., New York.
30. Remenyik, C. J., Dombi, G. W., and Halsall, H. B. (1980). *Arch. Biochem. Biophys.* **201**, 500.
31. Bøyum, A. (1968). *Scand. J. Clin. Invest.* **21 (Suppl. 97)**, 77.
32. Bøyum, A., Berg, T., and Blomhoff, R. (1984). In *Iodinated density gradient media: a practical approach* (ed. D. Rickwood) IRL Press, Oxford.
33. Ford, T. C. and Rickwood, D. (1990). *J. Immunol. Methods* **134**, 237.
34. McKnight, B., Ford, T. C., and Rickwood, D. (1982). *Develop. Comp. Immunol.* **6**, 381.
35. Lutz, M. P., Gaedicke, G., and Hartmann, W. (1992). *Anal. Biochem.* **200**, 376.

Appendix A: Estimation of the flow rate and rotor speed necessary to separate cells by centrifugal elutriation

The information given below has been extracted from the Beckman Instruction Manual for the JE-6B Elutriation System and Rotor.

Flow rate and rotor speed

Figure 1 is provided for the operator who wishes to determine the approximate flow rate and rotor speed for the separation of specific particles. This nomogram has been generated from Equation 1:

$$F = XD^2\left(\frac{\text{r.p.m.}}{1000}\right)^2 \quad [1]$$

where F is the flow rate at the pump in ml/min; X is a constant which is equal to 0.0378 for the Sanderson chamber and 0.0511 for the standard chamber; D is the diameter of the particles in micrometres (µm) and, r.p.m. is the rotor speed in revolutions/min. The characteristics of the standard and Sanderson chambers are given in *Table 1*.

Equation 1 is an expression relating to the conditions existing at the elutriation boundary. It is derived by setting the velocity of a particle sedimenting in a gravitational force field (from Stokes' Law) equal to the flow velocity at the elutriation boundary, where flow velocity (V_f) is equal to flow rate (ml/min) divided by cross-sectional area (cm^2). Mathematically, expressed as:

$$V_f = \frac{F}{A} = \frac{D^2(\rho_p - \rho_m)}{18\eta}\omega^2 r \quad [2]$$

Solving Equation 2 for F, one arrives at Equation 1, where X is a constant which incorporates the cross-sectional area of the chamber at the elutriation boundary and the distance of the elutriation boundary from the axis of rotation, $\Delta\rho$, is 0.05 g/ml (a reasonable density difference when working with cells), η is 1.002 mPas (the viscosity of pure water at 20 °C), and factors which convert r.p.m. to radians per second.

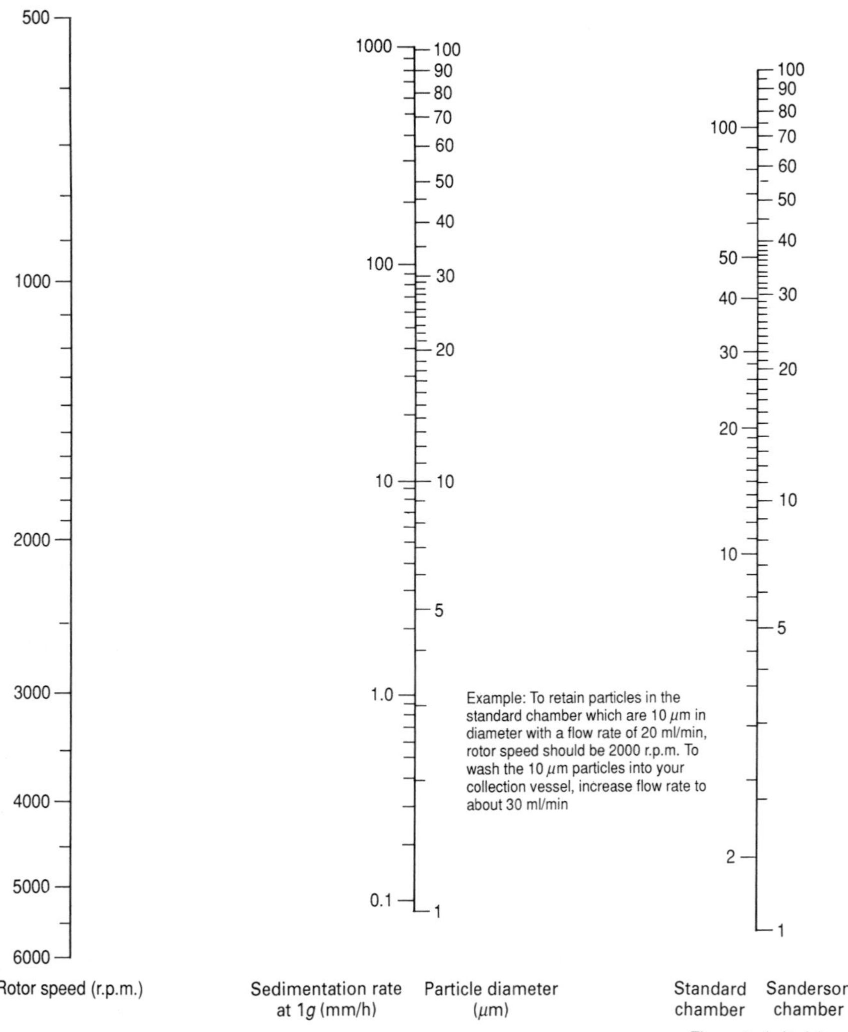

Figure 1. Flow rate and rotor speed for the separation of specific particles

Table 1. Specifications of the Sanderson and standard chambers

Chamber	Separable particle size (min and max approx.)	Particle diameter variation within an eluted population[a]		Number of particles to fill chamber		Chamber volume (ml)
		Population of small particles	Population of large particles	Min	Max	
Standard	5–50 μm	±2.5 μm	±5 μm	$\pm 10^7$	$\pm 10^9$	4.2
Sanderson	5–50 μm	±1.5 μm	±2.5 μm	$\pm 10^5$	$\pm 10^7$	5.9

[a] If particles are the same size, separation can be made using either chamber if population densities differ by at least 0.025 g/cm^3.

Before using the nomogram, the reader should note that if the sedimentation rate of the particles has been determined in a medium other than the elutriation medium (or if the diameter was calculated in a different medium, since, with the indicated assumptions at unit gravity the sedimentation rate is proportional to D^2), a new experimental sedimentation rate (S_E) must be calculated before the nomogram is used. This can be calculated using the following equation:

$$S_E = S_P \left(\frac{\eta}{\Delta\rho}\right) P \left(\frac{\Delta\rho}{\eta}\right) E \qquad [3]$$

where E is the elutriation medium and P is the medium previously used for measuring the sedimentation rate (S_P).

After using the nomogram, if viscosities other than that of pure water at 20 °C (at 4 °C, for example, the viscosity of water is 1.567 mPa.s) and liquid/particle differences other than 0.05 g/ml are required, then the parameters read from the nomogram must be adjusted:

$$F_{\text{new}} = F_{\text{nomogram}} \left(\frac{\Delta\rho}{0.05\,\eta}\right) \text{new} \qquad [4]$$

Once the magnitude of the adjustments required is known, use the nomogram as follows. Mark the known parameters on two of the scales in *Figure 1*. Then, using a ruler, draw a line through these two points so that the line intersects the third scale. Flow rate, particle size, and rotor speed are read where the ruler intersects these scales. With the adjustments of Equations 3 and 4, these parameters should usefully approximate the elution parameters for the particles in which you are interested.

Adriaan Brouwer, Henk F. J. Hendriks, Terry Ford, and Dick L. Knook

Appendix B: Isolation of liver cells

Isolation of parenchymal liver cells

The isolation of parenchymal liver cells was carried out as described elsewere (1). The method is a modification of an earlier procedure (2).

Media

The standard medium used for the isolation of parenchymal cells consists of 0.9 mM glutamine, 0.1 mM aspartic acid, 0.2 mM threonine, 0.3 mM serine, 0.5 mM glycine, 0.9 mM glutamic acid, 0.6 mM alanine, 10.0 mM lactate, 1.0 mM sodium pyruvate, 3.0 mM KCl, 0.5 mM NaH_2PO_4, 20.0 mM glucose, 25.0 mM Hepes, 0.5 mM $MgCl_2$, 24.0 mM $NaHCO_3$. All components except $NaHCO_3$ are dissolved, pH is adjusted to pH 7.4 and then $NaHCO_3$ is added. Osmolarity is adjusted to 308 mOsm with NaCl. The dissociation medium used during some steps of the isolation procedure consists of the standard medium containing, in addition, 0.05% (w/v) collagenase type I (Sigma Chemical Co.) and 1 mM $CaCl_2$ to activate the collagenase.

During the isolation procedure, perfusates are constantly saturated through an air-stone with a mixture of oxygen and carbon dioxide (95:5 v/v) to meet oxygen needs of the metabolically active parenchymal cells. The temperature of the perfusates is maintained at 37 °C for optimal collagenase activity.

Preparation of the parenchymal cell suspension

Under ether anaesthesia, the liver, portal vein, and inferior *vena cava* are dissected free and left *in situ*. A cannula (G18 Luer Braunüle, Melsungen, FRG) is inserted into the portal vein. An initial linear perfusion of the liver is done for 2 min with dissociation medium, to rinse blood from the liver. The perfusate is discarded via an incision in the abdominal aorta. It is important to perfuse the liver with a pressure of not more than 40 Torr, because parenchymal cells are highly sensitive to hydrostatic pressure. In *BN/BiRij* rats with an average liver weight of about 4 g, a rate of about 10–15 ml/min can be achieved without exceeding this pressure.

During the first perfusion, a cannula is inserted into the thoracic portion of the inferior *vena cava* to transport the effluent, and the abdominal aorta is then clamped off. A second linear perfusion is done using 225 ml of the standard medium at a rate of 10–15 ml/min to eliminate calcium ions from the liver; this dissociates the junctions between parenchymal cells. Then, a third, recirculation perfusion is done with dissociation medium for 30 min. The perfusate is returned to the medium reservoir. After this recirculating perfusion, the liver is excised and the Glisson's capsule is disrupted. At this stage the liver should be transformed into a thick paste, which is transferred to a flask and agitated for 10–20 min at 37 °C with 50 ml of dissociation medium, in which an oxygen tension of at least 50×10^3 Pa is maintained. The cell suspension obtained is filtered through five layers of gauze and through a 50 μm Nylon filter.

Cell clumping may be reduced by adding 2% albumin to the medium. The resulting total liver cell suspension contains about 100×10^6 parenchymal cells, 6×10^6 Kupffer cells, 15×10^6 endothelial cells, and less than 1×10^6 fat-storing cells per gram liver.

Parenchymal cells can be harvested, and cellular debris and non-parenchymal cells are largely removed, by centrifugation at 50*g* for 1–4 min. Sinusoidal cells can be harvested from the supernatant by subsequent centrifugation at 450*g* for 10 min. However, the parenchymal cell preparation still may contain substantial numbers of sinusoidal cells, and the yield of sinusoidal cells is low for Kupffer and fat-storing cells.

Preparation of sinusoidal liver cell suspensions

Media

The composition of the isolation medium (GBSS) is given in Section 5.2.1.

Preparation of sinusoidal cell suspensions

The liver is first perfused linearly *in situ* through the portal vein with Gey's balanced salt solution (GBSS) for 5 min, followed by a perfusion of 6 min with 0.225% pronase E (Merck) dissolved in GBSS. During these perfusions, the cannula is fixed in the portal vein using two ligatures, and the liver is carefully dissected free and placed on a sieve. Once placed on the sieve, dehydration is prevented by covering the liver with a tissue, temperature is maintained by placing the liver under a heat lamp. The excised liver is then connected to a circulation perfusion system containing 0.055% collagenase type I (Sigma Chemical Co.) and 0.055% pronase E (Merck) dissolved in 60 ml GBSS and perfused for 30 min. All media should be kept at 37 °C and perfused at a flow rate of 10 ml/min.

Following perfusion, Glisson's capsule is removed and the main vessels are sectioned. The paste-like substance is then stirred in 100 ml GBSS containing 0.055% collagenase type I and 0.025% pronase E at 37 °C for 30 min. The pH of this mixture is monitored and kept at pH 7.4 by the addition of 1 M NaOH. The suspension is filtered through nylon gauze. The resulting filtrate is then centrifuged at 450*g* for 10 min to pellet the sinusoidal cells.

Sinusoidal liver cells have been isolated from rats of different ages, gender, and genetic background (3, 4). This isolation procedure has an average yield of 7×10^6 Kupffer cells, 19×10^6 endothelial cells, and up to $5\text{–}10 \times 10^6$ fat-storing cells per gram liver (4).

References

1. Van Bezooijen, C. F. A., Grell, T., and Knook, D. L. (1977). *Mech. Age. Dev.* **6**, 293.
2. Van Bezooijen, C. F. A., Van Noord, M. J., and Knook, D. L. (1974). *Mech. Age. Dev.* **3**, 107.
3. De Leeuw, A. M., Barelds, R. J., De Zanger, R., and Knook, D. L. (1982). *Cell Tiss. Res.* **223**, 201.
4. Hendriks, H. F. J., Brouwer, A., and Knook, D. L. (1990). *Methods in enzymology*, Vol. 190, (ed. L. Packer), p. 49. Academic Press Inc, San Diego CA, USA.

8

Separations in zonal rotors

JOHN GRAHAM

1. Introduction

This chapter is primarily concerned with the general design and operation of zonal rotors, but it also illustrates the advantages of zonal rotors by quoting some specific separations. The name 'zonal' rotor is unfortunate inasmuch as it is sometimes confused with 'rate-zonal' centrifugation. Rate-zonal centrifugation refers to the fractionation of a mixture of different sized particles as they move down (or up) through a gradient under the centrifugal force generated by the centrifuge; each population of particles moving independently of the others as a discrete band or zone at a rate roughly proportional to their size. Zonal rotors were originally designed to maximize the resolution of this process but they can be used for any type of centrifugation except, usually, simple differential pelleting. They also offer the capacity which makes the handling of large volumes of sample relatively easy.

There are two types of zonal rotor, batch-type and continuous-flow. Batch-type rotors permit the scaling up of the gradient fractionations of subcellular particles, viruses, and macromolecules which are normally carried out in tubes. Of the batch-type rotors there are those which are loaded and unloaded while spinning (i.e. dynamically) and those (reorienting rotors) which are loaded and unloaded while the rotor is stationary. Some reorienting zonal rotors can be loaded dynamically. Any particular gradient and separation method used with a dynamically-loaded rotor can, in general, also be used with the appropriate reorienting rotor.

Continuous-flow rotors can only be loaded in a dynamic mode, although they may be unloaded either dynamically or statically. Their primary function is for harvesting and purifying bacteria and virus particles from large volumes of culture fluid. Some of these are not strictly zonal rotors for they only permit the collection of particles by pelleting rather than banding within a density gradient, but they are included for the sake of completeness.

The only zonal rotors which are currently commercially available are for use in high-speed and ultracentrifuges. So-called A-type rotors which were capable of low-speed separations of cells are no longer available. Only those rotors which are available commercially will be described. Although cell elutriation is described in Chapter 7, this technique is discussed briefly along with the Field Flow method, since like continuous-flow rotors, they both involve liquid flow in a centrifugal field.

2. Zonal rotor design and operation theory

This section is concerned with the general principles of centrifugation in zonal rotors and their modes of operation. During the development of zonal rotors a coded categorization system which described the type of zonal rotor was introduced. It is still in use today when describing the mode of operation of the rotor but modern rotors are given non-standard codes by manufacturers. Individual rotors differ in the detail of some of their construction but their overall designs are very similar.

This section is divided into B14 batch-type, B29 batch-type, Reograd and continuous-flow zonal rotors. The design and operational strategy is given for each type of rotor, followed by a description of the commercially-available models. A detailed account of each model is beyond the scope of this chapter, but some of the specific design features and some of the more important operational points are brought to the readers' attention.

2.1 Dynamically loaded and unloaded batch-type rotors (B14 type)

2.1.1 Design

The body of the rotor is in two parts which fit together by a screw thread and are sealed by an O-ring; the bottom part supports a central core which also carries the septa assembly. The septa assembly, usually made of Noryl, comprises a cylinder carrying four (or sometimes six) septa or vanes which fits over the core. In many models the core and septa assembly are a single unit. The septa extend to the wall of the rotor effectively dividing the internal space of the rotor into four (or more) sector-shaped compartments (*Figure 1*).

The gradient fills the entire enclosed space and it is introduced into a dynamically-loaded rotor while the rotor is spinning at a low speed (around 2000 r.p.m.). To provide access to the top of the gradient (core of the rotor) a central channel within the core exits at the surface of the core just above the septa. To provide access to the bottom of the gradient (wall of the rotor) there is a second channel within the core, concentric with the central one, which is continuous with a radial or diagonal channel within each vane or septum (*Figures 1* and *2*).

Access from the outside to the interior of the spinning rotor is provided by the fluid seal (feed head), which facilitates transfer from a static to a rotating system. It comprises two main elements (*Figure 2*); a rotating seal which is situated at the top of the core, and a stationary seal, restricted from rotation by a restraining device (of one form or another) within the centrifuge chamber,

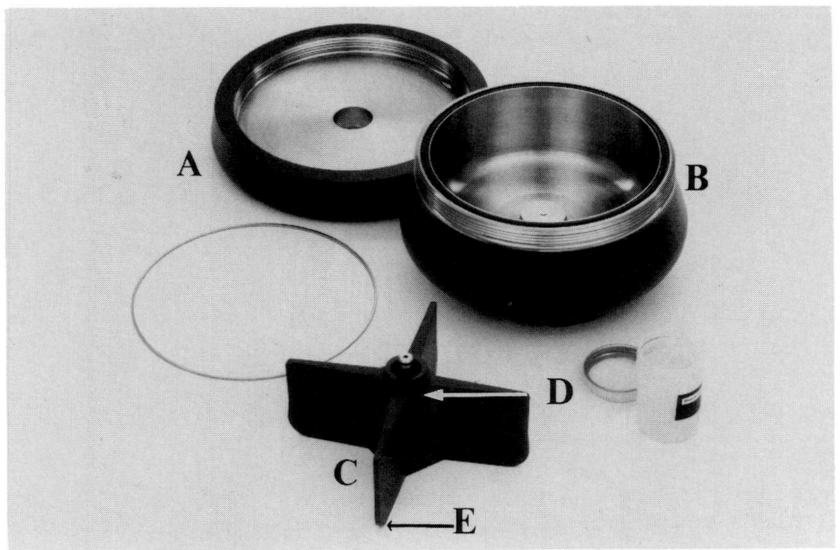

Figure 1. Disassembled B14 batch-type zonal rotor. A, top; B, bottom; C, septa assembly; D, exit to core surface; E, exit to edge of rotor. (Courtesy of Beckman Instruments Ltd.)

which is in contact, under pressure, with the rotating seal. Both seals contain the two concentric channels; one is normally made of polished stainless steel and the other of Teflon or Rulon (a filled fluorocarbon) so as to reduce the heat of friction between the two surfaces. In the stationary seal the two channels exit distally as two tubes; two other tubes may circulate cold water to the face of the seal to provide cooling at the fluid seal (*Figure 2*). The rotating seal may be contained within a bearing housing.

2.1.2 Operational strategy

To demonstrate the standard procedure for operating a dynamically-loaded (and unloaded) batch-type rotor, the loading of a rotor, capacity 650 ml, with a discontinuous 500-ml sucrose gradient (100 ml each of 10%, 20%, 30%, 40%, and 50% w/w) and a sample of 50 ml will be described.

The customary approach is to load the rotor, via the wall line, with the low density solution first. With the rotor spinning, the 10% sucrose will form a shell at the periphery of the rotor (*Figure 3*). Successively denser solutions will displace this lightest solution towards the core. The useful gradient is followed by a sufficient amount of a 'cushion' of 60% (w/w) sucrose solution so that the 10% sucrose reaches the core and is displaced out through the central channel.

The flow of liquid is then reversed and the sample is pumped into the rotor via the central channel, on top of the 10% sucrose (*Figure 3*). The 50-ml sample (normally in a solution marginally less dense than the top of the gradient, say

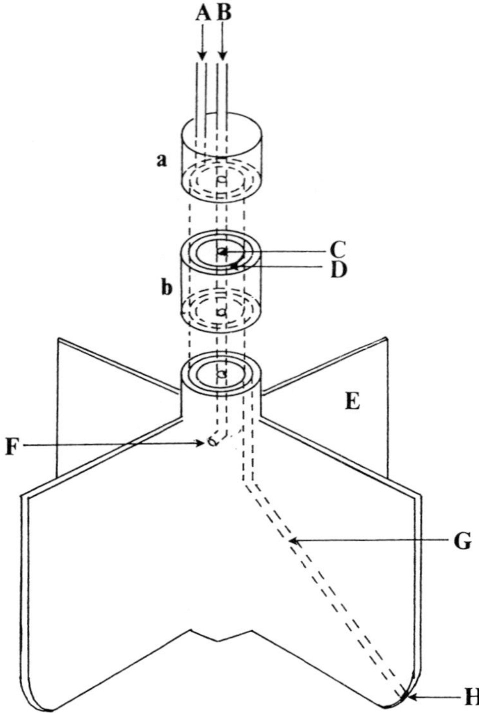

Figure 2. Fluid channels within feed head and septa assembly of B14 batch-type zonal rotor. A, edge feed tube of stationary seal (a); B, centre feed tube of stationary seal (a); C, central channel of rotating seal (b); D, annular (edge) channel in rotating seal (b); E, septa assembly; F, exit of central channel at core surface; G, diagonal channel through vane connecting annular channel to edge exit (H).

8% sucrose) will displace an equal volume of dense sucrose via the wall line. To clear the core lines of sample and to move it further out into the centrifugal field, 50 ml of 'overlay' solution (a low density solution, say 5% sucrose) is also pumped down the line to the core, which will displace a further 50 ml of cushion, leaving 50 ml of this dense solution still remaining in the rotor (*Figure 3*).

The function of the cushion of dense sucrose solution is twofold; first, it enables the useful gradient to be moved radially within the rotor without loss of any part of it, and second, so long as it is of a sufficiently high density, it will prevent sedimenting particles from reaching the wall of the rotor. If a sediment builds up on the wall of the rotor, the channels to the rotor edge may become blocked; it is important that this never happens.

Once the rotor is loaded, the fluid seal is removed, the rotor capped off, and accelerated to the centrifugation speed required. During unloading (also at low

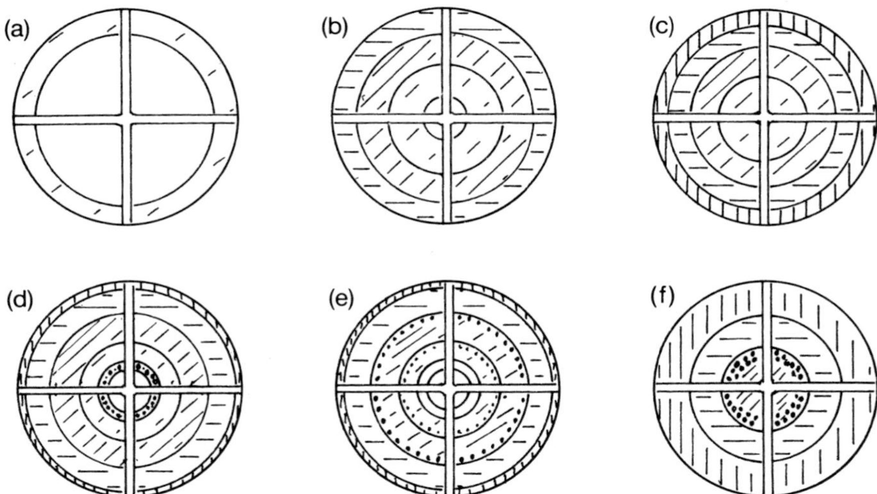

Figure 3. Dynamic loading and unloading of a batch-type rotor. (a) and (b); rotor filled via the edge with gradient, light-end first ▨ ▩ ▤ ; (c), dense cushion fed to edge ▥ displaces lightest gradient ▨ from centre. (d); sample ▦ and overlay □ fed to centre, some of cushion displaced. (e); separation phase. (f); unloading by displacement from centre by feeding dense cushion ▥ to edge.

speed in dynamically-unloaded rotors) the fluid seal is re-installed. More of the very dense solution (cushion) is then pumped to the wall of the rotor, displacing the gradient from the core, low-density end first. This is continued until all of the useful gradient has all been collected (*Figure 3*).

2.1.3 Resolution of gradients

Figure 4 shows one of the major reasons for the high resolution achievable with a gradient in a zonal rotor, compared to gradients in tubes (either in a swing-out or vertical rotor). Because the centrifugal field is radial, particles within the sample will travel down through the gradient radially. Only in the sector-shaped compartment of a zonal rotor will all particles travel unimpeded directly down through the gradient. In tubes this is true only for particles in the centre of the sample; the further a particle is from the centre the more it is directed towards the wall of the tube rather than straight through the gradient. The resolution of material in zonal rotors is greater than in any tube (see Section 3.2.1 of Chapter 1).

Collection of the material banded within the gradient is also ideal in that it moves through the conical section of the sector-shaped compartments prior to displacement out of the rotor. Within the central channel of the core each part of the gradient is supported and moved upwards by progressively denser solution. The shape of the core which slopes towards the centre, from the bottom to the top, also aids the effective collection of the gradient by directing it to the channel at the core surface above the septa.

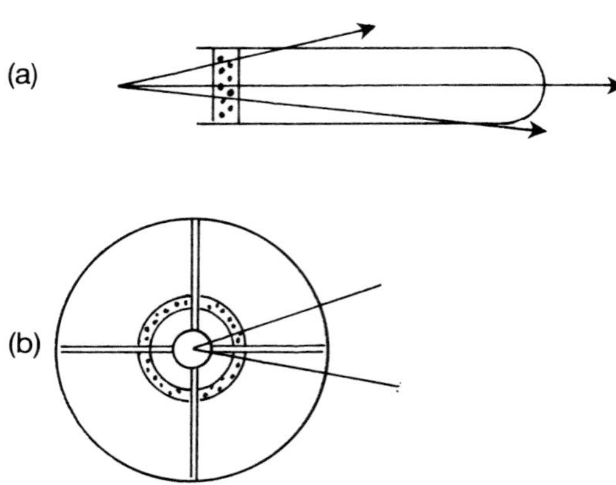

Figure 4. Comparison of sedimentation paths in radial centrifugal field in a swing-out rotor (a) and in a zonal rotor (b).

The gradient and sample in all rotors other than dynamically-loaded and unloaded zonal rotors experience acceleration and deceleration phases from, and to, rest. These reorientations of the bucket of a swing-out rotor or of the gradient within the tube of a vertical or fixed-angle rotor do not cause any serious disturbance to the gradients themselves; nevertheless, the complete lack of such potential disturbances with dynamically-loaded zonal rotors means that resolution is never compromised in such rotors.

2.1.4 Sample size

In separations of particles based on their sedimentation rate, the resolution is inversely proportional to the radial distance occupied by the sample. The tubes of swing-out rotors severely limit the amount of sample which can be accommodated on top of a gradient for these separations; its volume should not exceed 5% of that of the gradient. In a vertical rotor the volume of sample has similar restrictions but the resolution of the gradient is superior because of the small radial thickness occupied by the sample after reorientation. The size and geometry of a zonal rotor also produce small radial sample zones. The huge advantage of the zonal rotor over any tube system is that large volumes of sample can be conveniently handled even for rate-zonal separations.

Because the sedimenting chamber is sector-shaped, the radial distance occupied by the sample will decrease the further it is from the core of the rotor. In a 650-ml zonal rotor, for example, a 50-ml sample will occupy a radial distance of only 0.4 cm with a 50-ml overlay; with a 200-ml overlay it will be 0.33 cm and with a 400-ml overlay it will be 0.22 cm.

Likewise the useful gradient can also be manipulated so that it has a long pathlength (near the core of the rotor) or a short pathlength (near the periphery). The centrifugal force in the gradient will of course be modified as a result of these manipulations.

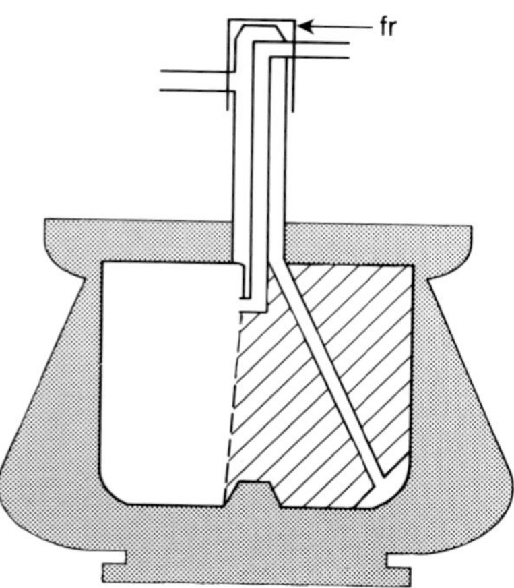

Figure 5. Diagram of JCF-Z rotor (standard zonal core). For full description see text (fr = fitting retainer).

2.1.5 Commercial models

i. High-speed rotors

The only model is the Beckman Instruments' JCF-Z with the standard zonal core. Its feed head is unlike other B14 type rotors. As shown in *Figure 5*, it comprises the seal housing which contains the familiar two lines (a central channel and annulus), which emerge as two horizontal tubes on either side of the housing. These two tubes are positioned on to the seal housing by the fitting retainer. The bearing assembly, rotating seal, seal housing, and fitting retainer are assembled on top of the rotor before transferring the rotor into the chamber. The annulus and central channel lead to the wall and the centre of the rotor as usual. The tubing connected to the fitting retainer provides the only restriction to rotation.

Before accelerating to the speed required, the delivery tubes and the fitting retainer are removed, but the sealing housing remains in place. Because the centrifuge does not operate under vacuum the rotor is not capped off. For unloading, the rotation of the seal housing is arrested manually and the fitting retainer and tubes relocated.

ii. Ultracentrifuge rotors

All the Beckman Instrument rotors (Al-14, Ti-14, Al-15, Ti-15, Z-60) conform to the basic design as described in Section 2.1.1. The removable feed head (*Figure 6*) contains the stainless steel stationary seal and a rotating Rulon seal within a lubricated bearing assembly which fits over the top of the rotor core. The feed head is placed in position when the rotor is at the loading speed. It is no longer considered necessary to feed cooling water to the face of the stationary seal.

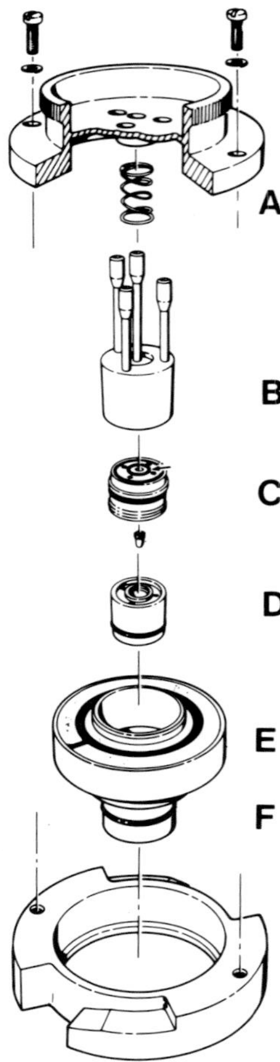

Figure 6. Exploded view of B14/15 feed head. A; upper plate and spring to provide pressure at the seal interface. B; distal end of stationary seal with tubes to centre and edge of rotor (and two cooling water tubes). C; stationary seal with central channel and annulus (edge line). D; rotating seal with central channel and annulus. E; bearing assembly for rotating seal. F; rotating extension to rotor. (Courtesy of Beckman Instruments Ltd.)

The restraining device used to prevent rotation of the stationary seal is rather more sophisticated in ultracentrifuges than it is in high-speed centrifuges. A clear plastic guard tray which is located above the rotor helps prevent any spillage from entering the rotor chamber and reduces condensation within it. This is very important since any liquid inside the chamber will interfere with the generation of the vacuum. The tray also supports the locking device by which the stationary seal is restrained.

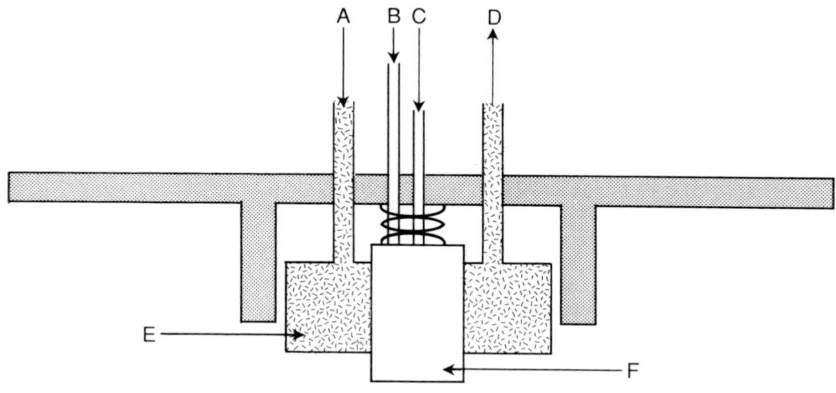

Figure 7. Kontron TZT48 stationary seal assembly. A; cooling water in. B; annular channel. C; core channel. D; cooling water out. E; cooling jacket. F; stationary seal. Note the spring discs for maintaining the pressure at the seal interface above the stationary seal.

Kontron zonal rotors also conform to the basic design and operational patterns described above. The feed head, however, is radically different (*Figure 7*). The rotating seal is the stainless steel top to the rotor core. The stationary seal is made from Teflon and it is surrounded by a cold water cooling jacket. It is located within a Perspex holder which is applied to the spinning rotor and restrained from rotation within the guard tray. There is no lubricated bearing assembly; this reduces any vibration which might arise from wear and irregularities of the bearing.

A series of inserts are available for the Kontron TZT 48 rotor which convert its 650-ml capacity to rotors of smaller capacities (325 and 160 ml); in both adaptions the pathlength is unchanged. A third adapter converts it to a Beaufay-type of zonal rotor in which only the edge portion of the enclosed space is used; it is used for rapid fractionations in short pathlength gradients.

The characteristics of both high-speed and ultracentrifuge commercial zonal rotors are summarized in *Table 1*.

2.2 B29 batch-type zonal rotors

2.2.1 Design

The standard B14 design described in the previous section is suitable neither for loading a sample at the edge of the rotor for flotation, nor for collecting gradients from the edge. At the wall of the rotor a fixed volume of liquid will have its greatest surface area and smallest radial thickness. If the rotor was being unloaded by pumping low density solution to the core, then material banded

Table 1. Characteristics of dynamically-operated batch-type zonal rotors

Manufacturer	Rotor	Capacity (ml)	Max RCF × 1000	Pathlength (cm)
Beckman	JCF-Z	1900	40	6.5 (av)
	Ti-14	665	172	5.4
	Ti-15	1675	102	7.6
	Z-60	330	256	5.5
Kontron	TZT 48.650	650	172	5.2
	TZT 40.325[a]	325	120	5.2
	TZT 40.160[a]	160	120	5.2
	TZT 40.140[b]	140	120	1.0
	TZT 32.1650	1650	102	7.4

[a]Adapted from TZT 48.650.
[b]Adapted from TZT 48.650, Beaufay-type rotor.

within the gradient would end up spread over the entire rotor wall and it would then have to be collected into the four channels within the septa. Considerable loss of resolution and recovery would occur.

In the B29 version, the wall of the rotor is modified by an insert which makes the diameter of the rotor larger at the top of the septa than at the bottom. The septa have the same shape; that is, they are wider at the top than at the bottom and the channels within each septum exit at the top (*Figure 8*). Consequently, the centrifugal force at the top of the septa is greater than at their base. These modifications enable material within the gradient to be directed towards the septa channels during unloading from the edge. They also stabilize small-volume samples loaded at the edge; flotation fractionations can be carried out with no loss of resolution.

2.2.2 Operational strategy

If the sample is going to be loaded at the centre, then the same procedure is followed for introducing both gradient and sample as for the B14 type rotor. If the sample is going to be loaded at the edge, then the gradient should be preceded by an overlay of about 50 ml, whose density will prevent the flotation of particles as far as the core of the rotor. The sample (say 50 ml) should be in a solution whose density is slightly higher than that of the densest part of the gradient, and followed by a cushion of dense sucrose solution to fill the rotor. It is not normal procedure to introduce a large volume of cushion, since it is customary to take advantage of the high centrifugal force at the periphery of the rotor.

Unloading the rotor can be accomplished in the normal manner by feeding a dense solution to the wall of the rotor, or by feeding water to the centre of the rotor. Which strategy will be adopted depends upon the position of the

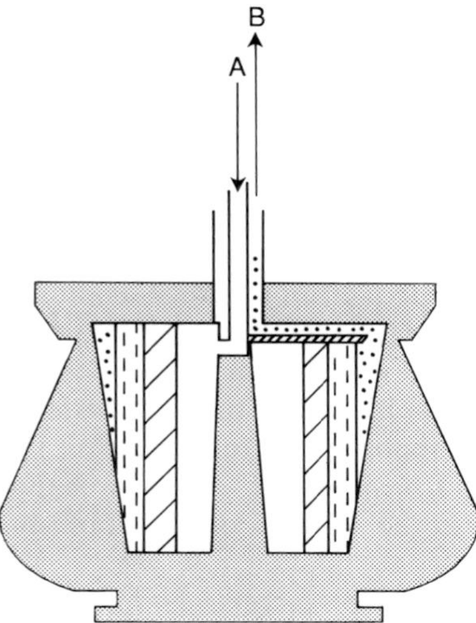

Figure 8. Unloading a B29 type zonal rotor at edge. Water ☐ is pumped to the centre A, displacing the sample ⚄ from the edge B. ▨, ◫ discontinuous gradient.

banded material of interest. If it bands close to the core then the former technique would be used, but if it bands close to the wall, then the latter.

2.2.3 Resolution of gradients

As with the B14 type rotors, the B29 rotors permit ideal sedimentation (or flotation) of particles in sector-shaped compartments. However, because unloading from the edge does not occur through an ideal conical section, some loss of resolution may occur at this stage. On the other hand, if the material of interest is banded close to the wall, centre unloading would require displacement of the bulk of the rotor contents by a dense solution, a process which itself causes some mixing. Edge-unloading can save time in such cases and the use of water rather than a viscous dense solution as the unloading medium is a practical advantage.

2.2.4 Commercial models

Beckman Instruments is the only company which manufactures this type of rotor. Adapters for the Ti-14, Al-14, Ti-15, and Al-15 convert these basic B14 type rotors to B-29 types. The adapter reduces the sedimentation pathlength slightly and the maximum centrifugal force is also slightly lower.

2.3 Reorienting (Reograd) batch-type rotors

2.3.1 Design

The 'weak point' of dynamically loaded and unloaded zonal rotors is the contact between the static and rotating faces of the fluid seal which must be kept highly polished and free from any scratches. Any damage to these faces will result in liquid within either of the two channels leaking radially across the seal under the influence of the centrifugal force at the rotating surface.

Reorienting (Reograd) rotors can be loaded with the gradient while stationary and then accelerated very slowly and smoothly to about 1000 r.p.m., before acceleration to the desired running speed. During the acceleration up to 1000 r.p.m. the gradient reorients (*Figure 9*) from a vertical to a horizontal one and vice versa during the deceleration, which again must be slow and smooth below 1000 r.p.m. to permit reorientation back to a vertical gradient prior to unloading when the rotor is stationary. Excellent linear gradients can be recovered from such rotors, in spite of the two reorientations. Moreover, these rotors do not require any sophisticated rotor seals for their operation. Therefore their feed heads are much simplified and the channel within the septa of Reograd rotors exists at the bottom of the rotor bowl rather than at the edge.

Reograd rotors are used for large-scale phase partitioning, simple two component separations, samples with heavy or light components which may obstruct the wall or centre lines of dynamically-loaded rotors, or samples which may be pathogenic. The lack of a rotating seal prevents the formation of any aerosols which can be a very effective means of transmitting infection.

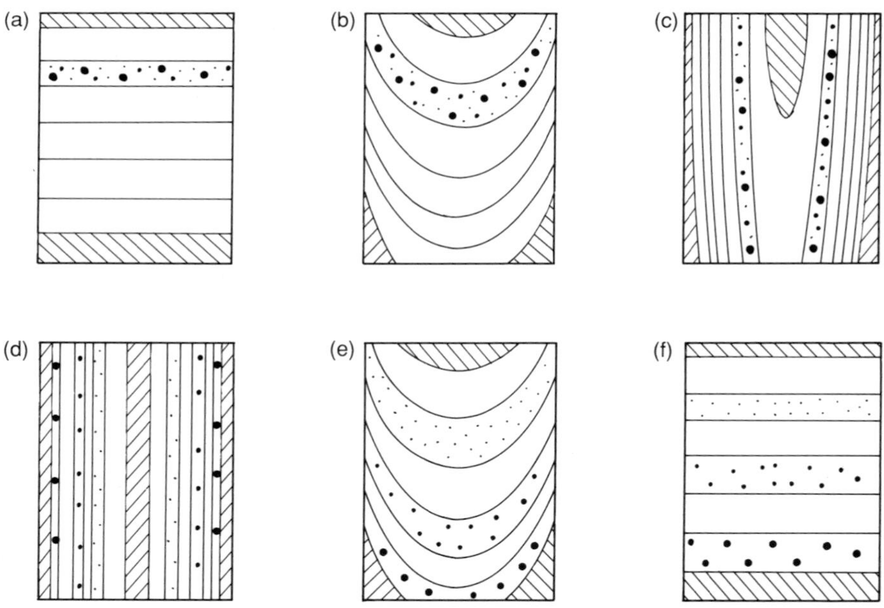

Figure 9. Schematic diagram of fractionation in a reorienting (Reograd) zonal rotor. (a) rotor at rest. (b) acceleration (early phase). (c) acceleration (later phase). (d) at speed. (e) deceleration. (f) at rest.

2.3.2 Operational strategy

These rotors must all be unloaded in a static mode, but, depending on the rotor type, they may be loaded statically or dynamically. This is discussed in more detail in Section 2.3.5.

2.3.3 Gradient design in Reograd rotors

During reorientation, the maximum shear occurs where the greatest changes in area occur, that is, those zones which are at the top and bottom of the stationary rotor, which are displaced to the centre and the edge of the spinning rotor. Shearing forces therefore decrease towards the centre of the gradient. Thus it is advantageous to include relatively large volumes of overlay and cushion solutions in these rotors to minimize the amount of shear experienced by the sample. This is particularly important of course with samples which are sensitive to shearing forces such as DNA.

Reorientation of gradients and sample zones is also important in vertical rotors where the dimensions of the tube ensure that in the spinning rotor, the sample zone is much thinner and the sedimentation pathlength much shorter than in an equivalent swing-out rotor, irrespective of where the sample is placed in the tube. In a Reograd zonal rotor, the depth of the sedimenting chamber is more or less the same as the radial length of the septa. If the sample is placed close to the top of the gradient in the stationary rotor, in the spinning rotor it will be close to the core and its radial thickness will be approximately the same as its original vertical thickness. Only if the sample is placed close to the bottom of the rotor, where it will reorient to a position close to the wall of the spinning rotor, will its thickness significantly decrease (by one half to a third).

2.3.4 Resolution

Unloading dense-end first is not ideal because within the core channel the gradient will be inverted (i.e. the density will decrease from top to bottom); this will lead to loss of resolution due to mixing. Passage through the peristaltic pump will also cause mixing. Otherwise the use of reorienting rotors does not significantly reduce resolution.

2.3.5 Commercial models

The DuPont Sorvall TZ-28 is specifically manufactured for reorienting gradient work; it can operate in both high-speed and ultracentrifuges in a variety of modes. The Beckman Instruments JCF-Z (high-speed) and Kontron TZT 32 and 48 (ultracentrifuge) rotors can be adapted to Reograd operation by the use of special cores and simplified feed heads.

i. Beckman Instruments JCF-Z (Reograd core)

In rotors which are adapted from dynamically loaded ones, gradient, cushion, sample, and overlay are introduced into the stationary rotor in a similar way to that described in Section 2.1.2. that is low-density end first to the bottom of the rotor via the central vertical line, followed by sample and overlay to the top, via the side annular line. After the sedimentation phase, the rotor is decelerated to rest and the gradient is unloaded from the bottom, usually with a peristaltic pump attached to the vertical exit line from the bottom of the rotor.

Compared to the dynamically loaded JCF-Z rotor (see *Table 1*) its capacity and pathlength are slightly reduced, but its maximum speed is the same.

ii. Du Pont Sorvall TZ-28 rotor

The feed head is replaced by a simpler adapter (sometimes called a distributor) which separates the pathways to the bottom of the rotor and to the core surface. In this zonal rotor the line to the bottom of the rotor septa emerges as the central channel in the distributor which fits on top of the rotor. The outlet ports at the surface of the core, however, are immediately adjacent to each vane and connect either with six holes or with an annular channel in the distributor (*Figure 10*).

This rotor can be loaded with gradient and cushion in the same manner as the Beckman Instruments JCF-Z, but the amount of dense solution used as a cushion should be just sufficient so that it forms a layer at the bottom of the rotor, not enough to cause the lightest part of the gradient to emerge from the rotor (*Figure 10*). This rotor is not completely filled with liquid before introducing the sample; it is only full after sample and overlay (if needed) have been added.

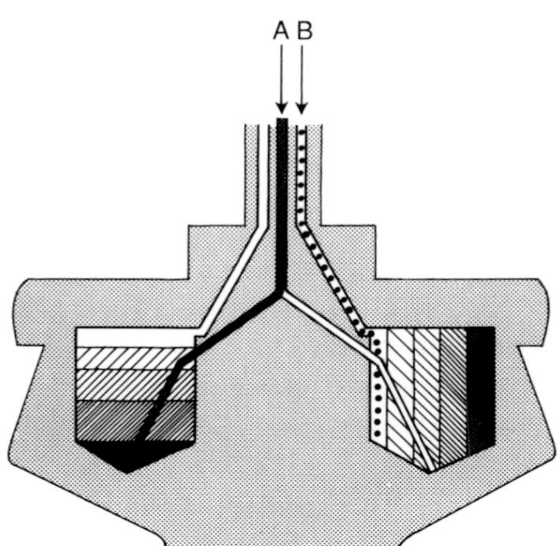

Figure 10. Loading of Sorvall TZ-28 Reograd rotor (in a high-speed centrifuge). Left hand sector: gradient loaded light-end first when the rotor is stationary through the centre line A. Right hand sector: at 1000 r.p.m., after reorientation, sample is fed through the annular channel B in distributor to the top of the gradient.

If the rotor is to be sealed during the separation run (which must be the case if it is being run in an ultracentrifuge), the sample (and overlay) must be injected into the six chambers separately; the rotor is then sealed prior to acceleration. If the rotor can be run unsealed (in a high-speed centrifuge) the sample can be applied in a dynamic mode, that is while the rotor is spinning at about 1000 r.p.m. In this mode the sample is introduced into an annular channel in the distributor; as the sample emerges at the surface of the core, the centrifugal force causes it to be directed along the surface of each vane and on to the top of the gradient. An overlay can be introduced in the same way to fill the rotor (*Figure 10*).

Indeed the entire gradient (dense end first), followed by sample and overlay, can be introduced via the annulus into the spinning rotor, the vanes directing the flow of liquid to the wall of the rotor. The only way to unload this rotor is to connect a line to the central channel of the distributor and to use a peristaltic pump to draw the gradient out via the bottom of the rotor, dense-end first, when it is stationary (*Figure 10*).

The maximum speed of the rotor is 28 000 r.p.m. (20 000 r.p.m. in a high-speed centrifuge); its capacity is 1350 ml and pathlength 5.9 cm.

iii. Kontron TZT rotors

Both the TZT-32 and TZT-48 can be used with a six-vane Reograd core and a much simplified feed head. Like the Beckman Instruments JCF-Z it must be loaded with gradient in a stationary mode light-density end first, followed by the sample to the top. It can be unloaded, however, by displacement of the gradient upwards with dense solution, or from the bottom, dense-end first. The characteristics of the adapted rotors are similar to that of the rotors in their dynamic mode (see *Table 1*).

2.4 Continuous-flow rotors

2.4.1 Design

Like zonal rotors, continuous-flow rotors are designed to cope with large volumes of sample, but unlike zonal rotors the amount of sample is not restricted to the volume of the space within the rotor. Some continuous-flow rotors are not zonal, that is they cannot be used to band material in a gradient; instead they are used to pellet the particles in the sample against the wall of the rotor. They will, however, be considered here because their designs and the concept of their use are rather similar.

In all cases a feed head is used to convert from the stationary environment of the centrifuge to the spinning rotor, that is a device similar to that on a dynamically-loaded batch-type zonal rotor, except that it is in place at all times, during the loading of the rotor with gradient (if used), during the separation run, and during unloading (if possible). Particularly in ultracentrifuges, the faces of the stationary and rotating seals can generate a great deal of heat and so very efficient cooling is needed.

2.4.2 Operational strategy

There are two parameters which have to be considered in continuous-flow centrifugation; the centrifugal force, and the flow-rate of the culture fluid passing through the rotor. As the fluid flows up through the rotor chamber or over the surface of the core, the particles of interest must be able to sediment out of the fluid path, either to the wall of the rotor or into the gradient. This will occur more efficiently the higher the centrifugal force and the lower the sample flow-rate.

2.4.3 Clean-out and resolution

The concept of clean-out is important in continuous-flow operation. It is the percentage recovery of particles within the rotor from the fluid. Clean-outs of about 90–95% should be aimed for. Often the increased time required for a reduced flow-rate to produce more than a 95% clean-out is not worthwhile.

Moreover, if the aim of the operation is not only to harvest, say bacteria, from culture medium, but also to achieve some purification from smaller contaminants in continuous-flow pelleting, then a faster flow-rate may be beneficial. The slower the flow-rate, the longer the residency time of the fluid within the chamber, so the smaller is the particle which can be sedimented.

The same considerations apply to zonal continuous-flow rotors, although here contaminating particles can be removed from the particles of interest by differential banding within the gradient. The flow of liquid over the core also tends to wash away progressively the least dense part of the gradient; it should therefore be arranged that the particles band close to the wall of the rotor. These rotors should not be used, however, to pellet material since this may cause the wall line to block.

2.4.4 Continuous-flow (non-zonal) commercial models

The Du Pont Sorvall TZ-28/GK and the Beckman Instruments JCF-Z (with either the small or large pellet core) can be used to sediment relatively large particles such as mitochondria, protozoa, bacteria, and some of the larger viruses. Their maximum speed of 20 000 r.p.m. develops about 40 000*g* at the rotor wall. *Table 2* summarizes some of the important characteristics of these rotors.

i. Du Pont Sorvall TZ-28/GK

This is designed primarily for the harvesting of particles from large volumes of culture fluid by sedimentation on to the wall of the rotor. It can only be used in the RC5 series of high-speed centrifuges, not in the RC-28S nor ultracentrifuges. The input sample is fed to the bottom of the rotor chamber via the channels in the septa, while the rotor is spinning. As the fluid is displaced upwards, particles sediment out towards the wall of the rotor; the clarified liquid then passes out of the rotor via the channels on the core surface.

Table 2. Characteristics of continuous-flow rotors

Manufacturer	Rotor type	Capacity (ml)	Flow rate (litres/h)	Max RCF × 1000	Pathlength (cm)
Beckman	JCF-Z(LP)[a]	1250	45/100[c]	40	3.6 (av)
	JCF-Z (SP)[a]	6 × 40	45/100[c]	40	2.9
Sorvall	TZ-28/GK[a]	1350	36	42	5.9
Beckman	JCF-Z(st)[b]	660	45/100[c]	40	2.3
	CF-32Ti[b]	430	9	102	1.2

[a]For pelleting only.
[b]Zonal type for sedimentation on to density barrier.
[c]Flow rate depends on type of feed head.
LP: Large pellet core.
SP: Small pellet core.

ii. Beckman Instruments JCF-Z with large and small pellet cores

These two cores are also designed for sedimenting particles from culture fluid on to the wall of the rotor.

The small pellet core is rather unusual in that at the operating speed, the sample is pumped through the central channel to the bottom of the rotor from where it is directed upwards into six canoe-shaped containers which collect the pelleted material (*Figure 11*). The clarified supernatant flows out through six radial channels at the top of the core to the annular line.

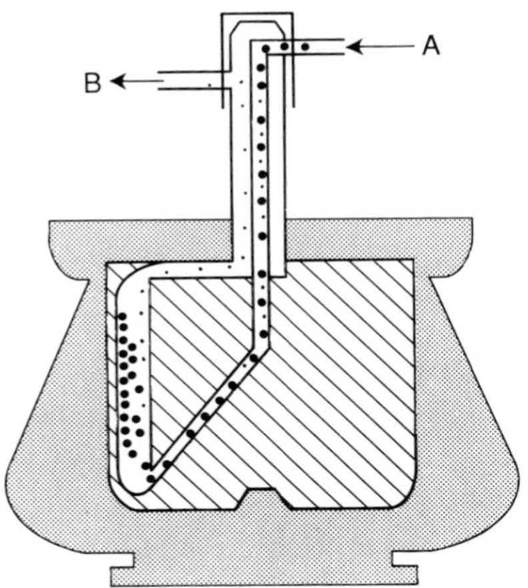

Figure 11. JCF-Z (small pellet core) continuous-flow rotor. At speed sample ▪ is fed in at A. Large particles sediment to the wall of each canoe-shaped chamber and the clarified supernatant ▫ exits the rotor at B.

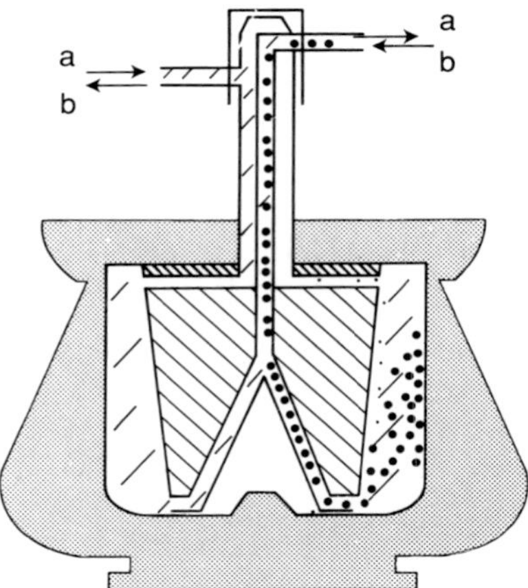

Figure 12. JCF-Z (large pellet core) continuous-flow rotor. Left hand sector: at low-speed, rotor is filled with buffer or a density barrier ▨ via the edge line (a), to emerge through the core channel (a) from the bottom of the rotor. Right-hand sector: at the operating speed, sample ▣ is passed to bottom of the core via the central channel (b); it passes up over the rotor core and the clarified supernatant emerges from the edge line (b). Rotor core ▧

The large pellet core is of a more conventional design. At 2000 r.p.m. the rotor is filled with buffer via the wall line at the top of the core; it exits via the central channel from the bottom at the core surface (*Figure 12*). Sample is then fed in the reverse direction to the bottom of the core. As it passes up over the core, particles sediment to the wall and the clarified supernatant exits at the top of the core.

Rather than fill the rotor with buffer, it can be advantageous to fill it with a medium of higher density than that of the sample. The latter would then flow over the core surface and again the particles of interest will pellet to the wall of the rotor. In this case, however, any less-dense contaminants may be held up in the higher-density medium and some degree of purification is thus achieved.

2.4.5 Continuous-flow zonal rotors

i. Beckman Instruments CF-32Ti and JCF-Z (with standard pellet core)
Both of these rotors are designed primarily for banding particles either within a gradient or on to a density cushion, from large volumes of culture fluid. Their characteristics are summarized in *Table 2*.

The design of these two rotors is shown in *Figure 13*. There are again two channels within the feed head and core. The central one exits at the base of the tapering core. The annulus is the common path for two channels, one of which exits at the top of the rotor core and one which travels over the top of the core to the wall of the rotor.

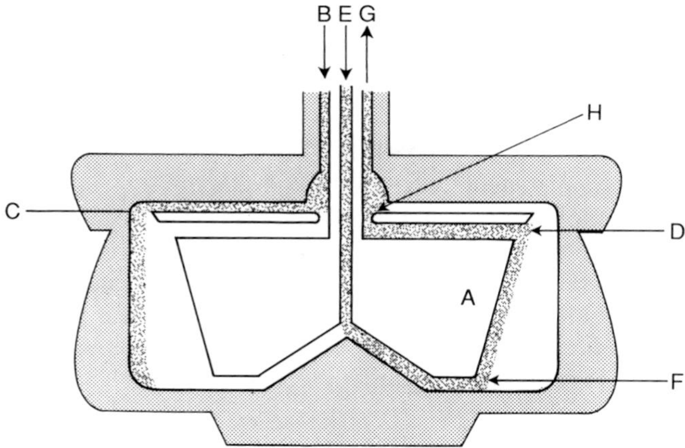

Figure 13. Continuous-flow operation of the CF-32Ti and JCF-Z (standard pellet core) zonal rotors. A; rotor core and septa. The annular channel B forms the common external path to the edge of the rotor C and the top of the core D. Gradient is fed to the wall of the rotor via B and C. Sample passes down the central channel E, up over the surface of the core between F and D; the clarified supernatant exiting at G. To unload the rotor an air block at H ensures the passage of dense sucrose to the wall from B to C and the effluent is collected at E.

The feed head for the JCF-Z (standard pellet core) is the same as for the other cores. That for the CF-32Ti is much more complex and it resembles much more closely the batch-type ultracentrifuge zonal rotors, except that a very complex seal is required in the lid of the ultracentrifuge for it to operate at high speed in a vacuum.

The rotor is filled with gradient (normally a discontinuous one or simply a density barrier), low-density end first while it is spinning at about 2000 r.p.m., in the same manner as a batch-type zonal rotor; the gradient enters the annular line within the core and the centrifugal force directs it across the top of the core to the edge (*Figure 13*). When the rotor is full, the flow is reversed and buffer is fed down the central channel to the bottom of the rotor. It then passes up over the core, flushing out part of the lightest solution, to exit in the annulus (*Figure 13*). While this flow of buffer continues, the rotor is accelerated to the running speed and then the buffer replaced with the sample. When all of the material has been fed into the rotor, the loading speed is maintained for a further period of time to permit banding of the last particles to enter the rotor.

The rotor is unloaded by deceleration back to 2000 r.p.m. and pumping dense solution to the wall of the rotor, to displace the gradient low-density end first from the core. Since the annular line is both the line to the wall of the rotor and to the top of the core, the latter must be blocked off. This is achieved simply by introducing a small bubble of air ahead of the unloading solution; the air will remain at the centre effectively blocking the line to the top of the core.

2.5 Other continuous-flow operations

For the sake of completeness, two other techniques deserve mention at this point since, like continuous-flow zonal centrifugation, they combine the effects of a centrifugal field and a flow. One, centrifugal elutriation, is a large-scale technique for fractionating cells, and experimental methodology is covered in Section 5.4 of Chapter 7. The other is sedimentation field flow fractionation which is for micropreparation or analytical work and whose detailed methodology is beyond the remit of this chapter.

2.5.1 Elutriation

In continuous-flow zonal centrifugation, the flow of liquid through the rotor occurs in a vertical direction (y) which is perpendicular to the centrifugal field (x). Small particles which do not sediment out of the flow path during their residency in the rotor are eluted from the rotor, whereas larger particles can sediment out into the gradient.

In centrifugal elutriation, the flow of liquid and the centrifugal field are in opposition to each other along the x axis. The kite-shaped chamber is first filled with the sample, suspended in a suitable buffer (*Figure 14*). The rate of buffer flow and the magnitude of the centrifugal field are chosen so that the tendency of the cells to sediment is balanced by the liquid flow. Within the chamber, the cells will band according to their size. When the buffer flow rate is increased the smallest cells will be eluted from the chamber and successively larger populations are eluted by increasing the flow rate further (*Figure 14*). A similar

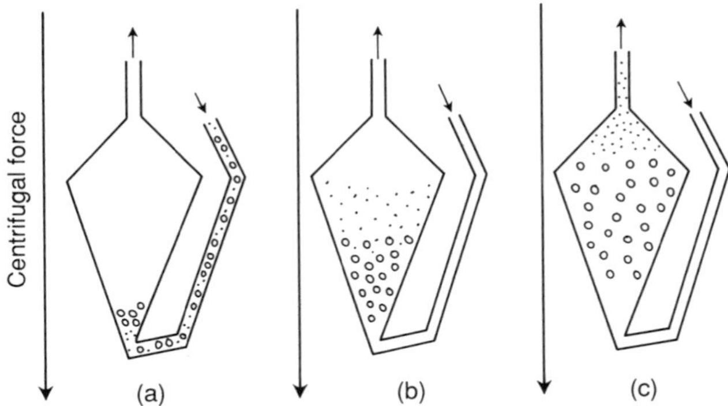

Figure 14. Elutriation. (a) sample is fed into the chamber. (b) flow rate and centrifugal field balance out so that particles remain in the chamber, smaller particles banding above the larger ones. (c) flow rate is raised to elute the smaller particles from the chamber.

effect could be obtained by successive reductions in the centrifugal force (i.e. rotor speed) but this is less frequently used.

The method has been used for fractionating liver cells (1), testes cells (2), and for the synchronization of Chinese hamster ovary cells (3). The methodology and descriptions of applications are discussed in greater detail in Section 5.4 of Chapter 7.

2.5.2 Sedimentation field flow fractionation

Like elutriation, sedimentation field flow fractionation uses a radial centrifugal field to sediment particles and a liquid flow to elute them sequentially. In this case the liquid flow is along the z axis perpendicular to the centrifugal field (x axis) and the variable force is a reducing centrifugal field rather than an increasing flow.

The flow of liquid occurs through a circular ribbon-like chamber (*Figure 15*) towards the periphery of the rotor. Typically the dimensions of the chamber are 90 cm long, 2.25 cm depth, and 0.025 cm radial width. So the sedimentation pathlength is very small (only 0.025 cm).

The chamber is first filled with carrier solution and the sample is injected as a uniform suspension: the void volume of the chamber is typically only 5.15 ml and routinely sample volumes of 400–600 µl are used. So, unlike zonal rotor centrifugation and elutriation, the technique is for micropreparation or analytical work only. The carrier flow is then stopped as the speed of the rotor is increased to give a centrifugal force of g_0 to achieve sedimentation equilibrium, when

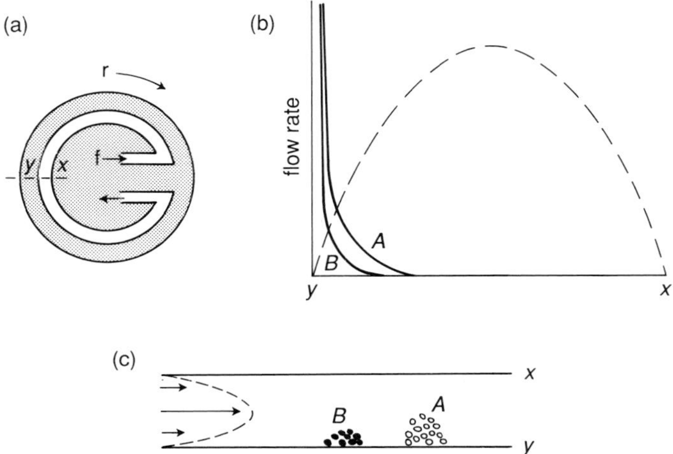

Figure 15. Sedimentation field flow fractionation. (a) the rotor; r, direction of rotation; f, buffer flow. (b) discontinuous line shows velocity of flow across the separating chamber (in the plane x–y). The solid lines show the distribution in the same plane of two particles, A and the faster sedimenting B. (c) shows the separation of A and B due to the buffer flow and the decrease in centrifugal force. (Reproduced from reference 4 with permission.)

the opposing forces of the gravitational field and diffusion are balanced and the concentration of each particle is maximal at the outer wall of the chamber, falling exponentially towards the centre (*Figure 15*). The time taken for this to occur is called the relaxation time (5–20 min) at 500–1500 r.p.m.

How far particles, on average, are from the peripheral wall depends on their size, density, and on the centrifugal force. Consider two populations of particles, *A* and *B*, which are distributed in the chamber as shown in *Figure 15* at equilibrium. When liquid flow is initiated, because of its velocity profile *A* particles will tend to move at a faster average velocity than *B* particles. Because the radial width of the chamber is small diffusion of the particles will be significant, so that over a substantive time all the particles of a given population will be moving longitudinally with the same velocity.

After this time therefore, *A* particles will tend to move ahead of *B* particles in the liquid flow. The centrifugal force, initially at g_0, for about 15 min, is then allowed to decrease (5–15 min), so that particles move away from the wall and so different populations of particles are eluted, smallest and lightest first. The technique has been used for purifying mitochondrial and microsomal fractions from corn roots (4), cells (5), liposomes (6), and macromolecules (7). It is used most effectively for fractionating material which has already been partially purified by differential centrifugation.

3. Operation of a B14 batch-type zonal rotor

The equipment required for operating a B14 zonal rotor is as follows:

- Silicone or polypropylene tubing (internal diameter 2.5–3.0 mm) for connections to the feed head.
- Peristaltic pump capable of delivering 10–100 ml/min.
- Plastic tubing connectors.
- Spencer–Wells artery forceps (haemostats) for clamping lines to the feed head.
- 50-ml plastic syringe.
- Recording spectrophotometer with a flow cell to monitor effluent from the zonal rotor. A special flow cell is required whose internal diameter is not less than 3 mm, otherwise a back pressure will build up at the fluid seal.

The assembly and operation of most zonal rotors is covered adequately in instruction manuals. However, some of the rather complex arrangements of tubes to the feed head, featuring a number of T-piece junctions, should be avoided. Here the simplest system which minimizes the length of tubing and hence lessens any mixing of the gradient is described. *Figure 16* describes an arrangement which allows for the greatest flexibility.

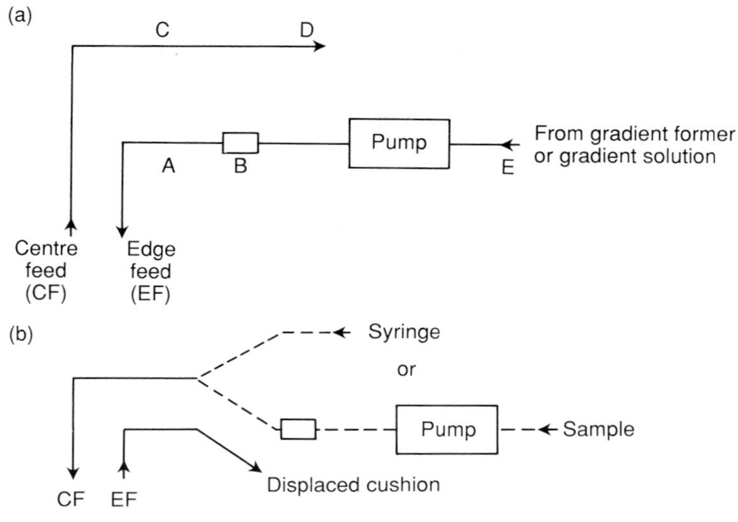

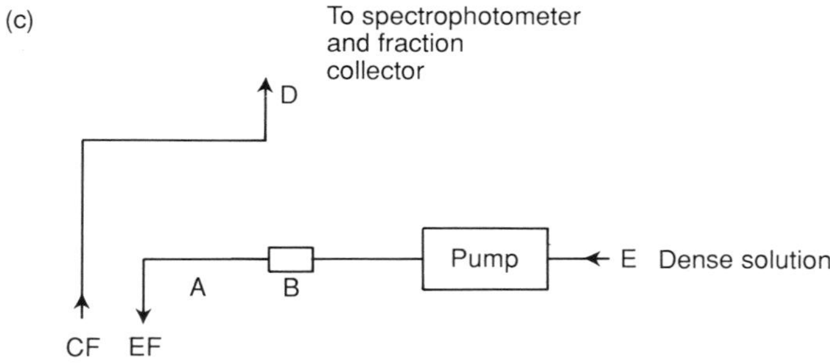

Figure 16. Simple feed head connections for operation of a B14 batch-type zonal rotor. (a) loading gradient. (b) loading sample. (c) unloading.

Protocol 1. Operation of a B14 batch-type zonal rotor

1. For loading the gradient, low-density solution first via the wall of the rotor, connect the fluid line from the wall (annulus) through tubing connector B (*Figure 16*) and a peristaltic pump to either a gradient former (continuous gradients) or to a vessel containing the lightest gradient solution (discontinuous gradients). For convenience, include a tubing connector between the pump and the gradient former and a bubble trap between A and the pump. Keep the tubing lengths minimal, but they should not be so short

as to impose any sideways pressure on the feed head, nor render manipulation of the connectors difficult. Make the tubing from the central channel long enough to reach connector B, to allow easy fixing to a syringe and to dip below the surface of about 50 ml of water placed on top of the guard tray. A steady stream of air bubbles emerging from the tube means that the gradient is entering the rotor correctly.

2. Should bubbles inadvertently enter the tubing, remove them by clamping the tubing at A and diverting the liquid flow into the bubble trap, before re-establishing the flow to the feed head.
3. When the low density solution emerges at D, turn off the pump and clamp the lines at A and C.
4. Apply the sample either by a syringe at D, or through the pump. If the latter, then disconnect the edge line at B; clean out the residual gradient medium and prime the tubing with sample before connecting to D.
5. Unclamp the edge tube slowly at A prior to loading the sample. Pressure often builds up in the edge line and this should be relieved gently.
6. After the sample and overlay have been loaded, clamp the tubes again at A and C as close to the feed head as possible to minimize any spillage of liquid from these lines when the feed head is removed.
7. Before applying the cap to the rotor core, after removing the feed head, always clean the top of the spinning core with a damp tissue to remove any residual gradient solution.
8. Prior to unloading the rotor by passing dense unloading solution to the wall of the rotor, remove any air which may have become trapped in the edge line during centrifugation. So before relocating the feed head on to the rotor core, connect a 50-ml syringe filled with water at D.
9. Prime the centre feed with water and clamp at C.
10. Pump unloading solution to the connector B which should not yet be linked to A.
11. Replace the feed head and unclamp at C while maintaining a slight pressure on the syringe barrel, then inject water to the centre until the edge line at A is full of the dense unloading solution and free of air bubbles.
12. Re-clamp at C while A is connected to B and then start the pump.

3.1 Possible problems associated with operating the B14 zonal rotor

The only major problem which may be encountered during loading or unloading is cross-leakage at the fluid seal, either from the annulus outwards, or from the annulus to the central channel. In the Kontron zonal rotor, peripheral leakage is collected in a channel in the collar of the guard tray, which can be removed with a syringe, and such leakage is not detrimental to the gradient so long as it is not excessive (less than 0.3% of the total rotor capacity). Leakage inwards is more of a problem. During loading, small columns of liquid will be expelled from the centre channel by the displaced air within the rotor. During unloading, dense unloading solution will mix incompletely with the emerging lower density gradient, and refractive index discontinuities will interfere with any absorbance measurements on the rotor effluent.

Cross-leakage may be overcome by decreasing the flow-rate; if this is unsuccessful it may be necessary to remove the feed head and clean the fluid seal.

4. Gradient design

Since the gradient within a zonal rotor fills the space of an enclosed cylinder, in the spinning rotor, equal volumes of gradient occupy a decreasing radial thickness the further they are from the rotor core. The volume/radius relationship in a B14 type rotor is given in *Figure 17*. Thus, unlike tube gradients, a zonal gradient which is linear with volume will not be linear with radius; it will instead be concave. So to achieve a gradient which is linear with radius in a zonal rotor, the gradient must be convex with volume. The reader is referred to Sections 5.5 and 6.8 of Chapter 3 for equipment for the generation of convex and concave gradients. The gradient profile shown in *Figure 18* was recovered from a B14 type rotor using 400 ml of 45% (w/v) sucrose and 400 ml of 5% (w/v) sucrose in a constant volume mixer.

Steensgaard and colleagues have devised a number of data processing programs for ultracentrifugation (see Chapters 1 and 5). These include a program for the calculation of sedimentation coefficients from centrifugation runs in zonal rotors.

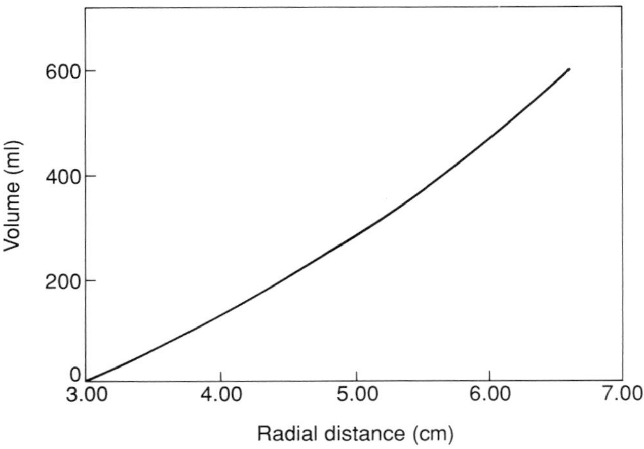

Figure 17. Volume–radius relationship for a B14 zonal rotor. For details see text.

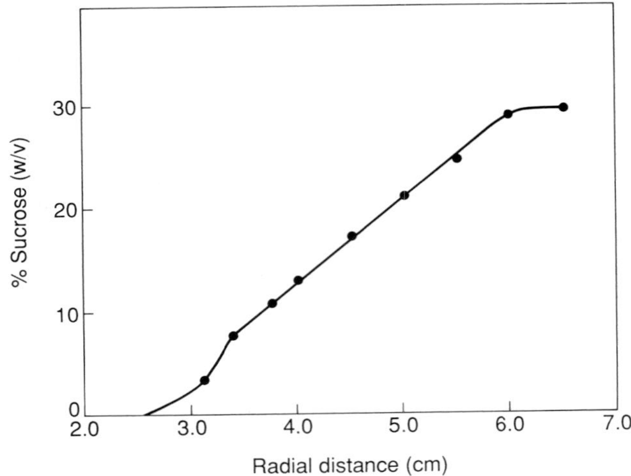

Figure 18. A 'linear-with-radius' gradient recovered from a B14 zonal rotor. For details see text.

The quantitative aspects of zonal centrifugation have also been discussed elsewhere in detail by Steensgaard *et al.* (8, 9).

5. Examples of separations with zonal rotors

This section includes only a few representative separations which show the zonal rotor system to advantage. Only those rotors which are currently commercially available are included. Any gradient system which has been worked out for a swing-out, vertical, or even fixed-angle rotor can be adapted to a zonal rotor, so long as the operator is aware of the difference in geometry of the zonal rotor. The zonal rotor will provide a fractionation at least as good as any tube gradient and often a superior separation will be obtained. A big advantage of the zonal system is that it saves using multiple fractionations with smaller-volume tube systems, and the production of a large amount of biological material in one separation has important consequences for the reproducibility of subsequent experimentation.

5.1 Separation of 40S and 60S ribosomal subunits

For the separation of ribosomal subunits on the basis of their sedimentation rate, it is important that the radial distance occupied by the sample is as small as possible to maximize the resolution of the gradient. For this reason it is inconvenient to use swing-out rotors for bulk preparation of these subunits. In the method described here, 35 ml of sample can be fractionated, and with an overlay volume of 65 ml the radial distance occupied by the sample is only 3 mm. To achieve a similar fractionation with a 38-ml tube in a Beckman SW28 swing-out rotor, 24 gradients would be required. A zonal rotor clearly provides a big advantage for this purpose, both in terms of time-saving and consistent quality of the product.

Rat-liver polysomes are prepared as described by Schrier and Staehelin (10) and incubated *in vitro* under conditions optimal for protein synthesis, in order to allow for the completion of nascent peptide chains and detachment of the ribosomes from endogenous mRNA. Those ribosomes which have released their nascent chains will dissociate into 40S and 60S subunits in 0.5 M KCl with full retention of biological activity on subsequent lowering of the salt concentration.

The following equipment and chemicals are required:

- B14 batch-type rotor (Beckman Ti-14, Kontron TZT-48) and the appropriate ultracentrifuge
- gradient maker with 300-ml capacity chambers
- recording spectrophotometer with zonal flow cell
- gradient solutions: 65 ml overlay containing 20 mM Tris–HCl (pH 7.6), 0.3 M KCl, 3 mM $MgCl_2$, 1 mM dithiothreitol (DTT); 250 ml of 15% (w/v) sucrose, 450 ml of 40% (w/v) sucrose, both solutions containing the same additions as the overlay.

Protocol 2. Separation of ribosomal subunits

1. Carry out all operations at 4 °C.
2. Introduce to the edge of the B14 rotor spinning at 2000 r.p.m. the following solutions: 500 ml of a 15%–40% (w/v) sucrose gradient (linear with volume) and 40% (w/v) sucrose to fill the rotor. Do this while the subunits are being incubated in the dissociating solution (10).
3. Feed the sample (35 ml) to the centre of the rotor; follow by 65 ml of overlay and centrifuge at 47 000 r.p.m. (115 000*g*) for 4 h.
4. Unload the rotor at 2000 r.p.m. by pumping 50% sucrose (w/v) to the edge and pass the effluent from the centre through a zonal flow cell in a spectrophotometer to monitor the absorbance (or transmission) at 280 nm and collect the effluent in 20-ml fractions.

The primary aim of this particular separation is to isolate purified 40S subunits and as shown in *Figure 19*, and the resolution of 40S from both 60S subunits and from the soluble protein is good; the majority of material is collected between 230 ml and 300 ml. This preparation is very active in binding initiation factors and methionyl-$tRNA_f$ (11). Clearly the 60S peak is heterogeneous and the leading edge of this material sediments almost to the wall of the rotor: this band probably also contains undissociated ribosomes, and to purify this further the rotor could be partially unloaded with a 40–60% (w/v) linear sucrose

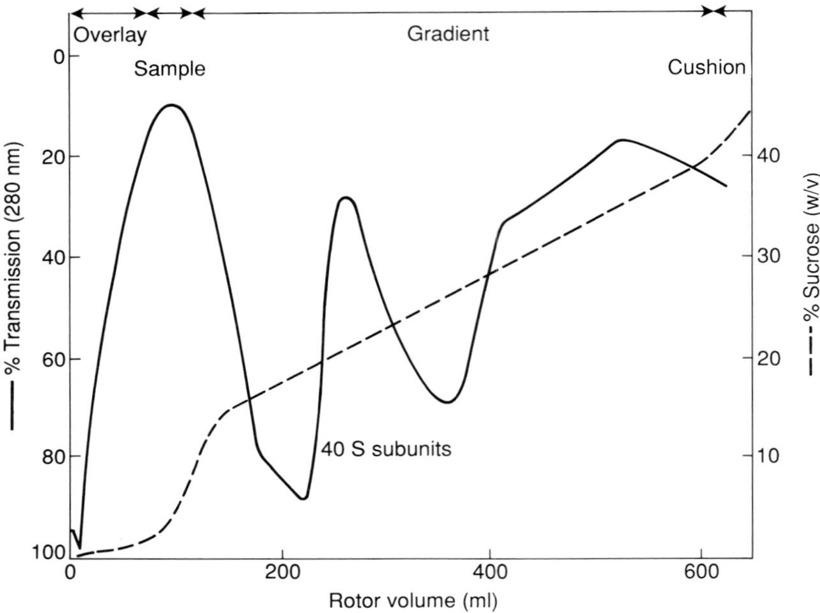

Figure 19. Separation of ribosomal subunits. — % transmission at 280 nm; ---- density in % sucrose (w/v). For experimental details see text.

gradient (400 ml) to recover all the 40S subunits and then recentrifuged. The separation of ribosomes, polysomes, and ribosomal subunits in zonal rotors has been discussed by Birnie *et al.* (12).

5.2 Fractionation of a post-nuclear supernatant

The initial partial fractionation of an homogenate from animal (or plant) tissue or from cultured cells can be very simply carried out in any batch-type zonal rotor. A series of differential centrifugation steps can be replaced by a single centrifugation in a zonal rotor for the initial separation of membrane fragments, mitochondria, lysosomes, and rough and smooth membrane vesicles. Homogenate from 10–20 g of liver or 10^9–10^{10} cultured cells can be fractionated at a time using this procedure; zonal rotor can cope with a sample volume of 50–200 ml (depending on its total capacity) quite easily. Carrying out a rate-zonal separation in a zonal rotor avoids the low resolution and poor recoveries associated with differential centrifugation and the inconvenience of using many tubes for conventional rate-zonal separations in tube rotors.

The mode of preparation of the homogenate will depend on the starting material (13) but the aim should be to suspend the sample in about 50 ml of 0.25 M sucrose containing a suitable buffer and any ionic additions necessary for the preservation of the material (14). The same buffer and ions should be in all gradient solutions. Centrifugation conditions will also be variable to some extent and so the conditions given should be viewed as suggestions.

The following equipment and chemicals are required:

- any dynamic or Reograd batch-type rotor (the Beckman JCF-Z with standard zonal core or the Sorvall TZ-28 are probably the most convenient) and an appropriate high-speed centrifuge
- recording spectrophotometer with zonal flow cell
- gradient solutions: 15%, 20%, 25%, 30%, 35%, 40%, 50%, and 55% (w/w) sucrose in the same buffer as that in the homogenate.
- overlay 0.1 M sucrose in the same buffer as that in the homogenate
- unloading solution: 60% (w/w) sucrose if you are using a rotor which can be unloaded by displacement from the centre

---

Protocol 3. Fractionation of cell membranes

1. Carry out all operations at 4 °C.
2. Introduce sequentially the following solutions; 50 ml each of 15%, 20%, 25%, and 30% sucrose, 100 ml of 35%, 50 ml of 40%, 100 ml of 50% sucrose for rotors of about 650 ml capacity. These volumes should be doubled or trebled for rotors of 1300 ml and 1900 ml capacity respectively.
3. The amount of 55% sucrose needed will depend on the type of rotor, but together with the sample and overlay should be enough to fill the rotor. With all rotors except the Sorvall TZ-28 introduce sufficient sucrose so that the top of the gradient emerges from the rotor.
4. Pump the sample in isotonic sucrose (50–200 ml depending on rotor type) followed by 30–100 ml of overlay on to the gradient. With the Sorvall TZ-28 use about 300 ml of 55% sucrose; accelerate the rotor slowly to 1000 r.p.m. and then apply 100-ml sample and overlay to fill the rotor.
5. Centrifuge at 10–16 000*g* for 60 min at 5 °C.
6. Unload dynamically-unloaded rotors by displacing the gradient with 60% (w/w) sucrose. Unload Reograd rotors, depending on type, in the same way, or by pumping the gradient out from the bottom of the rotor (dense-end first).
7. Monitor the effluent by passing it through the flow cell and collect in 20-ml, 50-ml, or 100-ml fractions, depending on the size of gradient.

Figure 20 shows the results obtained with a Lettre cell homogenate (15). All the solutions contained 1 mM $NaHCO_3$ and 0.2 mM $MgCl_2$; a B14 rotor was

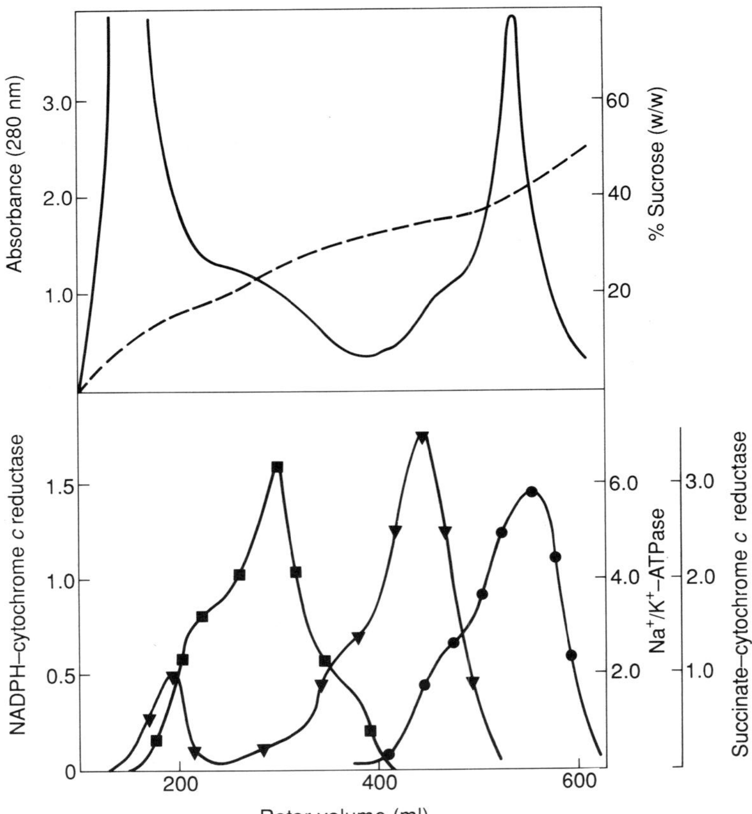

Figure 20. Fractionation of cell membranes from a Lettre cell post-nuclear supernatant. Top panel: — 280 nm absorbance, ---- % sucrose (w/w). Lower panel: ■ NADPH–cytochrome *c* reductase; ▼ Na^+/K^+-ATPase; ● succinate–cytochrome *c* reductase. For experimental details see text. (Derived from reference 15.)

centrifuged at 16 000*g*. The enzyme profiles show that plasma membrane fragments (containing Na^+/K^+-ATPase) sediment well ahead of the endoplasmic reticulum (NADPH–cytochrome *c* reductase), but just behind the mitochondria (succinate–cytochrome *c* reductase).

Very large fragments of plasma membrane formed during homogenization of, say, rat liver, may sediment faster than mitochondria (16) and it may be necessary to reduce the centrifugal force to as low as 4000*g*. The general problems of fractionation and analysis are beyond the scope of this chapter but are discussed elsewhere (13, 14).

5.3 Separation of smooth and rough endoplasmic reticulum in a Reograd rotor

Although Reograd rotors are recommended (rather than dynamic ones) if it is likely that a large pellet may be generated during centrifugation, it is only practical to allow this to happen if the pellet is firmly packed. A poorly-packed pellet will tend to fall back into the gradient medium as the latter passes over the pellet during unloading. In the separation of rough and smooth endoplasmic reticulum the pellet of rough endoplasmic reticulum is gelatinous and very well packed. This separation also highlights another use of Reograd rotors—for separations involving simple density barrier systems.

The following equipment and chemicals are required:

- Reograd zonal rotor: Beckman JCF-Z with Reograd core (in a high-speed centrifuge), Sorvall TZ-28 (in a high-speed or ultracentrifuge), Kontron TZT48 or TZT32 with Reograd core (in an ultracentrifuge)
- a post-mitochondrial supernatant (20 000*g* for 10 min) from a rat-liver homogenate (from up to six livers) in 0.25 M sucrose, 10 mM Tris–HCl (pH 7.6)
- density barrier solutions (volume will depend on rotor), 0.6 M sucrose, 15 mM CsCl and 1.3 M sucrose, 15 mM CsCl (both solutions buffered with 10 mM Tris–HCl, pH 7.6)

Protocol 4. Fractionation of endoplasmic reticulum

1. While stationary fill the rotor sequentially from the bottom with sample, 0.6 M sucrose, and 1.3 M sucrose using volumes in the ratio of 3:1:2.
2. Centrifuge at 100 000*g* for 1.5 h or 50 000*g* for 3 h.
3. Unload the rotor by pumping out the contents dense end first.

The smooth microsomes band very obviously at the interface between the two sucrose solutions, and the rough microsomes can be recovered from the wall of the rotor. The original method (17) used small tubes in a fixed-angle rotor. With large volumes of homogenate (say 400 ml produced from eight rat livers) the use of tubes is very time-consuming and the preparation using a Reograd rotor saves a lot of time.

5.4 Harvesting of virus from tissue culture fluid (1–3 litres)

The harvesting of virus from culture fluid by simple pelleting frequently causes inactivation of the virus. At the same time pelleting will not resolve the virus from some of the membrane contaminants and the volumes of culture fluid normally processed make it impractical to purify the virus by sucrose gradients in a swing-out rotor. The extra capacity available in batch-type rotors makes the harvesting of virus from up to one litre in a B14 type rotor or up to 2.8 litres in a B15 type rotor a practical proposition by using two loadings. If a B29 adapter is used, volumes of 800 ml, respectively, are practical. In the following, allantoic fluid containing Sendai virus will be purified, but any similar material can be used.

The following equipment and chemicals are required:

- refrigerated low-speed centrifuge
- B14 type rotor (Beckman Ti-14 or Kontron TZT48) or a B29 type rotor (Beckman Ti-14 with adapter) in the appropriate ultracentrifuge
- recording spectrophotometer with zonal flow cell
- phosphate-buffered saline (PBS), 30% (w/w) sucrose in PBS, 55% (w/w) sucrose in PBS, 60% (w/w) sucrose unloading solution (for B14 type rotors only

---

Protocol 5. Harvesting virus from tissue culture fluid using a batch-type zonal rotor

1. Carry out all operations at 4 °C.
2. Clarify the allantoic fluid by centrifugation at 3000*g* for 20 min.
3. Pump the following solutions into the rotor via the edge line; 25 ml PBS, 500 ml (B14) or 400 ml (B29) allantoic fluid, 50 ml of 30% sucrose, and 55% sucrose to fill the rotor.
4. Centrifuge at 45 000 r.p.m. (100 000*g*) for 90 min.
5. Using the spectrophotometer at 280 nm to monitor the effluent, harvest the virus which bands at the 30%/55% interface by pumping 60% sucrose to the edge (B14) or PBS to the centre (B29).

6a Repeat the process with the second batch of clarified allantoic fluid. With the B14 rotor, because it is full of 60% sucrose at the end of the first unloading, it is probably best to disassemble the rotor to discard the sucrose before re-loading.

6b. With the B29 adapter, continue pumping PBS to the centre after harvesting the virus to make sure all the 30% sucrose is removed (say an extra 100 ml); then make the second batch of fluid about 5% with respect to sucrose and pump this to the wall of the rotor. Make sure that about 25 ml of the PBS remains in the rotor after introducing the new 30% and 50% sucrose layers.

7. Pool the virus concentrate and dilute to about 20% (w/w) sucrose so that the total volume is less than 200 ml.

8a. Now set up the third run. In the cleaned-out B14 rotor load 50 ml PBS, 200 ml sample, 250 ml 30% sucrose, and 55% sucrose to fill the rotor.

8b. With the B29 type, as before, make sure the rotor is cleared of all residual sucrose and reverse the flow to introduce the virus suspension and new sucrose solutions at the edge.

9. Recentrifuge at the same speed for 2.5 h and recover the virus as before.

Figure 21 shows the 280 nm transmission profile of the effluent from the first run and the third run with a B14 rotor. The first two runs achieve concentration of the virus and some purification. Because the virus bands so close to the edge of the sample layer, it will still be contaminated with soluble proteins and some membrane material. The third run achieves additional concentration, but its main purpose is purification, by the insertion of a sucrose barrier between the edge of the sample and the virus banding position. The method achieves excellent recovery (at least 90%) of intact virus and the purification (increase in infectivity titre over starting material) is regularly 50–100 fold.

The aim of these fractionations should be to band the virus sharply in the steep gradient formed between two layers of sucrose, and in the final purification gradient, the largest possible volume of virus-free sucrose between the leading edge of the sample and the banding position, to maximize the separation from soluble protein and slowly-sedimenting membranes. Although the use of a B29 adapter reduces the capacity, it makes it possible to carry out all three runs without having to clean out the rotor. Reograd rotors can have an advantage if the virus is highly infective but they are not ideal if the virus bands too close to the edge of the rotor (see Section 2.3.3). The use of zonal rotors for virus harvesting has been discussed elsewhere in more detail (18).

5.5 Harvesting of virus from culture fluid (continuous-flow)

The use of a continuous-flow zonal rotor permits a more convenient approach to the harvesting of micro-organisms and viruses from large volumes of culture fluid. A Beckman JCF-Z with a large or small pellet core or a Sorvall TZ-28/GK can be used for pelleting both micro-organisms and viruses in high-speed centrifuges. For banding micro-organisms at a density barrier the Beckman JCF-Z with the standard pellet core can be used in a high-speed centrifuge; for viruses it is probably more convenient to use the Beckman CF-32Ti in an ultracentrifuge. The flow rates and centrifugation speeds used depend on the amount of material and the sedimentation coefficient of the particles. A brief description is provided below for the harvesting of Semliki Forest virus from 4 litres of culture fluid.

The following equipment and chemicals are required:

- Beckman CF-32Ti in the appropriate ultracentrifuge
- recording spectrophotometer with zonal flow cell
- phosphate-buffered saline (PBS), 4 litres of clarified culture fluid, 30% (w/w), and 55% (w/w) sucrose in PBS.

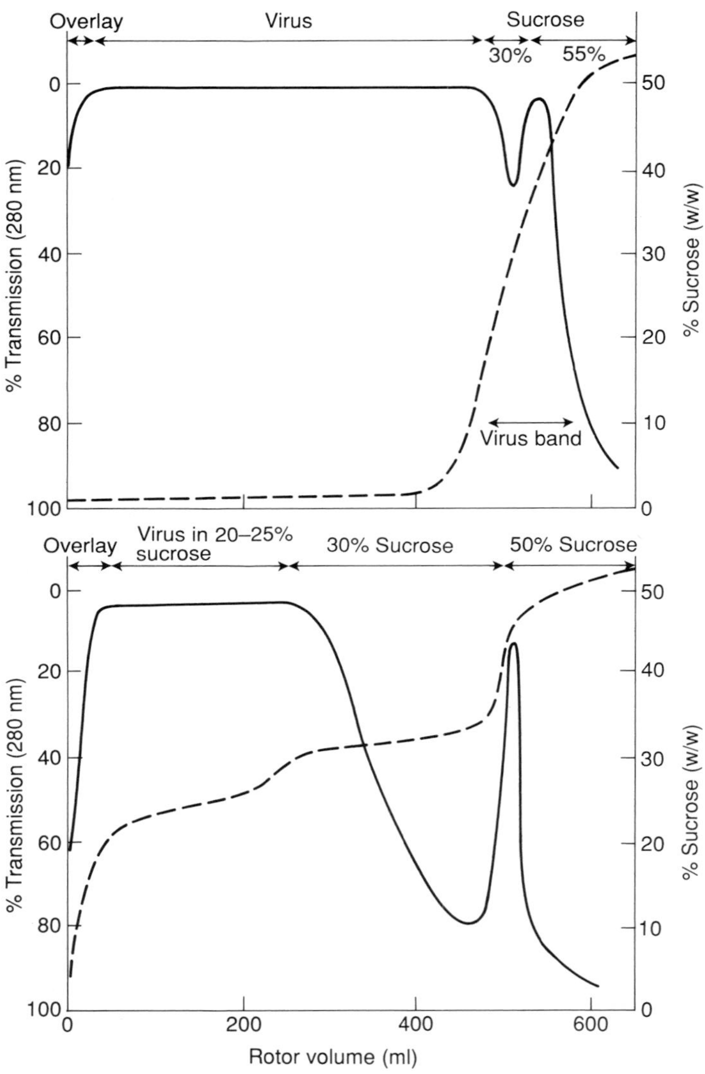

Figure 21. Concentration and purification of Sendai virus. Top panel: 1st centrifugation—concentration of virus from 450 ml of allantoic fluid. This was repeated with a second batch of allantoic fluid and the two virus bands combined together. Lower panel: 3rd centrifugation—concentration and purification of combined virus bands from 1st (and 2nd) centrifugation. In both panels: — % transmission at 280 nm, ---- % sucrose (w/w). For experimental details see text.

Protocol 6. Harvesting virus from tissue culture fluid using a continuous-flow zonal rotor.

1. While the rotor is spinning at 2000 r.p.m., load it, via the edge, with 200 ml of PBS, 150 ml 30% sucrose, and sufficient 55% sucrose to fill the rotor.
2. As the rotor accelerates to the operating speed of 32 000 r.p.m. (100 000*g*) pump PBS to the bottom of the rotor so that it passes up over the core surface.
3. At the operating speed, replace the PBS with the virus-containing fluid using a flow-rate of about 1 litre per hour.
4. When all the fluid has been processed, hold the rotor at 32 000 r.p.m. for a further hour to make sure all the virus is banded at the 30%/55% interface.
5. Decrease the speed to 2000 r.p.m. and (after introducing the air block) pump 55% sucrose to the wall and unload the rotor through a zonal flow cell in a spectrophotometer.

Approximately 95% clean out can be obtained under these conditions from 4 litres of culture fluid. With larger volumes of fluid it may be necessary to purify the virus in a second step using a batch type rotor. A more complete discussion of using continuous-flow rotors has been published (18).

5.6 Other fractionations

The A-type zonal rotor, for use in low-speed centrifuges, which is no longer commercially available, has been used for the synchronization of cells (19, 20); such separations could be carried out in high-speed zonal rotors. The important factors which influence the success of such fractionations are the use of relatively high sample volumes to minimize aggregative interactions between cells and a narrow sample band. The larger capacity high-speed zonal rotors should be capable of substituting for the A-type rotor. Methods of separating different cell types (21) in tube gradients should be readily adapted to zonal rotors.

Zonal rotors are usually used in a preparative mode but Steensgaard *et al.* (22) have used an ultracentrifuge zonal rotor in analytical mode to investigate the immune complex formation between monoclonal antibodies and human IgG in special isokinetic gradients (23). These workers were able to deduce the number of antigenic determinants and antibody-binding sites from the sedimentation profile of different mixtures of antigen and antibody.

Acknowledgements

I thank Dr A. Wyke for providing the Sendai virus samples and Dr M. Clemens and Miss V. Tilleray for the subribosomal unit preparation.

References

1. Knook, D. L. and Sleyster, E. C. (1976). *Exptl. Cell Res.* **99**, 444.
2. Grabske, R. J., Lake, S., Gledhill, B. L., and Meistrich, M. L. (1975). *J. Cell. Physiol.* **86**, 177.
3. Grabske, R. J., Lindl, P. A., Thompson, L. H., and Gray, J. (1975). *J. Cell Biol.* **67**, 142a.
4. Mozersky, S. M., Caldwell, K. D., Jones, S. B., Maleeff, B. E., and Barford, R. A. (1988). *Anal. Biochem.* **172**, 113.
5. Caldwell, K. D., Cheng, Z. Q., Hradecky, P., and Coddings, J. C. (1984). *Cell Biophys.* **6**, 233.
6. Dreger, R., Hawrot, E., Sartorelli, A. C., and Constantinides, P. P. (1988). *Anal. Biochem.* **175**, 433.
7. Schallinger, L. E., Yau, W. W., and Kirkland, J. J. (1984). *Science* **225**, 434.
8. Steensgaard, J. and Rickwood, D. (1985). In *Microcomputers in biology: a practical approach*, (ed. C. R. Ireland and S. P. Long), p. 241. IRL Press, Oxford.
9. Steensgaard, J., Moller, N. P. H., and Funding, L. (1978). In *Centrifugal separations in molecular and cell biology*, (ed. G. D. Birnie and D. Rickwood), p. 115. Butterworths, London.
10. Schrier, M. H. and Staehelin, T. (1977). *J. Mol. Biol.* **73**, 329.
11. Clemens, M. J., Echetebu, C. O., Tilleray, V. J., and Pain, V. M. (1980). *Biochem. Biophys. Res. Commun.* **92**, 60.
12. Birnie, G. D., Fox, S. M., and Harvey, D. R. (1972). In *Subcellular components preparations and fractionation* (ed. G. D. Birnie), p. 235. Butterworths, London.
13. Graham, J. M. (1975). In *New techniques in biophysics and cell biology* (ed. R. H. Pain and B. J. Smith), Vol. 2, p. 1. Wiley and Sons, London.
14. Graham, J. M. (1982). In *Cancer cell organelles* (ed. E. Reid, G. Cook, and D. J. Morre), p. 342. Ellis Harwood Ltd. Chichester, UK.
15. Graham, J. M. and Coffey, K. H. M. (1979). *Biochem. J.* **182**, 173.
16. Evans, W. H. (1970). *Biochem. J.* **166**, 833.
17. Bergstrand, A. and Dallner, G. (1969). *Anal. Biochem.* **29**, 351.
18. Graham, J. M. (1978). In *Centrifugal separations in molecular and cell biology* (ed. G. D. Birnie and D. Rickwood), p. 63. Butterworths, London.
19. Pasternak, C. A. and Warmsley, A. M. H. (1973). In *Methodological developments in biochemistry* (ed. E. Reid), Vol. 68, p. 249. Longmans, London.
20. Graham, J. M., Sumner, M. C. B., Curtis, D. H., and Pasternak, C. A. (1973). *Nature* **246**, 291.
21. Loos, D. M. and Roos, D. (1974). *Exptl. Cell Res.* **86**, 333.
22. Steensgaard, J., Jacobsen, C., Lowe, J., Ling, N. R., and Jefferis, R. (1982). *Immunology* **46**, 751.
23. Steensgaard, J. and Jacobsen, C. (1979). *J. Immunol. Methods* **29**, 173.

Appendix 1

Chemical resistance chart for tubes, adapters, and rotor materials

Key: S, satisfactory; M, marginal, test before using; U, unsatisfactory, not recommended; E, explosion hazard; —, not tested.

Reagent	Polycarbonate	Polysulfone	Polypropylene	Polyclear/Ultra-Clear[b]	Polyallomer	Cellulose nitrate	Celluose acetate butyrate	Nylon	Delrin	Noryl	Carbon fibre	Aluminium	Titanium
Acetaldehyde	U	—	M	M	M	U	U	—	—	—	S	S	S
Acetic acid (5%)	S	S	S	M	S	S	S	S	S	S	S	S	S
Acetic acid (60%)	U	S	S	U	S	U	U	M	S	S	S	S	S
Acetic acid (glacial)	U	M	M	U	S	U	U	—	S	—	U	S	S
Acetone[a]	U	U	M	U	M	U	U	M	S	—	U	S	S
Allyl alcohol	S	—	S	S	S	—	U	M	—	—	—	—	S
Aluminium chloride	S	—	S	S	S	S	S	S	S	—	U	—	S
Aluminium fluoride	U	—	S	S	S	—	—	S	S	—	—	—	S
Ammonium acetate	S	—	S	S	S	—	—	—	—	—	—	—	S
Ammonium carbonate	U	S	S	M	S	S	S	S	S	—	S	S	S
Ammonium hydroxide (10%)	U	S	S	U	S	U	U	S	U	S	S	—	S
Ammonium hydroxide (conc.)	U	—	S	U	S	U	U	S	U	U	U	—	S
Ammonium sulphide	U	—	S	—	S	—	—	—	S	—	—	—	—
Amyl alcohol	S	—	M	U	M	U	U	S	S	S	—	S	S
Aniline	U	U	M	U	U	U	U	S	S	U	U	S	S
Aqua Regia	U	—	U	U	U	U	U	—	S	—	U	U	S
Benzene[a]	U	U	U	U	M	S	S	S	S	—	U	S	S
Benzyl alcohol	U	—	U	U	U	S	U	M	S	—	—	—	S
N-Butyl alcohol	M	M	S	U	S	U	U	M	S	S	—	—	S
Caesium chloride	S	S	S	S	S	S	S	—	S	—	S	M	S
Caesium sulphate	S	S	S	S	S	S	S	S	S	S	M	M	S
Caesium trifluoroacetate	S	S	S	S	S	U	M	—	—	—	M	M	—
Calcium hypochlorite	M	S	S	S	S	—	—	S	S	—	U	M	S
Carbon tetrachloride	U	S	M	U	U	S	S	S	S	—	U	E	S

continued

Reagent	Polycarbonate	Polysulfone	Polypropylene	Polyclear/Ultra-Clear[b]	Polyallomer	Cellulose nitrate	Celluose actetate butyrate	Nylon	Delrin	Noryl	Carbon fibre	Aluminium	Titanium
Chlorobenzene	U	—	U	U	U	U	U	—	S	—	—	—	S
Chloroform	U	U	M	U	M	S	M	M	S	—	S	E	S
Chromic acid (10%)	M	U	S	S	S	U	U	—	S	S	U	M	S
Chromic acid (50%)	U	U	S	U	S	S	U	—	S	—	U	U	M
Citric acid (10%)	S	S	S	S	S	—	S	S	S	S	S	S	S
Cresol	U	—	S	S	S	—	—	U	S	—	—	S	S
Cyclohexyl alochol	M	—	S	U	S	—	U	S	S	—	—	S	S
Dextran	S	S	S	S	S	S	S	S	S	S	S	S	S
Diethyl ketone	U	—	M	M	U	U	U	M	—	—	—	S	S
Diethylpyrocarbonate	U	—	S	U	S	U	U	S	S	U	S	S	S
Dimethylformamide	U	—	S	U	S	—	—	—	S	U	—	S	S
Dimethylsulphoxide	U	—	S	—	S	—	—	—	S	—	S	S	S
Dioxane	U	—	M	U	M	—	U	—	M	U	—	S	S
Ether diethyl[a]	U	—	M	U	M	U	U	—	S	—	—	S	S
Ethyl acetate	U	U	M	—	M	U	U	M	S	U	—	M	S
Ethyl alcohol (50%)	M	S	S	U	S	S	S	M	S	S	S	S	S
Ethyl alcohol (95%)	U	S	S	U	S	U	U	M	S	S	S	S	S
Ethylene dichloride	U	—	U	U	M	U	U	S	—	S	—	S	S
Ethylene oxide	M	S	S	—	S	S	S	M	S	—	U	S	S
Ficoll	S	S	S	S	S	S	S	S	S	S	S	S	S
Formaldehyde (40%)	S	S	S	U	S	S	—	S	S	S	S	M	S
Formic acid (100%)	M	—	S	S	S	—	U	U	S	S	—	S	S
Glycerol	S	S	S	S	S	S	S	S	S	S	S	S	S
Guanidinium chloride	S	S	S	S	S	M	M	S	S	S	S	U	S
Hydrochloric acid (10%)	S	S	S	U	S	S	S	S	S	S	U	U	—
Hydrochloric acid (50%)	M	—	M	U	M	U	U	—	S	S	U	U	—
Hydrofluoric acid (10%)	M	S	S	—	S	M	M	S	S	—	U	U	U
Hydrofluoric acid (100%)	U	—	S	U	S	U	U	—	S	—	U	U	U
Hydroformic acid (100%)	—	—	S	U	S	—	—	—	—	—	U	U	S
Hydrogen peroxide (100%)	S	S	S	S	S	S	S	—	—	—	S	S	U
Isobutyl alcohol	—	—	S	U	S	—	U	M	—	—	—	—	S
Isopropyl alcohol	U	M	S	U	S	U	U	M	—	S	—	U	S
Lauryl alcohol	—	—	S	S	S	—	U	S	—	—	—	—	S
Lead acetate	—	—	S	S	S	—	S	—	S	S	—	M	S
Maleic acid	—	—	S	S	S	—	—	—	S	—	—	S	S
Magnesium hydroxide	U	—	S	S	S	—	U	—	S	S	—	U	S
Mercaptoethanol	S	U	S	—	S	—	S	—	S	—	—	M	S
Methyl alcohol	U	S	S	U	S	U	U	M	S	—	S	S	S
Methylene chloride	U	U	M	U	M	U	U	M	M	U	S	S	S
Metrizamide	S	S	S	S	S	S	S	S	S	S	S	S	S
Nitric acid (50%)	M	—	M	U	S	M	M	M	S	S	U	M	S
Nycodenz	S	S	S	S	S	S	S	S	S	S	S	S	S

Reagent	Polycarbonate	Polysulfone	Polypropylene	Polyclear/Ultra-Clear[b]	Polyallomer	Cellulose nitrate	Celluose actetate butyrate	Nylon	Delrin	Noryl	Carbon fibre	Aluminium	Titanium
Oleic acid	S	S	S	U	S	S	S	S	S	—	S	—	S
Oxalic acid	S	S	S	U	S	S	S	S	S	—	S	M	M
Perchloric acid (10%)	U	—	M	M	S	—	—	—	S	—	U	U	S
Percoll	M	S	S	S	S	S	S	S	S	S	S	M	S
Phenol (50%)	U	U	M	U	M	—	—	U	S	U	U	—	U
Phosphoric acid (10%)	S	S	S	—	S	S	S	—	S	S	S	—	—
Phosphoric acid (conc.)	M	S	M	U	M	M	M	—	S	—	M	—	M
Phosphorus trichloride	U	—	S	S	S	—	—	—	S	—	—	—	—
Potassium bromide	S	S	S	S	S	S	S	S	S	—	S	M	S
Potassium carbonate	U	—	S	S	S	S	S	S	S	S	S	M	S
Potassium chlorate	S	—	S	S	S	S	S	S	S	S	S	M	S
Potassium hydroxide (45%)	U	S	S	U	S	M	M	S	U	S	S	U	U
Potassium iodide	S	S	S	S	S	S	S	S	S	S	S	M	S
Potassium permanganate	—	—	S	S	S	—	—	S	S	—	—	—	—
Rubidium bromide	S	S	S	S	S	S	S	S	S	S	S	M	S
Rubidium chloride	S	S	S	S	S	S	S	S	S	S	S	M	S
Sodium bromide	S	S	S	S	S	S	S	S	S	S	S	M	S
Sodium carbonate	U	—	S	S	S	S	S	S	S	S	S	M	S
Sodium chloride	S	S	S	S	S	S	S	—	S	S	S	M	S
Sodium dichromate	—	—	S	S	S	—	—	S	—	—	—	M	S
Sodium hydroxide (1%)	U	S	S	—	S	S	S	S	M	S	S	U	S
Sodium hydroxide (10%)	U	S	S	U	S	U	U	S	U	S	S	U	S
Sodium hydroxide (conc.)	U	—	M	U	M	U	U	S	U	S	S	U	M
Sodium hypochlorite (5%)		S	S	—	S	S	S	S	S	—		M	S
Sodium iodide	S	S	S	S	S	S	S	S	S	S	S	S	S
Sodium peroxide	—	—	S	S	S	—	—	S	S	S	S	—	S
Sodium sulphide	U	—	S	S	S	—	S	S	S	—	S	S	M
Sodium thiosulphate	S	—	S	S	S	—	—	S	S	S	S	M	S
Sucrose	S	S	S	S	S	S	S	S	S	S	S	S	S
Sulphuric acid (50%)	S	S	S	U	S	U	U	M	U	U	U	U	M
Sulphuric acid (conc.)	U	U	S	U	S	U	U	U	U	U	U	U	U
Toluene	U	U	U	U	U	S	S	M	S	U	S	—	S
Trichloroacetic acid	M	—	S	U	S	—	—	—	U	S	—	—	S
Trichlorethylene	U	U	U	U	U	—	—	M	S	—	U	S	S
Trichloroethane	U	M	U	U	U	—	S	S	S	—	—	S	S
Turpentine	U	—	M	U	M	—	U	S	S	—	—	S	S
Urea	S	S	S	S	S	S	S	S	—	—	S	M	S
Xylene	U	U	U	U	U	—	S	U	S	U	S	S	S

[a]Flammability hazard.
[b]Polyclear and Ultra-Clear are registered names of Seton Scientific and Beckman Instruments Inc., respectively, for polyethyleneterephthalate (PET).

Appendix 2

Specifications of ultracentrifuge rotors

This appendix is an attempt to produce a comprehensive list of rotors manufactured for ultracentrifuges. Because scientific journals often do not insist on authors giving full details of the rotors used in centrifugal separations, this appendix is also useful if one is trying to reproduce separations obtained by other workers using a make of centrifuge other than the type available to the reader.

Most manufacturers of full-size ultracentrifuges have standardized on a drive shaft and overspeed sensor control system which are identical to those manufactured by Beckman Instruments Inc. Hence it is possible to use rotors from one manufacturer on an ultracentrifuge of another; some manufacturers support this practice while other manufacturers state that such practices will invalidate the warranty. In contrast, rotors for micro-ultracentrifuges (e.g. the Beckman Optima TLX) which are marked by †, are not usually interchangeable. Ultracentrifuge rotors are made from alloys of titanium or aluminium and these are marked in this appendix as Ti and Al, respectively. Some rotors are made from more than one type of alloy; for example, some swing-out rotors have an aluminium yoke and titanium buckets. Rotors of carbon fibre composite material (Cf) are also manufactured but usually these rotors also contain some metallic components (e.g. tube holders).

In the case of swing-out rotors, as described in Section 4.3.1 of Chapter 2, there are four different methods of attaching the buckets to the central yoke of the ultracentrifuge rotors. The original hinge pin (H/P) method which only allowed up to three buckets to be attached is now only found on older rotors. The two commonest designs for attaching buckets are the hook on (H/O) and the ball and socket (B/S) method; the former is much more common than the latter. Both the hook-on and ball-and-socket designs have the advantage that up to six buckets can be attached to a rotor. The other method, originally developed from the ball and socket method, is the top loading (T/L) of the bucket through the top of the rotor and in this case rotors with up to four buckets are available.

Acknowledgements

The assistance of Jens Steensgaard (Aarhus) and individual centrifuge rotor manufacturers in the preparation of these tables is gratefully acknowledged.

Small volume fixed-angle rotors

Supplier	Rotor designation	Tube number and volume (ml)	Rotor material	Tube angle	Maximum speed (r.p.m.)	R_{min} (cm)	R_{max} (cm)	RCF_{max} (g)	k-factor
Beckman Instruments Inc.	†TLA-120.1	14 × 0.5	Ti	30	120 000	2.5	3.9	626 000	8
	†TLA-120.2	10 × 2.0	Ti	30	120 000	2.5	3.9	626 000	8
	†TLN-100	8 × 3.9	Ti	9	100 000	2.3	4.0	449 000	14
	†TLA-100	20 × 0.2	Ti	30	100 000	3.0	3.9	435 000	7
	†TLA-100.1	14 × 0.5	Ti	30	100 000	2.5	3.9	435 000	12
	†TLA-100.2	10 × 1.0	Ti	30	100 000	2.5	3.9	435 000	12
	†TLA-100.3	6 × 3.5	Ti	30	100 000	2.5	4.8	540 000	16
	†TLA-100.4	8 × 5.1	Ti	28	100 000	2.6	4.9	542 000	16
	NVT90	8 × 5.1	Ti	8	90 000	5.2	7.7	698 000	10
	Type 90 Ti	8 × 13.5	Ti	25	90 000	3.4	7.7	698 000	25
	Type 80 Ti	8 × 13.5	Ti	26	80 000	4.1	8.4	602 000	28
	Type 70.1Ti	12 × 13.5	Ti	24	70 000	4.1	8.2	449 200	36
	NVT65	8 × 13.5	Ti	8	65 000	6.0	8.5	402 000	21
	Type 65	8 × 13.5	Al	24	65 000	3.7	7.8	368 000	45
	Type 50Ti	12 × 13.5	Ti	26	50 000	3.7	8.1	226 400	79
	Type 50	10 × 10	Al	20	50 000	3.7	7.0	195 600	65
	Type 50.3Ti	18 × 6.5	Ti	20	50 000	4.9	8.0	223 600	50
	Type 50.4Ti	44 × 6.5	Ti	20	50 000	6.6	9.6	270 000	38
						8.1	11.2	312 000	33
	†TLA-45	12 × 1.5	Al	45	45 000	2.5	5.5	125 000	99
	Type 42.2Ti	72 × 0.2	Ti	30	42 000	10.4	11.3	222 900	12
	Type 40	12 × 13.5	Al	26	40 000	3.7	8.1	144 900	124
	Type 25	100 × 1.0	Al	25	25 000	8.1	10.0	69 900	85
						9.7	11.6	82 000	72
						11.3	13.2	92 200	62

Du Pont–Sorvall'	†RP120AT		8 × 1.0	Ti	30	120 000	2.2	3.7	601 000	9
	†RP100AT		8 × 1.0	Ti	30	100 000	2.5	3.9	436 000	12
	†RP100AT2		14 × 0.5	Ti	30	100 000	2.5	3.9	436 000	12
	†RP100AT3		20 × 0.2	Ti	30	100 000	3.0	3.9	436 000	7
	†RP100AT4		6 × 3.0	Ti	30	100 000	2.5	4.8	541 000	16
	†RP80AT		8 × 4.0	Ti	30	80 000	2.9	5.6	401 000	26
	†RP80AT2		30 × 0.5	Ti	30	80 000	2.8	4.3	304 000	17
							3.6	5.0	358 000	14
	TFT 80.2		12 × 2.0	Ti	25	80 000	3.4	6.0	429 200	22
	TFT 80.4		10 × 4.4	Ti	25	80 000	3.7	6.6	468 600	23
	T-880		8 × 12.5	Ti	26	80 000	4.4	8.5	608 100	26
	T-875		8 × 12.5	Ti	24	75 000	4.6	8.7	547 300	29
	T-1270		12 × 12.5	Ti	24	70 000	4.0	8.2	448 800	37
	†RP70AT		20 × 0.5	Ti	30	70 000	3.0	5.6	307 000	31
	T-865.1		8 × 12.5	Ti	24	65 000	4.6	8.7	410 900	38
	T-1256		12 × 12.5	Al	24	56 000	4.0	8.2	287 200	58
	†RP45A		12 × 1.5	Al	45	45 000	3.2	5.5	125 000	67
	TFT-45.6		40 × 6.5	Ti	20	45 000	6.1	9.0	204 300	49
							7.6	10.6	238 900	42
Hitachi	†S120AT	(RP120AT)	8 × 1.0	Ti	30	120 000	2.2	3.7	601 000	9
	†S120NT	(RP120NT)	8 × 2.0	Ti	8	120 000	2.2	3.6	586 000	9
	†S100AT	(RP100AT)	8 × 1.0	Ti	30	100 000	2.4	3.9	436 000	12
	†S100AT2	(RP100AT2)	14 × 0.5	Ti	30	100 000	2.5	3.9	436 000	12
	†S100AT3	(RP100AT3)	20 × 0.2	Ti	30	100 000	3.0	3.9	436 000	7
	†S100AT4	(RP100AT4)	6 × 3.0	Ti	30	100 000	2.6	4.8	541 000	16
	P100AT	—	8 × 6.5	Ti	26	100 000	3.5	7.2	800 000	18
	†S100NT	(RP100NT)	8 × 4.0	Ti	8	100 000	2.6	4.3	479 000	12
	P90AT	—	8 × 12	Ti	26	90 000	3.4	7.7	700 000	25
	P90NT	—	8 × 5.0	Ti	10	90 000	5.1	7.1	646 000	10
	†S80AT	(RP80AT)	8 × 4.0	Ti	30	80 000	2.9	5.6	401 000	26
	†S80AT2	(RP80AT2)	30 × 0.5	Ti	30	80 000	2.8	4.3	304 000	17
							3.6	5.0	358 000	14
	†S70AT	(RP70AT)	20 × 0.5	Ti	30	70 000	3.1	5.6	307 000	31
	P70AT2	(SRP70AT)	12 × 12	Ti	24	70 000	4.1	8.3	454 000	36
	P65NT	(RP65NT)	10 × 12	Ti	10	65 000	5.8	8.5	402 100	23

Small volume fixed-angle rotors *(continued)*

Supplier	Rotor designation	Tube number and volume (ml)	Rotor material	Tube angle	Maximum speed (r.p.m.)	R_{min} (cm)	R_{max} (cm)	RCF_{max} (g)	k-factor
Hitachi *(cont)*									
	P65AT (RP65T)	10 × 12	Ti	26	65 000	3.5	7.8	368 400	48
	P65A (RP65)	10 × 12	Al	26	65 000	3.5	7.8	368 400	48
	P55AT (RP55T)	12 × 12	Ti	26	55 000	3.8	8.1	274 000	63
	P50AT4 (RP50AT4)	44 × 6.5	Ti	20	50 000	6.7	9.8	274 600	38
						8.2	11.3	316 000	32
	†S45A (RP45A)	12 × 1.5	Al	45	45 000	3.2	5.5	125 000	67
	P42AT (RPL42T)	72 × 0.23	Ti	30	42 000	10.4	11.3	222 800	12
Kontron	TFT 80.490	10 × 4.4	Ti	25	90 000	3.7	6.6	597 600	18
	TFT 80.2	12 × 2.0	Ti	25	80 000	3.4	6.0	429 259	22
	TFT 80.4	10 × 4.4	Ti	25	80 000	3.7	6.6	468 600	23
	TFT 80.13	8 × 13.5	Ti	26	80 000	4.4	8.5	605 300	26
	TFT 75.13	8 × 13.5	Ti	26	75 000	4.0	8.2	514 700	32
	TFT 65.13	12 × 13.5	Ti	26	65 000	4.0	8.2	386 600	43
	TFT 70.13	12 × 13.5	Ti	26	70 000	4.1	8.2	449 600	36
	TFT 60.13	12 × 13.5	Ti	26	60 000	4.1	8.2	330 700	50
	TFT 50.13	12 × 13.5	Ti	26	50 000	4.0	8.2	228 700	73
	TFT 45.6	40 × 6.5	Ti	20	45 000	6.1	9.0	204 300	49
						7.6	10.6	238 900	42
	TFT 32.13	32 × 13.5	Ti	30	32 000	4.4	9.3	106 900	185

Medium volume fixed-angle rotors

Supplier	Rotor designation	Tube number and volume (ml)	Rotor material	Tube angle	Maximum speed (r.p.m.)	R_{min} (cm)	R_{max} (cm)	RCF_{max} (*g*)	*k*-factor
Beckman Instruments Inc.	Type 70Ti	8 × 38.5	Ti	23	70 000	4.0	9.2	504 000	44
	Type 60Ti	8 × 38.5	Ti	24	60 000	3.7	9.0	362 200	63
	Type 55.2Ti	10 × 38.5	Ti	24	55 000	4.7	10.0	338 200	63
	Type 50.2Ti	12 × 38.5	Ti	24	50 000	5.4	10.8	301 900	70
	Type 42.1	8 × 38.5	Al	30	42 000	3.9	9.9	195 200	134
	Type 30	12 × 38.5	Al	26	30 000	4.9	10.5	105 700	214
	Type 28	8 × 50	Al	34	28 000	3.2	10.8	94 800	393
Du Pont–Sorvall	T-865	8 × 36	Ti	24	65 000	3.8	9.1	429 800	52
	T-1250	12 × 38.5	Ti	24	50 000	5.7	10.8	301 800	65
	A-841	8 × 36	Al	24	41 000	3.8	9.1	171 000	131
Hitachi	P70AT (RP70T)	8 × 40	Ti	23	70 000	3.9	9.2	504 000	46
	P50AT2 (RP50T-2)	12 × 40	Ti	24	50 000	5.4	10.8	303 300	71
	P50A2 (RP50-2)	8 × 40	Al	30	50 000	3.9	10.0	276 700	94
	P30A2 (RP30-2)	12 × 40	Al	26	30 000	4.9	10.6	106 700	217
Kontron	TFT 70.38	8 × 38.5	Ti	20	70 000	4.3	9.2	504 000	39
	TFT 65.38	8 × 38.5	Ti	20	65 000	4.3	9.2	434 500	46
	TFT 55.38	12 × 38.5	Ti	24	55 000	5.8	10.8	364 600	51
	TFT 50.38	12 × 38.5	Ti	24	50 000	5.8	10.8	301 800	63

Large volume fixed-angle rotors

Supplier	Rotor designation	Tube number and volume (ml)	Rotor material	Tube angle	Maximum speed (r.p.m.)	R_{min} (cm)	R_{max} (cm)	RCF_{max} (g)	k-factor
Beckman Instruments Inc.	Type 45Ti	6 × 94	Ti	24	45 000	3.6	10.6	235 400	133
	Type 35	6 × 94	Al	25	35 000	3.5	10.4	142 400	225
	Type 21	10 × 94	Al	18	21 000	6.0	12.2	60 100	407
	Type 19	6 × 250	Al	25	19 000	3.4	13.3	55 600	951
	Type 16	6 × 250	Al	25	16 000	3.5	13.7	39 300	1350
Du Pont–Sorvall	T-647.5	6 × 100	Ti	23	47 500	3.4	9.9	249 600	120
	A.641	6 × 98	Al	23	41 000	3.4	9.9	186 000	161
	A-621	6 × 250	Al	23	21 000	3.7	13.6	67 000	748
Hitachi	P45AT (RP45T)	6 × 94	Ti	24	45 000	3.7	10.4	237 700	130
	P42A (RP42)	6 × 94	Al	26	42 000	3.6	10.4	205 100	152
	P19A (RP19)	6 × 230	Al	26	19 000	4.7	13.6	54 900	746
Kontron	TFT 45.94	6 × 94	Ti	23	45 000	4.0	10.4	235 400	119
	TFA 20.250	6 × 250	Al	23	20 000	5.0	13.6	60 800	632

Vertical rotors

Supplier	Rotor designation	Tube number and volume (ml)	Rotor material	Maximum speed (r.p.m.)	R_{min} (cm)	R_{max} (cm)	RCF_{max} (g)	k-factor	k*
Beckman Instruments Inc.	[†]TLV-100	8 × 2.0	Ti	100 000	2.5	3.6	400 000	9	25
	VTi90	8 × 5.1	Ti	90 000	5.8	7.1	645 000	6	17
	VTi65	8 × 5.1	Ti	65 000	7.2	8.5	404 000	10	27
	VTi65.1	8 × 13.5	Ti	65 000	6.9	8.5	401 700	13	34
	VTi65.2	16 × 5.1	Ti	65 000	7.5	8.8	416 000	10	34
	VC53	8 × 39	Cf/Ti	53 000	5.3	7.9	248 000	36	96
	VAC50	10 × 39	Cf/Al	50 000	6.1	8.7	242 000	36	96
	VTi50	8 × 39	Ti	50 000	6.1	8.7	242 000	36	96
Du Pont–Sorvall	[†]RP120VT	8 × 2.0	Ti	120 000	2.0	3.1	500 000	8	22
	[†]RP100VT	8 × 2.0	Ti	100 000	2.5	3.6	400 000	9	25
	TV-1665	16 × 6	Ti	65 000	7.5	8.8	415 500	10	26
	TV-865	8 × 6	Ti	65 000	7.2	8.5	401 500	10	27
	TV-865B	8 × 19	Ti	65 000	5.9	8.5	401 500	22	58
	TV-860	8 × 36	Ti	60 000	5.9	8.5	342 000	26	69
Hitachi [†]S120VT	(RP120VT)	8 × 2.0	Ti	120 000	2.0	3.1	500 000	8	22
[†]S100VT	(RP100VT)	8 × 2.0	Ti	100 000	2.5	3.6	400 000	9	25
P100VT	—	8 × 5	Ti	100 000	4.9	6.3	700 000	6	17
P83VT	(SRP83VT)	8 × 5	Ti	83 000	5.8	7.2	549 100	8	19
P65VF	(RP65VF)	8 × 5	Cf/Al	65 000	7.2	8.6	404 300	10	27
P65VT2	(RP65VT2)	16 × 5	Ti	65 000	7.5	8.8	416 100	10	27
P65VT3	(RP65VT3)	10 × 12	Ti	65 000	6.9	8.5	402 100	13	34
P55VF2	(RP55VF2)	12 × 5	Cf/Al	55 000	7.4	8.7	243 200	14	37
P50VT	(RPV50T)	8 × 40	Ti	50 000	6.1	8.7	243 000	36	89
Kontron	TVF70.4	10 × 4.4	Cf/Al	70 000	6.6	7.7	421 770	7	19
	TVF65.13	10 × 13.5	Cf/Al	65 000	6.9	8.4	397 900	12	32
	TVF50.4	18 × 4.4	Cf/Al	50 000	10.7	11.7	327 000	9	25
	TVF50.6	18 × 6.5	Cf/Al	50 000	10.6	11.8	330 000	11	29
	TVF50.5	18 × 5.1	Cf/Al	50 000	10.6	11.8	330 000	11	29

Small volume swing-out rotors

Supplier	Rotor designation	Tube number and volume (ml)	Rotor material	Bucket system	Maximum speed (r.p.m.)	R_{min} (cm)	R_{max} (cm)	RCF_{max} (g)	k-factor	k*
Beckman Instruments Inc.	SW 65Ti	3 × 5	Ti	H/P	65 000	4.1	8.9	420 300	46	122
	SW 60Ti	6 × 4.4	Ti	H/O	60 000	6.3	12.0	482 900	45	120
	SW 55Ti	6 × 5	Ti	H/O	55 000	6.1	10.9	368 600	49	129
	†TLS 55	4 × 2.2	Al	T/L	55 000	4.2	7.6	259 000	50	130
	SW 50.1	6 × 5	Al/Ti	H/O	50 000	6.0	10.7	299 000	59	156
	SW 30.1	6 × 8	Al/Ti	H/O	30 000	7.5	12.3	123 700	139	373
Du Pont–Sorvall	TST 60.4	6 × 4.4	Ti	B/S	60 000	6.4	12.1	491 000	45	120
	†RPS 55S	4 × 2.2	Al/Ti	H/O	55 000	4.5	7.7	259 000	44	119
	AH 650	6 × 5	Al/Ti	H/O	50 000	6.0	10.7	299 000	59	156
Hitachi	P65ST(RPS65T)	3 × 5	Ti	H/P	65 000	4.0	8.9	420 400	48	123
	P56ST(RPS56T)	6 × 4	Ti	H/O	56 000	6.0	11.7	406 700	55	131
	P55ST2(RPS55T-2)	6 × 5	Ti	H/O	55 000	6.0	10.8	368 600	49	117
	†S55S(RPS55S)	4 × 2.2	Al/Ti	H/O	55 000	4.5	7.7	259 000	44	119
	P50S2(RPS50-2)	6 × 5	Al/Ti	H/O	50 000	5.9	10.7	301 800	60	144
Kontron	TST 60.2	6 × 2.0	Ti	B/S	60 000	6.4	11.2	451 000	39	104
	TST 60.4	6 × 4.4	Ti	B/S	60 000	6.4	12.1	486 900	45	120
	TST 55.5	6 × 5	Ti	B/S	55 000	6.6	11.3	382 100	45	120

Medium volume swing-out rotors

Supplier	Rotor designation	Tube number and volume (ml)	Rotor material	Bucket system	Maximum speed (r.p.m.)	R_{min} (cm)	R_{max} (cm)	RCF_{max} (g)	*k*-factor	*k**
Beckman Instruments Inc.	SW 41Ti	6 × 13.2	Ti	H/O	41 000	6.7	15.3	287 500	136	355
	SW 40Ti	6 × 14	Ti	H/O	40 000	6.9	15.9	284 400	132	346
	SW 28.1	6 × 17	Al/Ti	H/O	28 000	7.3	17.1	149 900	275	720
Du Pont–Sorvall	TH-641	6 × 14	Ti	H/O	41 000	6.8	15.2	285 600	121	318
	AH-629 (17 ml)	6 × 17	Al/Ti	H/O	29 000	6.7	16.6	156 000	273	705
	AH-629 (20 ml)	6 × 20	Al/Ti	H/O	29 000	8.1	13.0	122 000	142	382
Hitachi	P40ST(RPS40T)	6 × 13	Ti	H/O	40 000	6.6	15.9	284 400	139	352
	P28S2(SRP28SA1)	6 × 16	Al/Ti	H/O	28 000	7.0	16.6	145 600	278	719
Kontron	TST 41.14	6 × 14	Ti	H/O	41 000	6.8	16.1	302 500	130	340
	TST 28.17	6 × 17	Ti	H/O	28 000	7.4	17.2	150 700	273	714

Large volume swing-out rotors

Supplier	Rotor designation	Tube number and volume (ml)	Rotor material	Bucket system	Maximum speed (r.p.m.)	R_{min} (cm)	R_{max} (cm)	RCF_{max} (g)	k-factor	k*
Beckman Instruments Inc.	SW 30	6 × 20	Al/Ti	H/O	30 000	7.5	12.3	124 100	138	368
	SW 28	6 × 38.5	Al/Ti	H/O	28 000	7.5	16.1	141 200	247	650
	SW 25.1	3 × 34	Al	H/P	25 000	5.6	12.9	90 100	338	886
Du Pont-Sorvall	AH-629 (36 ml)	6 × 36	Al/Ti	H/O	29 000	7.7	16.1	151 300	222	586
Hitachi	P28S(SRP28SA)	6 × 40	Al/Ti	H/O	28 000	7.4	16.1	141 100	252	582
Kontron	TST 28.38	6 × 38.5	Ti	H/O	28 000	7.8	16.1	141 100	234	618

Zonal and continuous-flow rotors

Supplier	Rotor designation	Rotor volume (ml)	Rotor material	Maximum speed (r.p.m.)	Maximum pathlength (cm)	Depth of rotor (cm)	RCF_{max} (g)
Beckman Instruments Inc.	Z-60	330	Ti	60 000	5.2	3.1	256 000
	Ti-14	665	Ti	48 000	5.4	5.4	171 800
	Ti-15	1675	Ti	32 000	7.6	7.6	102 000
	CF-32Ti	430	Ti	32 000			102 000
Du Pont–Sorvall	TZ-28	1330	Ti	28 000	5.9	9.8	83 500
Hitachi	P48ZT(RPZ48T)	660	Ti	48 000	5.2	5.5	172 000
	P35ZT(RPZ35T)	1690	Ti	35 000	7.4	7.6	121 800
	P32CT(RPC32T)	430	Ti	32 000	1.4	7.3	102 000
Kontron	TZT 42.650	650	Ti	42 000	5.2	5.4	131 500
	TZT 42.325	325	Ti	42 000	5.2	5.4	131 500
	TZT 42.160	160	Ti	42 000	5.2	5.4	131 500
	TZT 42.140	140	Ti	42 000	4.0	5.4	118 200
	TZT 32.1650	1650	Ti	32 000	7.4	7.6	101 800

Appendix 3

Equations relating the refractive index to the density of solutions

When solutes are dissolved in water the refractive index of the resulting solution differs from that of water. The increase in the refractive index is proportional to the concentration of solute. In the case of gradient solutes the density of the solution is directly related to the solute concentration as well as the refractive index. This relationship can be expressed in terms of the following equation:

$$\rho = a\eta - b$$

The tables in this appendix list the coefficients *a* and *b* for a number of ionic and non-ionic gradient media. However, before applying the equations it is essential that allowance is made for the presence of other solutes (e.g. EDTA and buffers). The following equation should be used:

$$\eta_{\text{corrected}} = \eta_{\text{observed}} - (\eta_{\text{buffer}} - \eta_{\text{water}})$$

The refractive index of solutions is also temperature dependent and so it is important to make allowance for this when reading the refractive index. The refractive indices of water at various densities are: 15 °C, 1.3334; 20 °C, 1.3330; 25 °C, 1.3325. It is important to measure the refractive index of solutions to four decimal places as otherwise the calculated density will be inaccurate; this should be possible for most types of refractometers.

Note that this method will not work for colloidal gradient media such as Percoll. In the case of mixed solute gradients (e.g. CsCl–Cs_2SO_4 and Nycodenz–glucose gradients) then it will be necessary to determine the exact relationship between the density and refractive indices of solutions. Even so, the differences in the distributions of the solutes will vary depending on their rates of sedimentation and diffusion during centrifugation and so the accuracy of density measurements using refractive index is rather limited. For mixed solute gradients, unless the time of centrifugation is short, it is better to use an alternative method for determining density of mixed solute gradients (e.g. pycnometry or a density meter).

Ionic Gradient Media

Gradient solute	Temperature (°C) for		Coefficients		Valid density range (g/ml)
	η	ρ	***a***	***b***	
CsBr	25	25	9.9667	15.166	1.2–1.4
CsCl	20	20	10.9276	13.593	1.2–1.9
	25	25	10.8601	13.497	1.3–1.9
Cs_2SO_4	25	25	12.1200	15.166	1.1–1.4
	25	25	13.6986	17.323	1.4–1.8
Cs(HCOO)	25	25	13.7363	17.429	1.7–1.8
	25	20	12.8760	16.209	1.8–2.3
CsTCA[a]	20	20	7.6232	9.1612	1.1–1.7
CsTFA[b]	20	20	23.041	29.759	1.2–1.8
NaBr	25	25	5.8880	6.852	1.0–1.5
NaI	20	20	5.3330	6.118	1.1–1.8
KBr	25	25	6.4786	7.643	1.0–1.4
KI	20	20	5.7317	6.645	1.0–1.4
	25	25	5.8356	6.786	1.1–1.7
RbBr	25	25	9.1750	11.241	1.1–1.7
RbCl	25	25	9.3282	11.456	1.0–1.4
RbTCA[a]			6.5869	7.7805	1.1–1.6

[a]Trichloroacetate.
[b]Trifluoracetate, CsTFA is a registered trademark of Pharmacia-Biosystems AB.

Non-ionic gradient media

Gradient solute	Temperature (°C) for		Coefficients	
	η	ρ	***a***	***b***
Sucrose	20	0	2.7329	2.6425
Ficoll	20	20	2.381	2.175
Metrizamide	20	5	3.453	3.601
Metrizamide/D_2O	25	25	3.0534	2.9541
Nycodenz	20	20	3.242	3.323
Metrizoate	25	5	3.839	4.117
Renografin	24	4	3.5419	3.7198
Iothalamate	25	25	3.904	4.201
Chloral hydrate	4	4	3.6765	3.9066
Bovine serum albumin	24	5	1.4129	0.8814

Appendix 4

Marker enzymes and chemical assays for the analysis of subcellular fractions

1. Assays of marker enzymes

1.1 Aminopeptidase

Wachsmuth, E. D., Fritze, E., and Pfleiderer, G. (1966). *Biochemistry* **5**, 169.

Solutions:

- 25 mM leucine *p*-nitroanilide
- 50 mM sodium phosphate (pH 7.2)

Assay method:

1. Add 0.9 ml phosphate buffer, 100 µl leucine *p*-nitroanilide and 100 µl of the sample to a 1.0 ml cuvette.
2. Monitor the absorbance at 405 nm using a continuous chart recorder, with a full-scale deflection equivalent to 0.2 absorbance units.

Notes:
The amount of product of the reaction, *p*-nitrophenol, can be determined by constructing a standard curve (0–1 mM), or can be calculated based on its extinction coefficient at 405 nm of 9620.

1.2 Aryl sulphatase

Chang, P. L., Rosa, N. E., and Davidson, R. G. (1981). *Anal Biochem.* **117**, 382.

Solutions:

- 5 mM 4-methyl umbelliferyl sulphate (4-MUS)
- 30 mM lead acetate
- 0.1 M sodium acetate (pH 5.6)
- 1.0 M EDTA, 0.2 M glycine carbonate buffer (pH 10.3)

Assay method:

1. Calibrate the fluorimeter using 4-methyl umbelliferone (4-MU) in the glycine-carbonate buffer as a standard.
2. Incubate a mixture of 100 µl acetate buffer, 20 µl lead acetate, 20 µl of 4-MUS, 50 µl water, and 10 µl of sample (containing 10 µg protein) at 37 °C for 15 min.
3. Stop the reaction by addition of 1.0 ml of 0.2 M glycine-carbonate buffer.
4. Measure the fluorescence of the solution using an excitation wavelength of 387 nm and an emission wavelength of 470 nm.

The range of standard concentrations used will depend entirely upon the sensitivity of the fluorimeter.

Notes:
This method is a very sensitive fluorimetric assay which measures the production of 4-methyl umbelliferone from 4-methyl umbelliferyl sulphate.

1.3 Catalase

Cohen, G., Dembiec, D., and Marcus, J. (1970). *Anal. Biochem.* **34**, 30.

Solutions:

- 6 mM H_2O_2 in 10 mM sodium phosphate buffer
- 10 mM sodium phosphate buffer (pH 7.0)
- 3 M H_2SO_4
- 2 mM $KMnO_4$

Assay method:

1. Set up the following tubes at 4 °C:
   i. test: 50 µl sample, 0.5 ml H_2O_2, vortex, and after 3 min add 100 µl of H_2SO_4 and vortex again.
   ii. sample blank: 50 µl of sample, 100 µl H_2SO_4, and 0.5 ml of H_2O_2 in that order.
   iii. standard: add 100 µl of H_2SO_4 to 0.5 ml of buffer.
   iv. reagent blank: mix 100 µl of H_2SO_4 with 0.5 ml H_2O_2 and 0.5 ml of buffer.
   v. spectrophotometer blank: add 100 µl of H_2SO_4 to 0.55 ml of buffer and 0.7 ml of distilled water.
2. To tubes *i–iv* add 0.7 ml of $KMnO_4$ and read the tubes against tube *v* at 480 nm within 60 sec.

1.4 Galactosyl transferase

Fleisher, B., Fleisher, S., and Osawa, H. (1969). *J. Cell Biol.* **43**, 59.

Solutions:

- 1.0 mg/ml uridine diphospho-[6-^{3}H]-galactose, 74 kBq/ml (2 μCi/ml)
- 40 mM Hepes–NaOH (pH 7.0)
- 30 mM 2-mercaptoethanol
- 0.4 M $MnCl_2$
- 0.4 M *N*-acetylglucosamine
- 0.3 M EDTA (pH 7.4)

Assay method:

1. For each sample prepare two assay tubes containing:
   20 μl 40 mM Hepes buffer
   10 μl 0.4 M $MnCl_2$
   10 μl 30 mM 2-mercaptoethanol
   10 μl 1 mg/ml [^{3}H]-uridine-diphosphogalactose
   10 μl sample (20–50 μg protein)
   to one tube add 10 μl of distilled water to the other add 10 μl of 0.4 M *N*-acetylglucosamine solution as an acceptor.
2. Set up blanks at the same time, as above but with no sample added.
3. Incubate all the tubes at 37 °C for 60 min and stop the reaction by the addition of 20 μl of EDTA solution and cool the tubes to 4 °C.
4. Prepare 2 cm Dowex-1 columns (chloride form) in Pasteur pipettes, washing them through with three volumes of water.
5. Apply each incubation mixture to separate columns and wash twice with 0.5 ml water: collect the eluant in scintillation vials and add a water-soluble scintillation fluid (e.g. Beckman Ready Solv. HP) and measure the radioactivity of each.

1.5 Glucose-6-phosphatase

Aronson, N. N. and Touster, O. (1974). *Methods Enzymol.* **31**, 90.

Solutions:

- Prepare the substrate solution by mixing 0.1 M glucose-6-phosphate (sodium salt), 35 mM histidine buffer (pH 6.5), and 10 mM EDTA in the ratio of 2:5:1.
- 8% (w/v) trichloroacetic acid (TCA)

- 2.5% ammonium molybdate in 2.5 M H_2SO_4
- Reducing solution; 0.5 g 1-amino-2-napthol-4-sulphonic acid, 1.0 g anhydrous Na_2HSO_4 in 200 ml $NaHSO_3$[a].

Assay method:

1. Add 50 µl of sample (about 1 mg/ml protein) to 0.45 ml of substrate solution and incubate at 37 °C for 30 min.
2. For each sample, prepare a blank containing no substrate, together with a reagent blank to determine the background hydrolysis of the substrate. At the same time prepare a set of standards of known phosphate concentration.
3. Stop the reaction by addition of 2.5 ml of cold TCA and keep the acidified sample for 20 min at 0 °C.
4. Centrifuge at 700*g* for 20 min, take 1.0 ml of the supernatant and add 1.15 ml distilled water.
5. Add 0.25 ml ammonium molybdate solution and 100 µl of the reducing solution.
6. After 10 min, read the absorbance at 820 nm.

[a]The reducing solution must be freshly prepared.

1.6 NADPH-cytochrome *c* reductase and NADH-cytochrome *c* reductase

Williams, C. H. and Kamin, H. (1962). *J. Biol. Chem.* **237**, 587.

Solutions:

1. NADPH–cytochrome *c* reductase
   - 2 mg/ml NADPH[a]
   - 25 mg/ml cytochrome *c*
   - 10 mM EDTA

All made up in 50 mM sodium phosphate buffer (pH 7.7).

2. NADH–cytochrome *c* reductase
   - 1 mg/ml NADH[a]
   - 25 mg/ml cytochrome *c*

Both solutions are made up in 50 mM sodium phosphate (pH 7.2).

Assay method:

1. In a 1.0 ml cuvette add 0.9 ml of the appropriate buffer, 50 µl cytochrome *c*, 10 µl EDTA (for NADPH–cytochrome *c* reductase only) and 10–20 µl of the sample.

2. Use a chart recorder (full-scale deflection equivalent to 0.2 absorbance units) to record absorbance at 552 nm until steady.
3. Add 100 µl NADPH or NADH as appropriate and measure the increase in the absorbance due to the reduction of cytochrome *c*.

The molar extinction coefficient of cytochrome *c* is 27 000 at 552 nm.

[a]NADPH and NADH solutions must be freshly prepared, protected from light, and kept at 0 °C until used.

1.7 5′-Nucleotidase

Avruch, J. and Wallach, D. F. H. (1971). *Biochim. Biophys. Acta* **233**, 334.

Solutions:

- 10 mM [U-^{14}C]AMP, 20 kBq/ml (0.5 µCi/ml)
- 1.8 mM $MgCl_2$
- 0.5 M Tris–HCl (pH 8.0)
- 0.3 N $ZnSO_4$[a]
- 0.3 N $Ba(OH)_2$[a]

Assay method:

1. Prepare assay tubes containing:
   100 µl 10 mM [^{14}C]AMP
   100 µl 1.8 mM $MgCl_2$
   100 µl 0.5 M Tris–HCl (pH 8.0)
2. Add 50 µl water and 50 µl of membrane suspension (0.1–1.0 mg/ml protein).
3. Prepare a blank (Steps 1 and 2), minus the sample, to measure the background rate of AMP hydrolysis.
4. Incubate the reaction mixtures at 37 °C for 60 min and stop the reaction by addition of 0.3 ml of each of the $ZnSO_4$ and $Ba(OH)_2$ solutions. (Unhydrolysed AMP is precipitated by $BaSO_4$).
5. Stand the tubes at 4 °C for 30 min with occasional shaking, followed by centrifugation at 700*g* for 20 min.
6. Transfer 0.5 ml of the supernatant to a scintillation vial, add a water-soluble scintillator (e.g. Beckman Ready Solv. HP) and measure the [U-^{14}C]-adenosine in a liquid-scintillation counter.

[a]It is difficult to prepare satisfactory solutions of $ZnSO_4$ and $Ba(OH)_2$; solutions of the latter often tend to be cloudy. Hence it is recommended that both of these are purchased as ready-made solutions from Sigma Chemical Co.

1.8 Ouabain-sensitive Na^+/K^+-ATPase

Avruch, J. and Wallach, D. F. H. (1971). *Biochim. Biophys. Acta* **233**, 334.

Solutions:

- 10 mM [γ-^{32}P]ATP, 40 kBq/ml (1 μCi/ml)
- 1.5 M NaCl
- 10 mM $MgCl_2$
- 0.3 M KCl
- 0.2 M Tris–HCl (pH 7.4)
- 10 mM ouabain
- 10 mM HCl, 1 mM sodium phosphate
- Activated charcoal suspension (4% Norit A, 0.1 M HCl, 0.2 mg/ml bovine serum albumin, 1 mM sodium phosphate, 1 mM sodium pyrophosphate).

Assay method:

1. Prepare two assay tubes for each sample containing:

   100 μl 10 mM ATP
   100 μl 1.5 M NaCl
   100 μl 10 mM $MgCl_2$
   100 μl 0.3 M KCl
   100 μl 0.2 M Tris–HCl (pH 7.4)
   0.45 ml distilled water
2. Add 50 μl of sample suspension (0.2–1.0 mg/ml protein) to each tube, in the presence or absence of 10 μl of ouabain solution.
3. Set up a blank assay as in Steps 1 and 2, but minus sample, to measure the background rate of ATP hydrolysis.
4. Incubate the mixtures for 60 min at 37 °C and stop the reaction by the addition of 3.0 ml of the charcoal suspension. Unhydrolysed ATP is then adsorbed onto the activated charcoal.
5. Stand the tubes at 4 °C for 30 min with occasional shaking and filter the solution through a 2.5 cm Whatman glass-fibre disc (GF/C), in a Millipore microanalysis filter-holder, directly into a scintillation vial. Wash the residue twice with 3.0 ml of 0.01 M HCl containing 1.0 mM phosphate.
6. Measure the non-adsorbed [^{32}P]-phosphate by Cerenkov counting in a liquid-scintillation counter.

1.9 Alkaline and acid phosphatase

Engstrom, L. (1961). *Biochim. Biophys. Acta* **52**, 36.

Alkaline phosphatase

Solutions:

- 16 mM *p*-nitrophenol phosphate
- 50 mM sodium borate buffer (pH 9.8)
- 1.0 M $MgCl_2$
- 0.25 M NaOH

Assay method:

1. Prepare a stock assay mixture of 5.0 ml *p*-nitrophenol phosphate, 5.0 ml sodium borate buffer, and 20 µl $MgCl_2$.
2. Add 50 µl of sample (0.1–1.0 mg/ml protein) to 0.2 ml assay mixture.
3. Incubate at 37 °C for 60 min and stop the reaction by addition of 0.6 ml of NaOH solution.
4. Centrifuge at 700*g* for 20 min and measure the absorbance of the supernatant at 405 nm.
5. Calculate the amount of product formed either by calculation as described in Section 1.1 or from a standard curve of *p*-nitrophenol (0–1 mM).

Acid phosphatase

Carry out the alkaline phosphatase assay substituting 180 mM sodium acetate buffer (pH 5.0) in place of the borate buffer and omitting the $MgCl_2$.

1.10 Succinate–cytochrome *c* reductase

Method 1

Mackler, P., Collip, P. J., Duncan, H. M., Rao, N. A., and Heunnekens, F. M. (1962). *J. Biol. Chem.* **237**, 2968.

Solutions:

- 50 mM sodium phosphate buffer (pH 7.4)
- 0.66 M sodium succinate in phosphate buffer[a]
- 12.5 mg/ml of cytochrome *c* in phosphate buffer[b]
- 0.1 M KCN in phosphate buffer (*CAUTION: very poisonous*)[b]

Assay method:

1. Prepare an assay mixture, in a 1.0 ml cuvette, containing:

    0.9 ml 50 mM sodium phosphate buffer

    20 µl 0.1 M KCN

    50 µl 12.5 mg/ml cytochrome *c*

    10–50 µl of the sample

2. Record absorbance at 552 nm, using a chart recorder with a full-scale deflection equivalent to 0.2 absorbance units.
3. When the trace is steady, add 0.1 ml of sodium succinate solution and record the increase in absorbance due to the reduction of cytochrome *c*.

The molar extinction coefficient of cytochrome *c* at 552 nm is 27 000.

Method 2

Graham, J. M., Ford, T., and Rickwood, D. (1990). *Anal. Biochem.* **187**, 318.

Solutions:

- 2.5 mg/ml 2-*p*-iodo-phenyl-3-*p*-nitrophenyl tetrazolium chloride (INT) dissolved in dimethyl formamide
- 0.2 M sodium succinate (pH 7.5)[a]
- 20 mM Tris–HCl (pH 7.4), 0.1 mM EDTA
- Ethyl acetate/ethanol/trichloroacetic acid (5/5/1 by vol)

Assay method:

1. Prepare the incubation mixture in 1.5 ml microcentrifuge tubes:

    0.25 ml 20 mM Tris–HCl (pH 7.4), 0.1 mM EDTA

    50 µl 0.2 M sodium succinate

    50 µl 2.5 INT

2. Start the reaction by the addition of 50 µl of the sample.
3. Incubate for 5 min at room temperature (20–23 °C) and stop the reaction by addition of 1 ml of ethyl acetate/ethanol/TCA.
4. Centrifuge for 2 min at top speed in a microcentrifuge.
5. Measure the absorbance of the supernatant at 500 nm.

[a]The sodium succinate solution can be stored indefinitely at −20 °C.
[b]The potassium cyanide and cytochrome *c* solutions must be freshly prepared. *CAUTION: Potassium cyanide is extremely poisonous.*

2. Chemical assays for nucleic acids

2.1 DAPI fluorescent assay for DNA

Brunk, C. F., Jones, K. C., and James, T. W. (1979). *Anal. Biochem.* **92**, 497.

Solutions:

- DAPI reagent: 0.1 μg/ml 4′,6-diamidino-2-phenylindole (DAPI) dissolved in 0.1 M NaCl, 10 mM EDTA, 10 mM Tris–HCl (pH 7.0)
- DNA standard solution (20 μg/ml)

Assay method:

1. Determine the fluorescence of the DAPI solution in the absence of DNA at an excitation wavelength of 360 nm and an emission wavelength of 450 nm.
2. Add 15 μl of DNA solution and remeasure fluorescence.
3. Add a further 15 μl of DNA and repeat measurement; repeat twice to obtain four points from which to draw a standard curve.
4. After the addition of each of four 15 μl aliquots of the DNA sample, take further fluorescence measurements.
5. Calculate the DNA content of the sample by comparison of the fluorescent yields of the standards and the sample material.

Notes:

When DAPI is complexed with DNA, the fluorescence of the complex is enhanced about 20-fold as compared with the fluorescence of the dye alone. The fluorescence yield is unaffected by pH between pH 5 and pH 10. Cations, especially divalent or heavy metal ions, cause significant quenching of the fluorescence and quenching is also observed at low ionic strengths (hence the composition of the buffer used). Quenching is found to increase with decreasing temperature and therefore it is necessary that all measurements are made at a uniform temperature. The binding of DAPI is highly specific for A–T base pairs, hence it is essential that the DNA of the standard and sample have the same base composition.

2.2 Diphenylamine assay for DNA

Schneider, W. C. (1957). *Methods Enzymol.* **3**, 680.

Solutions:

- Diphenylamine reagent: dissolve 1.0 g of diphenylamine in 100 ml of glacial acetic acid and add 2.75 ml of concentrated H_2SO_4.
- DNA standard solution: 500 μg/ml DNA ($A_{260} = 11$)
- 20% (w/v) trichloroacetic acid (TCA).

Assay method:

1. Dilute the DNA standard solution to give 1.0 ml sample volumes in 5% TCA containing 0–200 µg DNA for preparing a standard curve.
2. Hot-acid digest the samples as well as the DNA standards in 5% TCA in a water bath at 90 °C for 20 min to solubilize the DNA.
3. Add 2.0 ml diphenylamine reagent to each 1.0 ml sample and incubate at 100 °C in a water bath.
4. Cool to room temperature and read the optical density at 595 nm, using a blank containing 2.0 ml of diphenylamine reagent and 1.0 ml 5% TCA.

Notes:
When the DNA sample has been purified on sucrose, Ficoll, or metrizamide gradients, it is necessary to remove the gradient medium by acid precipitation before proceeding, in order to prevent these media interfering with the assay. The absence of a sugar group on the Nycodenz molecule allows the assay to be carried out without interference by the medium. To precipitate the DNA, add an equal volume of cold 20% (w/v) TCA to the sample and centrifuge at 2000*g* for 10 min to pellet the DNA. Wash the pellet twice with cold 10% TCA. This assay cannot be used for Percoll or metrizoate gradient fractions. Do not store the diphenylamine reagent in the cold since it solidifies.

2.3 Ethidium bromide assay for DNA and RNA

Karsten, U. and Wollenberger, A. (1972). *Anal. Biochem.* **46**, 135.
Karsten, U. and Wollenberger, A. (1977). *Anal. Biochem.* **77**, 464.

Solutions:

- Phosphate buffered saline (PBS): contains 0.1 g $CaCl_2$, 0.2 g KCl, 0.2 g KH_2PO_4, 0.1 g $MgCl_2 6H_2O$, 8.0 g NaCl, 1.15 g Na_2HPO_4, adjusted to pH 7.5 and made up to 1 litre
- ethidium bromide (25 µg/ml) in PBS
- heparin (25 µg/ml) in PBS
- RNase (50 µg/ml) in PBS (heat to 100 °C for 10 min to destroy any DNase activity)
- DNA standard solution (25 µg/ml DNA) in PBS. Stock DNA solutions should be stored frozen. Before use dilute the solution by addition of four volumes of PBS.
- The sample material should be either in PBS or another appropriate buffer.

Assay method:

1. Make up the following assay mixtures in 3-ml cuvettes:
   (i) Standard: 0.5 ml DNA standard solution, 0.5 ml heparin solution, 1.0 ml PBS
   (ii) Blank I: 0.5 ml heparin solution, 1.5 ml PBS.
   (iii) Blank II: 2.5 ml PBS.
   (iv) Sample (DNA + RNA): 0.5 ml homogenate, 0.5 ml heparin solution, 1.0 ml PBS.
   (v) Sample (DNA only): 0.5 ml homogenate, 0.5 ml heparin solution, 0.5 ml RNase solution, 0.5 ml PBS.
   (vi) Sample (background correction): 0.5 ml homogenate, 2.0 ml PBS.
2. Incubate mixtures (1–6) in a waterbath at 37 °C for 20 min then add 0.5 ml of ethidium bromide solution to tubes (1), (2), (4), and (5). Full intensity of fluorescence is attained within 60 sec of the addition of ethidium bromide and remains constant for at least 60 min. Briefly stir the reaction mixture before taking any measurements.
3. Measure the fluorescence using an excitation wavelength of 360 nm and an emission wavelength of 580 nm. The fluorescence of the standard is set at 100. Temperature during the measurements is critical and should be kept constant to within ±0.5 °C.
4. Calculation of nucleic acid content:
   DNA:

$$A_{dna} = \frac{A_{std}(F_{(5)} - F_{(2)} - F_{(6)} + F_{(3)})}{F_{(1)} - F_{(2)}}$$

where A_{dna} = amount of DNA/mixture (μg), A_{std} = amount of standard DNA/mixture, F = fluorescence intensity (units).

RNA:

$$A_{rna} = \frac{A_{std}(F_{(4)} - F_{(5)})}{0.46(F_{(1)} - F_{(2)})}$$

The factor 0.46 is empirically derived from the ratio of the fluorescence yield of RNA and DNA with ethidium bromide. Alternatively, an RNA standard solution can be used.

Notes:

Up to 5 μg/ml DNA or RNA fluorescence increases linearly with concentration. Check the RNase solution for fluorescence and subtract from the reading of mixture (5).

2.4 Methyl green assay for DNA

Peters, D. L. and Dahmus, M. E. (1979). *Anal. Biochem.* **93**, 306.

Solution:

- Methyl green reagent: dissolve methyl green (Kodak-Eastman) in 0.1 M Tris–HCl (pH 7.9) to a final concentration of 0.01%. Remove any contaminating crystal violet in the solution by extraction with an equal volume of ethanol; usually two extractions are sufficient.
- Proteinase K solution: 1 mg/ml predigested at 37 °C for 30 min.
- DNA standard solution: 500 µg/ml DNA.

Assay method:

1. Prepare the standard curve samples by diluting the stock DNA solution to 0.2-ml aliquots containing 0–100 µg DNA. Prepare samples also as 0.2-ml aliquots.
2. Add 5 µl of the proteinase K solution to each DNA aliquot and incubate at about 20 °C for 60 min.
3. Add 1.0 ml of the methyl green reagent to each sample and incubate either overnight at 20 °C or 3 h at 45 °C.
4. Measure the optical density of the solutions at 640 nm, using a mixture of 1.0 ml reagent and 0.2 ml distilled water (treated in the same way as the sample fractions).

Notes:
In the absence of DNA, the blue colour of the reagent fades, but in the presence of DNA, the colour is retained in proportion to the amount of DNA present. The assay is compatible with samples containing Nycodenz, metrizamide, or metrizoate. Methyl green from some sources may not give consistent results.

2.5 Orcinol assay for RNA

Schneider, W. C. *Methods Enzymol.* **3**, 680.

Solutions:

- Orcinol reagent: dissolve 0.5 g of orcinol in 50 ml of concentrated HCl, then add and dissolve 0.25 g of $FeCl_3 \cdot 6H_2O$. Keep this reagent in ice until required.
- RNA standard solution: 250 µg/ml of RNA in distilled water. Alternatively, it is possible to use a ribose standard solution. 20% (w/v) tricholoracetic acid (TCA).

Assay method:

1. Dilute the RNA standard solution to give 0.5 ml volumes containing 0–250 μg RNA for the preparation of a standard curve. Dilute each of the samples to the same volume.
2. Add TCA to each sample to a final concentration of 5% (w/v) and incubate for 20 min at 90 °C.
3. Centrifuge for 2 min in a microcentrifuge, remove supernatant and add an equal volume of orcinol reagent. Incubate for 20 min in a boiling water-bath.
4. Cool the solutions and measure the optical density at 660 nm. Use 1.0 ml reagent and 1.0 ml 5% TCA as a blank.

Notes:
The yellow colour of the reagent becomes green in the presence of RNA. This assay is incompatible with the presence of Percoll and metrizoates; also, the orcinol reagent reacts with sucrose, Ficoll, and the glucosamide group of the metrizamide molecule. Metrizamide can be removed by acid precipitation. Proteins and formaldehyde also interfere with the orcinol assay.

3. Assays for proteins

3.1 Amido-black filter assay

Schaffner, W. and Weissman, C. (1973). *Anal. Biochem.* **56**, 502.

Solutions:

- Stain: dissolve 0.1 g of Amido black 10B in 100 ml of methanol/glacial acetic acid/water (45/10/45 by vol)
- Destain: methanol/glacial acetic acid/water (90/2/8 by vol)
- Eluant solution: 25 mM NaOH, 0.05 mM EDTA in 50% aqueous ethanol
- Standard protein solution: 150 μg/ml protein
- 60% (w/v) trichloroacetic acid (TCA) stock solution
- 1% SDS in 1 M Tris–HCl (pH 7.5)

Assay method:

1. Dilute the protein standard solution to 0.27-ml aliquots containing 0–150 μg protein for the preparation of a standard curve. Make up the samples to the same volume.
2. Add 0.03 ml of Tris–SDS solution and 0.6 ml of 60% TCA to each sample.
3. Filter each sample through a Millipore filter (0.45 μm pore size) and rinse the tube with 0.3 ml 6% TCA and wash filter with 2.0 ml 6% TCA.

4. Place each filter in stain for 2–3 min with gentle agitation and pass into a water rinse for 30 sec.
5. Pass each filter through three changes of destain, about 1 min per change rinse again with water and blot with a tissue.
6. Place filter in a test tube with 0.6 ml eluant for 10 min and vortex three times for 2–3 sec. Add a further 0.9 ml eluant and mix by vortexing.
7. Measure the optical density at 630 nm using the eluant solution as a blank.

Notes:
Using this assay, no interference is experienced from the presence of gradient solutes unless, like Percoll and metrizoates, they are not acid soluble. Sucrose, metrizamide and Nycodenz are acid soluble and thus such gradients may be analysed using this assay. Potassium ions must be excluded from the sample solutions since they precipitate the dodecyl sulphate ions.

3.2 Coomassie blue filter assay

McKnight, G. S. (1977). *Anal. Biochem.* **78**, 86.

Solutions:

- Stain: 0.25% Coomassie blue G-250, 7.5% acetic acid, 5% methanol in distilled water
- Destain: 7.5% acetic acid, 5% methanol
- Eluant solution: 0.12 M NaOH, 80% ethanol, 20% distilled water
- 3.0 M HCl
- 20% (w/v) trichloroacetic acid (TCA)

Assay method:

1. Use 25 mm Whatman glass-fibre filters (GF/C) and number each along the edge. Do not touch filters with bare fingers.
2. Pipette protein samples onto the centre of the discs and allow them to be completely absorbed (5–30 sec).
3. The sample volume to be used will depend upon the protein (see Notes).
4. Prepare blank filters using distilled water in place of protein sample.
5. After absorption, place filters in 20% TCA at 4 °C and swirl gently for 2–3 min.
6. Transfer the filters to the stain at 4 °C for 20 min.
7. Transfer to destain, swirl for several minutes, decant solution, and repeat twice with fresh destain.
8. Place filters on a filtering apparatus and wash several times with the destain solution until the blank filters are completely white.

9. Cut out the stained spots, using a 10-mm cork borer, resting the filters on several sheets of filter paper.
10. Put the spots into 1.5-ml microfuge tubes, add 0.6 ml of eluant solution, vortex, and stand at room temperature until the colour disappears from the filters (about 5 min).
11. Acidify the eluant with 0.03 ml 3 M HCl (Coomassie blue is colourless in basic solutions), vortex, and centrifuge for 2 min in a microcentrifuge to remove any pieces of fibre in the solution.
12. Remove supernatant and measure absorbance at 590 nm, using acidified eluant as a blank.

Notes:
This assay is not compatible with Percoll or metrizoates. The sample volume that can be pipetted onto the filter depends on the salt concentration used and on the particular protein. Some proteins, such as RNase and histones, do not spread, and volumes of up to 100 µl can be used to give concentrated spots, but BSA and ovalbumin, for example, diffuse into the filter and in such cases volumes need to be less than 50 µl. In high salt (e.g. 1 M NaCl), all proteins bind tightly to the filter and produce a concentrated spot. The sensitivity of this assay can be increased 3-fold by using 200 µl of eluant solution and 10 µl of 3 M HCl.

3.3 Coomassie blue solution (Bradford) assay

Bradford, M. (1976). *Anal. Biochem.* **72**, 248.

Solutions:

- Coomassie blue reagent: dissolve 0.1 g of Coomassie blue G-250 in 50 ml of 90% ethanol, add 100 ml of 85% (w/v) phosphoric acid and then make up to 200 ml with distilled water. This concentrated solution can be stored at 5 °C. Before use dilute this reagent by the addition of four volumes of distilled water.
- Standard protein solution: 100 µg/ml.

Assay method:

1. Dilute protein standard solution to 0.1-ml volumes containing 0–100 µg/ml and dilute samples to same volume.
2. Add 5 ml of reagent to each fraction and measure the optical density at 595 nm.
3. Use 5.0 ml reagent and 0.1 ml protein buffer as a blank.

Measurements should be carried out after 2 min but less than 60 min after addition of the reagent.

Notes:
This assay provides a swift and easy determination of protein distribution in a gradient and is compatible with sucrose, metrizamide, and Nycodenz, but not metrizoates. This assay can be used for Percoll gradient fractions if the Percoll is first removed by centrifugation at 12 000*g* for 15 min after precipitation with 0.025% Triton X-100 in 0.25 M NaOH, as described by R. Vincent and D. Nadeau (*Anal. Biochem.* **135**, 355).

3.4 Fluorescamine assay

Bohlen, P., Stein, S., Dairman, W., and Udenfriend, S. (1973). *Arch. Biochem. Biophys.* **155**, 213.

Solutions:

- Fluorescamine reagent: dissolve 30 mg of fluorescamine in 100 ml of dioxane (histological grade)
- 50 mM sodium phosphate buffer (pH 8.0).

Assay method:

1. Place the sample, containing 0.5–50 µg protein, into a tube and add phosphate buffer to give a total volume of 1.5 ml.
2. Add 0.5 ml of fluorescamine solution, with vortexing to ensure vigorous mixing. Rapid mixing is essential for reproducible results as the reagent is readily hydrolysed. Maintain the pH at pH 8–9. The reaction takes place at room temperature.
3. Measure the fluorescence with an excitation wavelength of 390 nm and an emission wavelength of 475 nm.
4. All values should be corrected using a blank prepared with 1.5 ml of the sample buffer.

More reproducible results are obtained by carrying out direct fluorescence measurements in the reaction tubes, using a cell adaptor, instead of transferring the mixtures to a cuvette.

Notes:
Fluorescamine reacts with amino groups (primary amines), thus Tris and other primary amine buffers cannot be used. Large amounts of secondary and tertiary amine buffers also interfere with the assay. Hence inorganic buffers, such as phosphate or borate, are most successful. A modification of this procedure uses gel filtration to remove contaminating low molecular weight amines in the sample which interfere with the assay (Bohlen *et al.*, *Anal. Biochem.* **58**, 559). Using this modification of the assay as described, it should be noted that contaminants with molecular weights in excess of 300 do interfere due to incomplete separation from the proteins on the Sephadex G-25 columns.

3.5 Folin–Ciocalteu (Lowry) protein assay

Lowry, O. H., Rosebrough, N. J., Farr, A. L., and Randall, R. J. (1951). *J. Biol. Chem.* **193**, 265.
Peterson, G. L. (1979). *Anal. Biochem.* **100**, 201.

Solutions:

- Folin–Ciocalteu reagent is available commercially.
- Solution A: dissolve 20 g Na_2CO_3, 4.0 g NaOH, and 0.2 g sodium potassium tartrate in distilled water and make the solution up to 1 litre
- Solution B: 0.5% $CuSO_4 \cdot 5H_2O$
- Solution C: mix 50 ml of Solution A and 1 ml of Solution B immediately before use.

Assay method:

1. Dilute samples to 1.0 ml, add 4.0 ml solution C, mix, and allow to stand for 10 min.
2. Add 1.5 ml of Folin–Ciocalteu reagent (diluted 1:9 immediately before use), mix again, and stand for 30 min in dark.
3. Measure the optical density at 660 nm, using 1.0 ml water in place of sample as the blank.

Notes:
This is the most widely used method for estimating proteins but over the years it has been extensively modified. Most modifications are aimed at improving sensitivity and in avoiding the many sources of interference in this assay. Gradient media that interfere with this assay include sucrose, glycerol, metrizamide, Nycodenz, and Percoll.

3.6 Microbiuret assay

Itzhaki, R. F. and Gill, D. M. (1964). *Anal. Biochem.* **9**, 401.

Solutions:

- Benedict's reagent: first dissolve 17.3 g trisodium citrate and 10.0 g Na_2CO_3 in warm distilled water, add to this 1.73 g of $CuSO_4$ dissolved in 10 ml of distilled water and make the solution up to 100 ml
- 1.0 M NaOH
- Standard protein solution: 250 μg/ml of protein

Assay method:

1. Prepare a standard curve, using the standard protein solution diluted to 1.0 ml volumes containing 0–250 μg protein.

2. Prepare 1.0 ml sample volumes and add 3.0 ml of NaOH solution and 0.2 ml of Benedict's reagent. Mix well and allow to stand at room temperature for 15 min.
3. Measure the optical density at 330 nm, using water in place of sample as the blank.

Notes:
Gradient media containing sugars, such as sucrose, Ficoll, and metrizamide, will interfere with this assay. Other media may be unstable in the hot acid conditions of the assay.

4. Assays for polysaccharides and sugars

4.1 Anthrone assay

Seifter, S., Dayton, S., Norvic, B., and Muntwyler, E. (1950). *Arch. Biochem. Biophys.* **25**, 191.

Solutions:

- Anthrone reagent: 0.2% anthrone dissolved in 95% H_2SO_4
- Glucose standard solution: 50 µg/ml glucose

Assay method:

1. Prepare a standard curve of 0–125 µg glucose using the standard glucose solution.
2. Place 2.5 ml of sample into a boiling tube, cool in ice-water, and add 5.0 ml of anthrone reagent from a fast-flowing burette, mixing thoroughly, still in the ice-water.
3. Cover the tube with a marble and heat for 10 min in a boiling water bath.
4. Cool quickly and measure the optical density at 620 nm. Use 2.5 ml of distilled water as the blank.

Use caution when boiling the H_2SO_4 solution.

4.2 Phenol–H_2SO_4 assay

Dubois, M., Gilles, K. A., Hamilton, J. K., Rebers, A., and Smith, F. (1956). *Anal. Chem.* **28**, 350.

Solutions:

- 80% (w/w) phenol
- Concentrated H_2SO_4 (95%)
- Standard sugar solution (100 µg/ml)

Assay method:

1. Using tubes of at least 10 mm diameter to allow good mixing and minimize heat dissipation, add 50 µl of phenol solution to 1.0 ml of the standard or sample solution.
2. Add 5.0 ml of H_2SO_4, allowing the acid to leave the pipette and fall on to the liquid surface quickly to ensure good mixing.
3. Let the tubes stand for 10 min, shake, and place into a waterbath at 25–30 °C for 10–15 min.
4. Measure the optical density at 490 nm for hexoses and at 480 nm in the case of pentoses and uronic acids. Prepare blanks using water instead of the sample solution.

Notes:
5% (w/w) phenol may be used in place of 80% phenol, in which case 1.0 ml of phenol solution is added, other volumes as given above. Most compounds that interfere with the anthrone assay also interfere with this assay.

Appendix 5

Names and addresses of suppliers of centrifuges and ancillary equipment

AB-Biox: PO Box 143, Jarfalla, Sweden.

Accurate Chemical and Scientific Corp: 300 Shames Drive, Westbury NY 11590, USA.

A.L.C. Apparecchi Per Laboratori Chimici Srl: via Jean Jaures 12, 20125 Milano, Italy.

Alfa laval Inc: 211, Linwood Ave, PO Box 1316, Fort Lee, NJ 07024, USA

Alfa Laval Separation AB: S-14780 Tumba, Sweden.

Alfa Laval Sharples: Doman Road, Camberly, Surrey GU15 3DN, UK.

Anderman and Co. Ltd: Central Avenue, East Molesey, Surrey KT8 0QZ, UK.

Anton Paar: see Paar K. G.

Artisan Industries: Waltham, Mass, USA.

Astra: Södertalje, Sweden.

Baird and Tatlock Ltd: Freshwater Road, Chadwell Heath, Romford, Essex, UK.

Baskerville and Lindsay Ltd: Barlow Moor Road, Manchester M21 2AX, UK.

BDH Chemicals Ltd: Broom Road, Poole, Dorset BH12 4NN, UK.

Beckman Instruments Spinco Division: 1117 California Ave, Palo Alto, CA 94304, USA.

Beckman Instruments International SA: 17, Rue des Pierres du Niton, PO Box 308, CH-1207 Geneva, Switzerland.

Beckman-RIIC Ltd: Turnpike Road, Cressex Industrial Estate, High Wycombe, Bucks HP12 3NR, UK.

Bellingham and Stanley Ltd: Longfield Road, North Farm Industrial Estate, Tunbridge Wells, Kent, UK.

Bio-rad Laboratories: 2200 Wright Ave, Richmond, CA 98404, USA.

Biorad Laboratories Ltd: Caxton Way, Watford, Herts, UK.

Biospec Inc: PO Box 722, Bartesville, OK 74005, USA.

Braun Melsungen International GmbH: Schwarzenbergerweg, D-3508 Melsungen, FRG.

Buchler Instruments Inc: 1327 Sixteenth Street, Fort Lee, NJ 07024, USA.

Burkhard: PO Box 55, Uxbridge, Middx UB8 1LA, UK.

Christ: Heraeus Sepatech.

Clandon Scientific Ltd: Lysons Ave., Ash Vale, Aldershot, Hants GU12 5QR, UK.

Corex: see Dupont–Sorvall and Dupont (UK) Ltd.
Corning Ltd: Stone, Staffs ST15 0BG, UK.
Damon: IEC see IEC.
Denley: Daux Road, Billingshurst, Sussex RH14 9SJ, UK.
Du Pont-de Nemours and Co. Inc: Wilmington, DE 19898, USA.
Du Pont–Sorvall: Pecks Lane, Newtown, CT 06470, USA.
Du Pont (UK) Ltd: Wedgwood Way, Stevenage, Herts SG1 4QN, UK.
Eastman Kodak Co: 343 State Street, Rochester, NY 14650, USA.
Edco Scientific Inc: PO Box 64c, Sherborn, MA01770, USA.
Electro Nucleonics Inc: 368 Passaic Ave, Fairfield, NY 07006, USA.
Eltex of Sweden Ltd: Lane Close Mills, Bartle Lane, Great Horton, Bradford, Yorks BD7 4QQ.
Esco (Rubber) Ltd: 14–16 Great Portland Street, London W1N 5AB, UK.
Fisons Scientific Ltd: Bishop Meadow Rd, Loughborough, Leics LE11 0RG, UK.
F.T. Scientific Ltd: Station Industrial Estate, Bredon, Tewkesbury, Glos GL20 7HH, UK.
Grant Instruments Ltd: Barrington, Cambridge CB2 5QZ, UK.
Hereus Sepatech GmbH: Am Kalkberg, PO Box 1220, D-3360 Osterode/Harz, FRG.
Hermle GmbH and Co: Industriestrasse 8-12, D-7209 Gosheim, FRG.
Hettich: Gartenstrasse 100, D-7200 Tuttlingen, FRG.
Hitachi Koki Co. Ltd: 1060 Takeda, Katsuta City, Ibaraki Pref 312, Japan.
A. R. Horwell Ltd: 2 Grangeway, Kilburn High Road, London NW6 NB, UK.
V. A. Howe and Co. Ltd: 88 Peterborough Road, London SW6 3EP, UK.
IEC: 300, 2nd Av, Needham Heights, MAS 02194, USA.
IEC (UK) Ltd: Unit 1, Lawrence Way, Brewers Hill Rd, Dunstable, Beds LU6 1BD, UK.
ISCO (Instrument Specialities): Box 5347, Lincoln, NE 68505, USA.
Janetzki: Leipzig, FRG.
Jouan SA: rue Bobby Sands, Case Postale 3203, F44800 St Herblain, France.
Jouan Inc: PO Box 2716, 148 A Rte 7, East Winchester, VA 22601, USA.
Jouan Ltd: 130, Western Road, Tring, Herts HP23 4BU, UK.
Kompspin Inc: 2068, B Walsh Avenue, Santa Clara, CA 95050, USA.
Kontron Instruments Spa: Via G Fantoli 16/15, 20138 Milano, Italy.
Kontron (UK) Ltd: Croxley Centre, Blackmoor Lane Watford, Herts WD1 8XQ, UK.
Kubota Corporation: 29-9 Hongo 3-chome, Bunkyo-ku, Tokyo 113, Japan.
Life Science Labs Ltd: Sedgewick Road, Luton, Beds LU4 9DT, UK.
Merck: Postfach 4119, D-6100 Darmstadt 1, FRG.
Monsanto Chemicals Limited: 10–18 Victoria Street, London SW1, UK.
MSE Scientific Instruments Ltd: See SANYO–Gallenkamp.
Nalge Company: 75, Panorama Creek Drive, PO Box 20365, Rochester, New York 14602-0365, USA.

Nissei Sangyo GmbH: Berliner Strasse 91, 4030 Ratingen 3, FRG.
Nyacol Inc: Megunco Road, Ashland, MA 01721, USA.
Nycomed-Pharma AS: Lillogaten 3, Postbox 4284, Torshov, N-0401 Oslo 4, Norway.
Nycomed (UK) Ltd: 2111 Coventry Road, Sheldon, Birmingham B26 3EA, UK.
Paar KG: Kamtnerstrasse 322, A-8054 Graz, Austria.
Paar Scientific Ltd: 594 Kingston Road, Raynes Park, London SW20 8DW, UK.
Pennwalt Ltd: Doman Road, Camberley, Surrey GU15 3DN, UK.
Pharmacia Biosystems Ltd: Davey Avenue, Knowhill, Milton Keynes MK5 8BH, UK.
Pharmacia Biosystems Inc: 800, Centennial Ave., Piscataway, NJ 08854, USA.
SANYO–Gallenkamp Ltd: Park House, Meridian East, Meridian Business Park, Leicester LE3 2UZ.
Schering, A.G.: 1000 Berlin 65, Müllerstrasse 170-172, FRG.
Schering Chemicals Ltd: Burgess Hill, Sussex, UK.
Scientific Supplies Co. Ltd: Scientific House, Vine Hill, London EC1 5EB, UK.
Searle Instrument Division, West Road, Temple Fields Industrial Estate, Harlow, Essex, UK.
Seton Scientific: PO Box 60548, Sunnyvale, CA 94088, USA.
Shandon Southern Ltd: Frimley Road, Camberley, Surrey, UK.
Sharples: see Pennwalt.
Sigma Chemical Co: 3050 Spruce Street, PO Box 14508, St Louis, MO 63178, USA.
SIGMA laborzentrifugen GmbH: PO Box 1727, D-3360 Osterode/Harz, FRG.
Sorvall: see Du Pont–Sorvall and Du Pont (UK).
E.R. Squibb and Sons: Regal House, Twickenham, Middlesex, UK.
Townson and Mercer Ltd: Chadwick Road, Astmoor, Runcorn, Cheshire WA7 1PR, UK.
VWR Scientific: PO Box 13645, Philadelphia, PA 19101-3645, USA.
Whitley Scientific Ltd: 14 Otley Road, Shipley, Yorks BD17 7SE, UK.
Wifug AB: Box 147, S-34300 Elmhult, Sweden.
Wifug Ltd: Lustra Works, Parry Lane, Bradford, Yorks BD4 8TQ, UK.
Winthrop Laboratories: Winthrop House, Surbiton, Surrey, UK.
Yamamoto Scientific Co: Minato-Ku, Tokyo 105, Japan.

Index

Index

Index